The Plastic
Deformation of Metals

The Plastic
Deformation of Metals

R. W. K. Honeycombe

Goldsmiths' Professor of Metallurgy
University of Cambridge

EDWARD ARNOLD

© R. W. K. Honeycombe 1984

First published 1968
by Edward Arnold (Publishers) Ltd
41 Bedford Square
London WC1B 3DQ

Edward Arnold
300 North Charles Street
Baltimore
Maryland 21201
U.S.A.

Edward Arnold (Australia) Pty Ltd
80 Waverley Road
Caulfield East
Victoria 3145
Australia

Reprinted 1971, 1974
First published as a Paperback 1975
Reprinted 1977
Second Edition 1984
Reprinted 1985

British Library Cataloguing in Publication Data

Honeycombe, R. W. K.
   The plastic deformation of metals. — 2nd ed.
   1. Metals — Plastic properties
   I. Title
   620.1′6′33   TA490

ISBN 0-7131-3468-2

To June, Juliet and Celia

Printed in Great Britain by Butler & Tanner Ltd.,
Frome and London

Preface

In the last twenty years, the fundamental basis of metallurgy has become much more firmly established. One of the most impressive developments has undoubtedly been the theory of dislocations which accounts for many of the characteristics of crystalline solids, in particular the behaviour during plastic deformation. Several comprehensive books have been written on this subject, but there seems to be a definite need for a text which describes systematically the actual behaviour of metals and alloys during various types of deformation, and attempts to explain this as far as possible in terms of dislocation theory.

The approach I have adopted echoes that of the classical *Kristallplastizitaet* by E. Schmid and W. Boas, in so far as the behaviour of single crystals is taken as a logical starting point, but the emphasis is on the large volume of post-war work which has been done in this field. The results of such investigations are then used to examine more complex deformation phenomena in polycrystalline aggregates, for example, textures, creep, fatigue and fracture.

The book is aimed at graduate and undergraduate students of metallurgy and materials science in Universities and Colleges of Technology who need an overall picture of the plastic deformation of metals, in which both the theory and behaviour of metals receive attention. It should also be useful in explaining to engineers the basic principles which determine the properties of the materials they use. The references included are not comprehensive, but are selected to provide a broad basis for further reading in the subject. An elementary working knowledge of metallography and crystallography has been assumed.

The book was mostly written while I was at the University of Sheffield, and I am very grateful to Professor A. G. Quarrell and colleagues in the Department of Metallurgy there for helpful discussions and encouragement. Mrs. Wendy Morton deserves special thanks for patiently deciphering my manuscript and for doing much of the final typing. I owe a particular debt to Professor E. O. Hall for reading the manuscript and for making many helpful comments. I must also thank Dr. Brian Ralph very much for raising numerous useful points at the proof stage and Mrs. Evelyn

Martin for her help in the preparation of the index. The sources of all figures have been acknowledged, and I would like to add my thanks to the various authors who have helped me in this way.

Finally, I would like to take this opportunity of expressing my deep gratitude to my old friend Dr. Walter Boas, who first introduced me to the study of the deformation of crystals.

R. W. K. Honeycombe

Cambridge,
1968

Preface to Second Edition

In this new edition I have taken the opportunity to switch to SI units. I have also included at the end of each chapter, additional general references to relevant books which have been published since 1968. The text is unchanged because I feel that the elements of the subject remain broadly the same today. However it is inevitable that, over the fifteen years since the book was first published, there have been many important developments in the subject, for example in the fields of creep, fatigue and fracture. To do justice to this work a new book would have to be written, but in the meantime I trust that this book, by emphasizing well established principles, will provide a springboard to later developments in the field.

Finally I would like to pay tribute to the late Dr Walter Boas who died earlier this year, and who will be remembered for his outstanding contributions to our understanding of the plastic deformation of metals.

R. W. K. Honeycombe

Cambridge
1982

Contents

Conversion of Units

$M\,Nm^{-2}$ to lb. in.$^{-2}$	$\times\ 145.038$
$M\,Nm^{-2}$ to kg. mm.$^{-2}$	$\times\ 0.10197$
lb in^{-2} to $M\,Nm^{-2}$	$\times\ \ 0.006895$
kg mm^{-2} to $M\,Nm^{-2}$	$\times\ \ 9.80665$
1 Angström unit (Å)	$=10^{-10}\,m$
1 Angström unit	$=10^{-1}$ nanometre (nm)
1 micron (μm)	$=10^{4}$ Angström units
1 micron	$=10^{-6}\,m$
1 Joule (J)	$=10^{7}$ erg
1 eV	$=1.602\times10^{-19}\,J$

Chapter 1

Introduction

Perhaps the most characteristic property of a metal is ductility or the ability to suffer much deformation without breaking. This property was extremely useful even to our remote ancestors when they discovered deposits of native gold and copper as well as occasional iron meteorites. Indeed, gold is the most ductile of metals, which for many centuries has been beaten into the thinnest of leaf. While practical knowledge of the forming of metals extends backwards for thousands of years, our understanding of the physical phenomena associated with deformation has only developed within the last forty years. Some of the basic principles have been elucidated, but many associated phenomena will require much further research before they are thoroughly understood. It is the aim in this book to provide a framework to the subject involving known facts and principles, but occasionally the temptation to indulge in some speculation will not be resisted.

If an increasing load is applied to a metal wire suspended from a fixed point, extension will occur which is completely recoverable when the load is removed. The metal is then said to be deformed elastically, and there is a linear relationship between stress σ and strain ϵ which Hooke's law defines thus, $E = \sigma/\epsilon$. At any given load the ratio of the stress σ to the elastic strain ϵ is a constant E, known as Young's modulus in the case of a uniaxial tensile stress. However, beyond a certain load, complete recovery of the strain does not occur on unloading because the metal has deformed plastically or permanently. We define the transition as the yield stress or the initial flow stress. The total elastic strain is extremely small, so the plastic strain accounts for the overwhelming proportion of the deformation. The engineer when designing machines tries to avoid stressing a part anywhere near the yield stress but, to form and fabricate metals, a metallurgist must work in the plastic range. In this book we shall explore the behaviour of metals from the yield stress to the point at which they break apart; we shall examine the basic mechanisms which allow such operations as rolling, forging, drawing and pressing to be successfully carried out.

In the early part of this century, Rosenhain and Ewing showed that plastic deformation of metals produced on the surface many parallel microscopic steps called slip bands, which suggested to them that the metal had sheared along the bands rather like cards in a shuffled pack. These early observations showed clearly that the shear occurred along well-defined crystallographic planes in the metal, the markings changing direction at the grain boundaries. However, detailed study was difficult in normal metal specimens because one grain might exhibit several different sets of markings, and the need for investigation of the behaviour of individual grains or crystals became apparent. Only in this way could the problem of plastic deformation be simplified.

A little later, in 1910, Andrade developed a technique of growing large individual crystals from the melt by a method later to be elaborated by Bridgman, who used it to prepare single crystals of many metals of uniform dimensions, covering a wide range of possible orientations relative to the specimen axis. The way thus lay open for the detailed study of plastic deformation of metal crystals, with the result that the period 1920–34 was rich in experimental investigations of crystal plasticity. Fundamental studies of the deformation of most of the common metals led to important generalizations, such as the critical resolved shear stress criterion, and allowed the principles of crystal geometry to be precisely stated. Behaviour of metals in the three main crystallographic groups, face-centred cubic, body-centred cubic and close-packed hexagonal, was compared and contrasted. Results of this period of extensive activity were summarized in the classical monograph on crystal plasticity by E. Schmid and W. Boas published in 1936. Chapter 2 of the present book will present the classical experimental results on the deformation of single crystals while in Chapter 4 the subject will be discussed in the light of more recent developments. Since 1945 much new work has been done, in some cases with metals of much higher purity which has led to the modification of earlier ideas.

In 1934 Orowan, Polanyi and Taylor independently introduced the concept of a dislocation, a crystal line defect which was necessary to account for the fact that the observed strength of metals generally was about a thousand times less than the theoretical estimates. It is no exaggeration to say that the dislocation has proved to be the most important discovery of metal physics in the last thirty years. What was at first an elegant theoretical concept has, in the post-war years, proved to be a triumphant reality, significant not only in the process of plastic deformation itself but also in crystal growth, recovery and other diverse phenomena. Development of the theory was interrupted by the war, but was resumed in 1945 to blossom in the following decade. The manifestations of the theory are now many, but the principles can be summarized so that they are of use to students of metallurgy. This has been the aim in Chapter 3, where simple types of dislocation are described and some of their properties outlined; this chapter also contains some of the direct evidence obtained for the existence of dislocations. It is interesting to reflect that after a decade of fruitful theory, the well-tried metallographic approach in the guise of thin-foil electron metallography has provided the final proof of many theoretical pronouncements.

Perhaps the most impressive of the plastic properties is work hardening or the ability of a metal to become stronger as it deforms. Work hardening has proved to be a very difficult problem to solve, so much so that it is impossible to present a concise and convincing quantitative theory. However, the dislocation theory has provided many useful ideas, some of which are considered in Chapter 5, and used in an attempt to explain the experimental results.

While work hardening contributes greatly to the strength of a metal, the addition of alloying elements is a much more effective way of increasing its resistance to deformation. The effect of even very small concentrations of solute atoms on the strength of a metal crystal can be very substantial, as shown in Chapter 6. The effect of alloying elements both in solid solution and present as a separate phase on the process of plastic deformation must be examined in an attempt to understand the behaviour of complex alloys (Chapter 7). While deformation by slip is widespread, it is not the only mechanism by which plastic strain is achieved. In Chapter 8 the important process of deformation twinning is introduced.

Many theories concerned with mechanical properties of metals are best tested on single crystals. However, there comes a stage when an attempt must be made to use single crystal behaviour to understand the deformation of polycrystalline aggregates. Clearly the properties of the individual crystals provide the key, but the deduction of polycrystalline behaviour from them is difficult and only limited progress has been made. In Chapter 9 the role of the grain boundaries is first outlined, followed by a discussion of the stress–strain curves of polycrystalline aggregates.

Atomic holes or vacant lattice sites are an important by-product of plastic deformation, which are often accompanied by the formation of interstitial atoms. Such point defects are of considerable significance not only in deformation processes, but in recovery and solid state diffusion. They are also a direct consequence of irradiation in crystalline solids, and are dominant factors in determining their mechanical properties (Chapter 10).

It has been known for thousands of years that a metal hardened by working can be restored to its original ductility by heating. There is a series of interesting processes by which this end is reached, which commences with the rearrangement of defects within the deformed crystals (recovery) and concludes with the replacement of the deformed grains by a new set of strain-free crystals (recrystallization). These phenomena are discussed in Chapter 11. While such processes may not at first glance appear to be relevant to a study of deformation, it should be appreciated that if the temperature is raised, these phenomena can occur during deformation.

The arrangement of the grains in a polycrystalline aggregate is often far from random. The various means of working metals induce characteristic preferred orientations or textures which are reflected in the variations of mechanical properties with direction. Textures also develop during re-crystallization, and influence markedly the working behaviour of annealed metals. In Chapter 12, these topics are discussed and, in addition, the anisotropy of magnetic and thermal properties.

The fact that under a constant stress there is a time-dependent com-ponent of the plastic deformation referred to as creep is now well known. The development of creep-resistant alloys is an important and difficult occupation which theory will take many years to supplement. However, a start has now been made, so that it is now possible to outline the basic characteristics of deformation by creep and to provide some theoretical bases for the observed behaviour (Chapter 13).

It has been estimated that at least 80 per cent of metal failures are fatigue failures, i.e. failures under repeated alternating loads. The fact that metals under these conditions cannot survive stresses much lower than their initial yield stress is sufficiently striking for this to be a subject of both practical and theoretical interest. In recent years structural changes during fatigue have been extensively studied, and in Chapter 14 the special features of this mode of deformation will be summarized. The ultimate end result of prolonged plastic deformation is fracture or rupture; in Chapter 15 the various modes of failure are considered as far as possible from a fundamental standpoint. The fact that a normally ductile metal such as iron can, in some circumstances, fracture disastrously after negligible plastic deformation, helps to emphasize the importance of brittle fracture. This phenomenon came into prominence during the last war when many Liberty ships were quickly built by extensive use of welded structures, but unfortunately several hundred failed in service by extensive brittle facture, which in many cases was a catastrophic process. The incidence of brittle failures of steel was so high, and the cost so great, that much research was initiated during and after the war. Some of the results of this work are dealt with in Chapter 15, and an effort is made to show how dislocation concepts have been applied to this formidable practical problem.

Chapter 2

The Deformation of Metal Crystals—
General Aspects

2.1 Introduction

Much of our present-day knowledge in physical metallurgy has stemmed from an appreciation of the properties of individual metal crystals. In the first place, these properties, whether they be mechanical or physical, determine the properties of the normal polycrystalline metal, and thus must be understood before we can try to explain the behaviour of poly-crystalline metals. Furthermore, some phenomena are difficult to study basically in polycrystalline specimens, because the results are complicated to interpret. This applies particularly to plastic deformation. The diffi-culties mainly arise from the fact that all mechanical properties, and in many cases physical properties such as thermal expansion and diffusion co-efficients, vary according to the direction in the crystal along which they are measured. In other words, the crystal is anisotropic with respect to the particular property.

Classical work on metal crystals during the period 1920–35 [1, 2] showed that many of the properties varied with direction in the crystals. For example, in all metals, the elastic and plastic properties are anisotropic. Some results are shown in Table 2.1 for typical cubic and non-cubic metals. While it can be seen that non-cubic metal crystals frequently show extreme varia-tions in properties, such as yield stress, elongation and U.T.S.,† the cubic metals too can exhibit substantial anisotropy. Likewise, the magnetic pro-perties show considerable variations with orientation in iron single crystals. The electrical and thermal properties of cubic metals do not vary with direction; however, with hexagonal metals such as zinc and cadmium marked anisotropy occurs.

Most of these variations in properties would normally have little effect on polycrystalline metals if the crystals were randomly arrayed, because then the property would be an average value, constant in all directions in the

† Ultimate Tensile Strength.

Table 2.1   Anisotrophy of some elastic and plastic properties of metal crystals at room temperature

Metal	Young's modulus, $MN\,m^{-2}$	Ultimate tensile strength $MN\,m^{-2}$	Elongation, per cent
Aluminium (f.c.c.*)	62 760–75 510	58·3–113·8	20–70
Copper (f.c.c.)	66 700–191 230	127·5–343·2	10–55
α-iron (b.c.c.†)	132 390–284 390	59·8–225·6	20–80
Cadmium (hexagonal)	28 240–80 900	2·5–10·3	10–700
Magnesium (hexagonal)	42 950–50 500	29·4–88·3	20–220
Zinc (hexagonal)	34 800–123 560	11·8–27·5	60–400

* f.c.c. = face-centred cubic.   † b.c.c. = body-centred cubic.

aggregate. However, metals are rarely completely random polycrystalline aggregates in practice, but instead they possess textures or preferred orientations which reflect to a greater or lesser degree the anisotropy of properties in the individual grains.

2.2   Preparation of metal single crystals

Several review articles[7] and books[4] have appeared on the growth of metal crystals in recent years so only a summary will be given.

A number of methods for the preparation of large single crystals of metals has been known for many years, although in the last decade many refinements have been made, as the importance of crystalline perfection and purity have been realized. The methods almost all fall into two main groups:

(1) Solidification from the melt.
(2) Grain growth in the solid state.

2.2.1.   Solidification from the melt

Most of the techniques in this category have been derived from the Czochralski and Bridgman methods. In the former, a seed crystal is gradually withdrawn vertically from the surface of a molten metal in such a way that a crystal rod is built up with the orientation of the seed. While the crystals produced can be somewhat irregular in cross-section, there is no contact with a mould and so contamination from this source is eliminated. The method has undergone a revival in recent years for the preparation of germanium and silicon crystals for transistors. This technique has distinct advantages for silicon, which is very reactive at temperatures in the vicinity of its melting point (1412°C). A modern apparatus is shown in Fig. 2.1 where the whole operation is carried out *in vacuo*. The

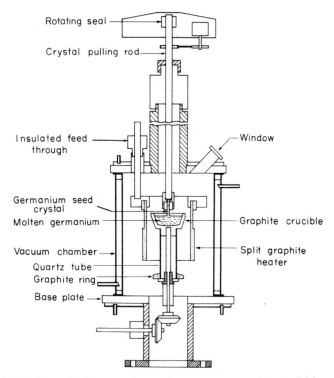

Fig. 2.1 Modern single crystal growing apparatus—Czochralski method. (After Lehovec *et al.*, 1953, *Rev. scient. Instrum.*, **24**, 652)

crystal rod is rotated during growth to give a more uniform cross-section and to disperse homogeneously any alloying additions. The success of the method depends on obtaining a delicate balance between the heat input to the system and the losses from crucible and crystal rod. The orientation of the resulting crystal can be readily controlled by choosing a seed crystal of the required orientation.

In the Bridgman method, the metal is contained in a mould pointed at one end, which is slowly lowered through a vertical furnace with a temperature gradient. The crystal is nucleated at the pointed end and grows upwards in the mould. Alternatively, the furnace can be moved relative to the mould, or both the mould and furnace can remain stationary while the current to the latter is gradually reduced. An important aspect of the method is the design of the mould and the choice of material for it. It is desirable though not essential to use split moulds from which the crystal is separated by dissembling the mould, not by moving the crystal. High purity graphite, sufficiently hard to take a good polish, is a satisfactory material for metals such as aluminium, copper, silver, nickel, but crystals of the lower melting metals such as tin, zinc and cadmium can be grown in precision-bore heat-resisting glass tubes provided they are adequately coated with dried aqua-dag.

For metals which react with carbon, alumina can be used. A useful variation is the soft mould technique, in which the metal rod to be melted is packed in powdered alumina in an alumina tube. The powder flows readily under pressure with the result that stresses are not set up in the crystal during and after solidification. In this way shouldered specimens can be made, which, if grown in a rigid mould, would become plastically deformed during cooling.

Single crystals can also be prepared from the melt by using horizontal travelling furnaces and moulds, the latter being in the form of graphite boats usually pointed at one end. The free surface of the resulting crystal is not always particularly flat but this may not matter for many applications. The method has the advantage that the process of solidification can be observed if the boat is held in a clear silica tube and the heat is provided by an induction coil. Single crystal wires can be made in the same way by putting the wire in loose-fitting glass or silica tubes coated with graphite.

One of the most important recent developments in the preparation of single crystals from the melt has been the application of the zone melting technique,[8] where a molten zone is moved from one end of a bar to the other. This is primarily a method of purification, but as this is gradually achieved the tendency for nucleation becomes less and less, and often the final result is a very pure single crystal. The method has been extensively used by the electrical industry in recent years to prepare single crystals of germanium containing only a few parts per million impurities. An apparatus is shown schematically in Fig. 2.2.

Crystals can also be prepared vertically by the floating zone method in which the metal is no longer in contact with a container.[8] The specimen rod is gripped top and bottom by water-cooled grips and a molten zone is produced either by high-frequency heating or electron bombardment from a filament, the process being usually carried out in a high-purity argon atmosphere or a good vacuum. The molten zone, which is passed slowly along the rod, can be made quite stable, the length of the zone being proportional to $(\sigma/\rho)^{1/2}$, where σ = surface tension and ρ is the density of the metal. Rods up to one inch in diameter can be melted, but a number of passes may be necessary before a single crystal is obtained. However, once a high level of purity is attained, many metals of high melting point can be obtained in single crystal form, e.g. tungsten, tantalum, molybdenum, vanadium, nickel, rhenium, etc.

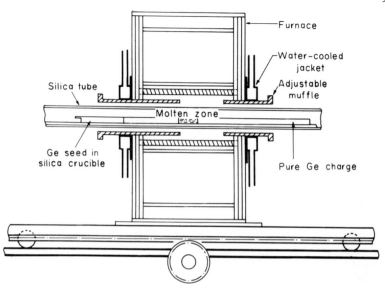

Fig. 2.2   Single crystal growth by zone melting. (After Cresswell and Powell, 1957, *Prog. Semicond.*, **2**, Heywood)

2.2.2. *Grain growth in the solid state*

The most widely used solid-state method is the strain-anneal technique first introduced by Carpenter and Elam in 1921 to prepare large aluminium crystals. A fine-grained annealed specimen is critically strained by 1–2 per cent elongation in tension, then annealed at a gradually increasing temperature. In this manner the nucleation and growth of a few, and often only one grain, takes place; at the maximum temperature this grain absorbs all the slightly strained matrix. For successful application, the method requires careful control of several variables, including the initial grain size, critical strain, rate of heating during the growth anneal and the maximum temperature reached. Once the optimum conditions are deter-mined, the resulting crystals are usually of high quality and of a shape and surface finish which can be accurately predetermined. The method is largely confined to metals and alloys which do not show a marked tendency to form annealing twins, e.g. aluminium. Metals which twin heavily on slight plastic deformation are unsuitable. The method is useful for pro-ducing crystals which cannot be grown from the melt because of the occurrence of phase changes in the solid state, e.g. titanium, α-iron.

In recent years, the technique has been improved by annealing the strained specimens in a moving gradient furnace so that a crystal nucleates at one end and grows along the specimen as the furnace moves. In this way, single crystal rods of aluminium and some of its alloys can be grown up to 50 cm long. The temperature gradient needed for pure aluminium need not be greater than $20°C\,cm^{-1}$, but for some aluminium alloys gradients of $100°C\,cm^{-1}$ are necessary, while gradients of $1000°C\,cm^{-1}$ have been used for the growth of silicon iron crystals.

Some metals, e.g. molybdenum, tungsten and niobium, can be converted into single crystals by grain growth without prior strain using a method developed by Andrade.[9] A long, thin (1 mm diam.) wire is heated uniformly by a current, then a small furnace is passed along this wire, establishing a temperature gradient in which one grain can be made to grow at the expense of the others.

Metals which undergo phase transformations can in some cases be prepared in a single crystal or coarse-grained form by slow cooling through the transformation, e.g. iron, zirconium, uranium, titanium. In some cases repeated cycling through the change point must be resorted to. It is likely that stresses resulting from the volume change associated with the phase change play a significant role in these methods.

The phenomenon of secondary recrystallization* can be used to produce very large grains in sheet form. Heavily straight-rolled copper can be recrystallized in the range 773–973 K. If this material is then annealed near the melting point, secondary recrystallization takes place and a few large grains or even one grain replaces the original recrystallized grains. Single crystals of iron–nickel and iron–silicon alloys have also been prepared in this way.

Summarizing, it can be said that melt methods are of wider application and can be used to produce large crystals in relatively simple mould shapes. Control of orientation by use of seed crystals is relatively easy. On the other hand, melt methods are not always very satisfactory, particularly for alloy crystals as segregation occurs. Solid-state methods eliminate segregation and the shape of the crystal can be predetermined; however, seeding is difficult although not impossible. The main limitation is that many metals for one reason or another do not lend themselves to solid-state methods of crystal preparation.

2.3 Crystallographic nature of plastic deformation

The surface of a metal crystal which has been polished, then plastically deformed, becomes covered with one or more sets of parallel fine lines called slip lines. The early work of Rosenhain and Ewing showed that these were steps on the surface resulting from microscopic shear movements along well-defined crystallographic planes. Such crystallographic planes are referred to as *slip* or *glide planes*, while the direction of shear in the plane is called a *slip* or *glide direction*. The combination of a particular slip plane and a slip direction in that plane is referred to as a slip or glide

* See Chapter 11.

system. Figure 2.3 shows some typical slip bands on the surface of an aluminium crystal deformed lightly at room temperature.

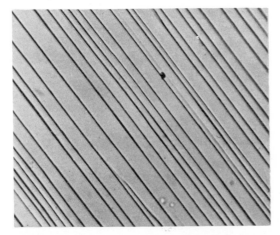

Fig. 2.3   Slip bands on an aluminium crystal deformed at room temperature.
× 250

In most metals the planes on which slip takes place are usually those with the closest atomic packing, while the slip direction is always the closest-packed direction in the slip plane. We can represent the closest-packed planes in a hexagonal or face-centred cubic crystal by a raft of uniform spherical balls in the closest packing arrangement (Fig. 2.4), in

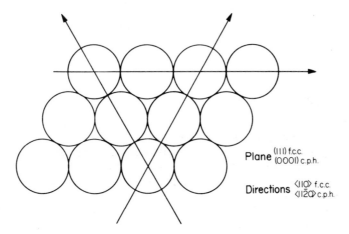

Plane   (111) f.c.c.
        (0001) c.p.h.

Directions   ⟨110⟩ f.c.c.
             ⟨112̄0⟩ c.p.h.

Fig. 2.4   The closest packed planes in face-centred cubic (f.c.c.) and close-packed hexagonal (c.p.h.) structures

which the three close-packed directions are apparent. The different symmetry of the two crystal structures arises from the order in which these close-packed planes are stacked. The close-packed hexagonal structure is created when the layers are arranged in the order *ABABAB* so that the third layer is in exactly the same relative position as the first. The face-centred cubic structure is formed when the layers are stacked in the order *ABCABCABC*, with the fourth plane in exactly the same relative position as the first.

Table 2.2 lists the common slip planes which have been detected in a number of familiar metals. In the close-packed hexagonal structure (Fig.

Table 2.2   Slip data for some metal crystals at room temperature

Metal	Structure	Slip plane	Slip direction	Critical resolved shear stress, $MN\,m^{-2}$ at RT (some typical values)	Purity, per cent
Aluminium	f.c.c.	{111}	$\langle 110 \rangle$	0·54–0·98	99·994
Copper	f.c.c.	{111}	$\langle 110 \rangle$	0·88–0·98	99·98
				0·34	99·999
Gold	f.c.c.	{111}	$\langle 110 \rangle$	0·49	99·999
Nickel	f.c.c.	{111}	$\langle 110 \rangle$	3·24–7·35	99·98
Silver	f.c.c.	{111}	$\langle 110 \rangle$	0·39–0·69	99·999
Cadmium	c.p.h.	(0001)	$\langle 11\bar{2}0 \rangle$	0·13	99·999
Magnesium	c.p.h.	(0001)	$\langle 11\bar{2}0 \rangle$	0·49	99·99
Zinc	c.p.h.	(0001)	$\langle 11\bar{2}0 \rangle$	0·29	99·999
Iron	b.c.c.	{110} {112} {123}	$\langle 111 \rangle$	14·71	99·96

2.5A) the basal plane (0001) is the closest-packed plane, and is the commonest slip plane in such hexagonal metals as zinc, cadmium and magnesium. Figure 2.5A illustrates the three close-packed slip directions $\langle 11\bar{2}0 \rangle$ in the basal plane, but for clarity the atoms are not shown touching. However, other slip systems can and do participate in the deformation of hexagonal metals, for example slip on pyramidal and prismatic planes has been observed frequently (see Chapter 4).

The face-centred cubic metals deform primarily on the close-packed octahedral {111} planes in the $\langle 110 \rangle$ close packed directions of which there are three in each {111} plane (Fig. 2.5B). As there are four different orientations of {111} planes, there is therefore a total of twelve possible slip

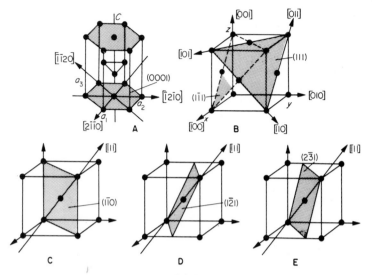

Fig. 2.5   A–E, Typical slip systems in close-packed hexagonal (A), face-centred cubic (B) and body-centred cubic structures (C–E)

Fig. 2.6   Slip lines in α-iron containing 3 per cent silicon deformed at room temperature, (1Ī0) face. (After Noble and Hull, 1965, *Phil. Mag.*, **12**, 777) × 100

systems which can take part in the deformation. All the familiar face-centred cubic metals behave in this way, and it is rare to find a slip plane other than one of the octahedral planes. Moreover, this behaviour persists in solid solutions of face-centred cubic structure. Slip on the cube plane has been observed occasionally in aluminium at high temperatures, and it also can occur at lower temperatures in some aluminium alloys.

The body-centred cubic metals present a somewhat more complicated picture, but in general they all slip in the closest packed *directions* $\langle 111 \rangle$. The choice of slip plane, however, varies greatly. Work on iron indicates that while the $\{110\}$ plane is a common slip plane, the $\{112\}$ and $\{123\}$ planes are also operative. Planes of these three types share a $\langle 111 \rangle$ direction as their zone axis; Fig. 2.5C–E illustrates one plane from each of the three systems, with a common $\langle 111 \rangle$ direction. The multiplicity of possible slip systems is sometimes reflected in the very wavy appearance of slip bands in a number of body-centred cubic metals (Fig. 2.6).

2.4   Anisotropy of plastic properties of crystals—geometry of slip

The orientation of a metal single crystal is an important variable. The principal axes of the crystal may make any angle with the artificially imposed surfaces, so that the axis of a crystal rod does not normally coincide with a crystallographic axis.

The usual way of designating the orientation of a crystal is to use the stereographic projection characteristic of the crystal structure of the metal concerned. This is a geometrical device to provide in two dimensions a plot in which the angular relationships between planes in the crystal are preserved and can be measured.* The normals from the various planes of a crystal are drawn to intercept a sphere, the surface of which could be used to measure angles, but in practice the sphere with its various intercepts is projected on to a plane surface to form a circular plot within which the various plane normals or poles fall. The angles between these are then measured with a circular stereographic net, which is subdivided into degrees and represents a projection of a sphere ruled to provide angular relationships.

The standard projection (Fig. 2.7) is of particular use in the representation of crystal structures and orientations. This type of projection is obtained by orienting the crystal with a plane of low indices in the plane of projection, e.g. a cube plane, so that the centre of the projection is then the normal to the cube plane, the [001] direction. In such a projection the symmetry of the crystal is fully revealed. In the case of a cubic crystal (simple cubic, face-centred cubic and body centred cubic alike), the projection is divided by intersecting great circles into 24 unit stereographic

* The stereographic projection is dealt with in most books on crystallography. The reader is referred to *The Structure of Metals*, 3rd edition by C. S. Barrett and T. B. Massalski for further details, McGraw-Hill, 1966.

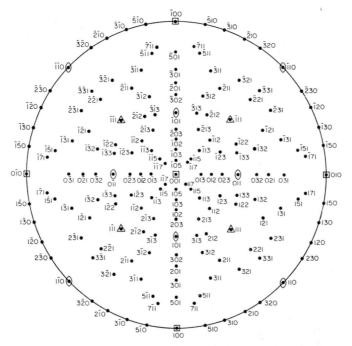

Fig. 2.7   Standard cubic projection. (After Barrett and Massalski, 1966, *The Structure of Metals*, McGraw-Hill, New York and Maidenhead)

triangles which are crystallographically identical (Fig. 2.8). In each case the three corners of the triangles represent equivalent directions, $\langle 001 \rangle$, $\langle 011 \rangle$ and $\langle 111 \rangle$ making always the same angles with each other. The triangles, of course, vary in shape because of the nature of the projection.

In the specification of a crystal orientation only one triangle need be used, and this is customarily the position of the [001], [011], [$\bar{1}$11] triangle in the centre of the projection. All possible orientations of crystals of cubic structure can then be specified by plotting the position of the specimen axis (e.g. tension or wire axis) within this triangle or along its boundaries. Consequently, in practice, if one wishes to represent the orientations of a number of single crystal rods, the angles between the rod axis and at least two of the [001], [011] and [$\bar{1}$11] directions are measured, and the axis plotted in the standard triangle using a stereographic net.

If crystals of a particular metal of widely differing orientations are deformed in tension, the stress–strain curves show large variations. The yield stress, ultimate tensile strength and elongation are all markedly anisotropic even in cubic crystals where some of the physical properties, e.g. electrical resistivity, are isotropic. For example, the yield stress in tension of zinc crystals varies by a factor of at least 6 as the orientation is changed; moreover, the crystals show elongations in a tensile test from a few per cent to several hundred per cent. While ordinary tensile tests reflect the inherent variation of plastic properties with crystallographic direction in metal crystals, it is obvious from the nature of the deformation process that a tensile stress is not the best way of measuring the stress in a process which occurs by shear, nor is the elongation in the direction of tension an appropriate measure for a strain which occurs along a well-defined set of crystallographic planes. The best solution to this dilemma would be to carry out all deformation tests in shear but, although this has

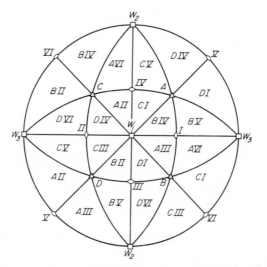

Fig. 2.8   Cubic stereographic projection showing 24 basic triangles. (After Schmid and Boas, 1935, *Kristallplastizität*, Springer Verlag, Berlin)

been done occasionally, it is usually much more convenient to do a tensile test. However, to enable the behaviour of crystals of different orientations to be compared, it is customary to resolve the tensile stress along the slip direction in the slip plane.

In Fig. 2.9 a crystal of cross-sectional area A has a tensile load of L imposed on it giving a tensile stress of σ_t. The slip plane is shown in which OX is the slip direction, while λ is the angle between the axis and the slip direction. The tensile axis makes an angle χ with the slip plane, so that the area of the slip plane is $A/\sin \chi$. Therefore, the tensile stress on the slip plane is

$$\frac{L}{A} \sin \chi = \sigma_t \sin \chi$$

and the shear stress on the slip plane *resolved* in the slip direction is

$$\tau = \sigma_t \sin \chi \cdot \cos \lambda = \sigma_t \cos \phi \cdot \cos \lambda \qquad \textbf{2.1}$$

where ϕ is the angle between the tension axis and the normal ON to the slip plane, and σ_t is the tensile stress. This equation shows that in some circumstances τ is zero, namely if the tension axis is normal to the slip plane so that $\lambda = 90°$, or if the tension axis is parallel to the slip plane when $\chi = 0°$. Thus deformation by slip would not be expected in these two extreme orientations, because the shear stress in the slip direction would be zero. On the other hand, the maximum shear stress is obtained when $\sin \chi \cos \lambda = 0.5$, that is when χ and λ are both $45°$. Then we have $\tau_{max} = 0.5 \sigma_t$.

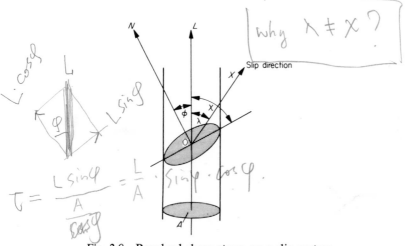

Fig. 2.9   Resolved shear stress on a slip system

2.5   Critical resolved shear stress for slip—Schmid's law

The early work of Schmid, Boas and collaborators[1] showed, particularly for the hexagonal metals such as cadmium, zinc and magnesium, that the tensile yield stress varied greatly with orientation. However, when the tensile yield stress is converted to a resolved shear stress using eqn. **2.1** above, it is found that the resulting shear stress τ_0 is a constant for a particular metal. In other words, crystals of a given metal will commence to deform plastically when the resolved shear stress on the slip plane in the slip direction reaches a constant critical value τ_0. This is usually referred to as Schmid's law. In Fig. 2.10 the theoretical curve for the equation

$$\tau_0 = \sigma_0 \sin \chi_0 \cos \lambda_0$$

where τ_0 = critical resolved shear stress

σ_0 = yield stress in tension

χ_0 = initial angle between slip plane and tension axis

λ_0 = initial angle between slip direction and tension axis,

is plotted, assuming a constant value of τ_0 for magnesium, giving an equilateral hyperbola with a minimum at $\sin \chi_0 \cos \lambda_0{}^* = 0.5$. Superimposed on this curve are the experimental points for the yield stress of magnesium crystals [10] as the function $\sin \chi_0 \cos \lambda_0$ is varied; the degree of

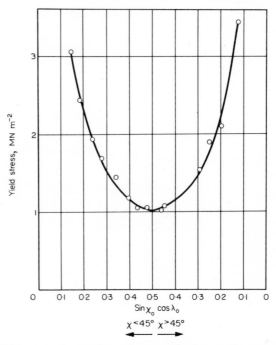

Fig. 2.10   Yield stress in tension of magnesium crystals as a function of orientation. (After Burke and Hibbard, 1952, *Trans. AIME*, **194**, 295)

correlation of the points with the curve shows how closely the Schmid law is obeyed. This correlation has been found for cadmium, zinc and magnesium by various workers,† and thus the concept of a constant critical resolved shear stress for slip is well established. In Fig. 2.11 is plotted data for 99.999 per cent zinc [11] where the curve is drawn for $\tau_0 = 0.18 \, \text{MN m}^{-2}$; again the experimental points are accurately on the curve.

* Usually referred to as the Schmid factor.
† The earlier work is summarized in General Reference 1.

The data for face-centred cubic and body-centred cubic metals is less conclusive, partly because the multiplicity of possible slip systems does not allow the law to be tested over a very wide range of orientations. Work[12] on copper has indicated that orientations towards the centre of the stereographic triangle give nearly constant values of τ_0, but this is no longer true

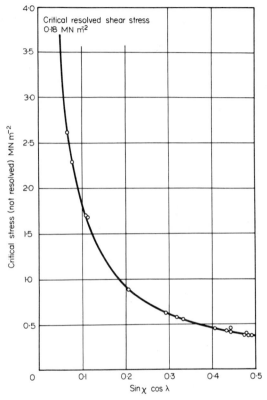

Fig. 2.11   Orientation dependence of yield stress of zinc crystals (99·999 per cent). (After Jillson, 1950, *Trans. AIME*, **188**, 1129)

for orientations approaching the boundaries of the triangle (Fig. 2.12) where the operation of other slip systems becomes more likely.

Despite these uncertainties, τ_0 represents a fundamental mechanical property, because it is so closely related to the basic mode of plastic deformation by shear along the slip planes. With a reproducible experimental technique, τ_0 remains more or less constant for crystals of one metal of given purity, and so it presents a starting point to determine the effect on mechanical properties of such variables as temperature, concentration of alloying element, rate of strain, etc. Some of the more recent values of τ_0

determined on metals of high purity are listed in Table 2.2 (p. 12) which illustrates the very low values which have been obtained. These results have posed the question whether in fact crystals of very high purity exhibit any elastic behaviour at all. However, work with even the highest purity metals now shows that there is always an elastic limit, although this can occur at extremely low stresses.

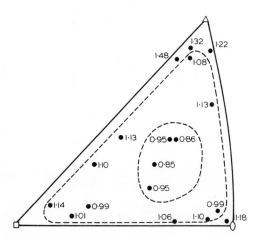

Fig. 2.12 Orientation dependence of τ_0 of copper crystals. Values in $MN\,m^{-2}$ (After Diehl, 1956, *Z. Metallk.*, **47,** 331)

2.6   Effect of variables on the critical shear stress

It has already been mentioned that τ_0 is very structure-sensitive. The most significant variable is undoubtedly the purity of the metal crystals used for the determination. Rosi[13] has determined τ_0 for silver crystals at room temperature, and found average values of 0·47, 0·72 and 1·28 $MN\,m^{-2}$ for 99·99 per cent, 99·97 per cent and 99·93 per cent silver respectively. Cadmium of 99·999 per cent purity has been shown to have a τ_0 of 0·1–0·17 $MN\,m^{-2}$, whereas material of 99·99 per cent purity has a τ_0 of about 0·59 $MN\,m^{-2}$. This large effect of impurities on τ_0 will be examined further when the strength of solid solutions is considered in Chapter 6.

The critical shear stress is also very dependent on the temperature at which it is measured, particularly if this temperature is low with respect to the melting point. Figure 2.13 shows two groups of results for magnesium crystals over the range 100–600 K in which there is no significant decrease of τ_0 above 300K, whereas a 100 per cent increase occurred between this temperature and 100K. The difference between the results of Schmid and Siebel[1] and those of Burke and Hibbard[10] can be attributed to different impurity levels. Aluminium crystals of 99·996 per cent purity show a marked dependence of τ_0 on temperature below 200 K, but not a very great dependence above this temperature, while similar results have been obtained on crystals of copper.[14]

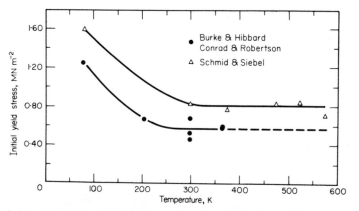

Fig. 2.13   Temperature dependence of τ_0 for magnesium. (After Conrad and Robertson, 1957, *Trans. AIME*, **209**, 503)

The rate of strain has also an influence on the value of τ_0, which increases with increasing strain rate. Cadmium crystals deformed at room temperature gave values of τ_0 from 0·20 MN m^{-2} at a strain rate of less than 10^{-2} s^{-1} to 0·44 MN m^{-2} at a strain rate of 10^{-1} s^{-1}. Rapid strain rates suppress time-dependent components of the deformation (creep), which would otherwise lead to plastic deformation at lower stresses.

2.7   Determination of shear strain

The shear strain is likewise a more precise measurement than the percentage elongation; it is defined as the relative displacement of two slip planes of unit distance apart. During the slip process a geometrical transformation occurs, with the result that a crystal originally of circular cross-section becomes oval as the crystal lengthens (Fig. 2.14A and B). This transformation results in the slip direction rotating towards the tensile axis, shown diagrammatically in Fig. 2.15, where the lattice position has been kept constant while the axis of the crystal has moved. This diagram permits several useful relationships to be calculated.

If l_0 and l_1 are the lengths of the crystal before and after a given deformation, and λ_0 and λ_1 are the angles between the axis and the slip direction before and after the deformation, then the following relation holds for triangle *ABB′*

$$\frac{l_1}{l_0} = \frac{\sin \lambda_0}{\sin \lambda_1} \qquad\qquad \textbf{2.2}$$

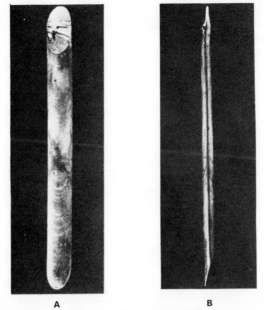

A   **B**

Fig. 2.14   Deformation of a zinc crystal in tension. Front and side view. (After Parker and Washburn)

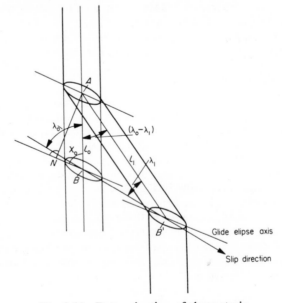

Fig. 2.15   Determination of shear strain

In the triangles ABN and $AB'N$

$$AN = l_0 \sin \chi_0 = l_1 \sin \chi_1 \qquad\qquad \textbf{2.3}$$

where χ_0 and χ_1 are the angles between the slip plane and the axis before and after deformation.

From triangle ABB'

$$BB' = \frac{l_1[\sin (\lambda_0 - \lambda_1)]}{\sin \lambda_0}$$

Now the *shear* or *glide* strain is $BB'/AN = \epsilon$, so

$$\epsilon = \frac{l_1}{l_0 \sin \chi_0} \cdot \frac{\sin (\lambda_0 - \lambda_1)}{\sin \lambda_0} \qquad\qquad \textbf{2.4}$$

eliminating λ_1 using **2.2** and **2.3** above.

$$\epsilon = \frac{1}{\sin \chi_0} \left(\sqrt{\left(\frac{l_1}{l_0}\right)^2 - \sin^2 \lambda_0} - \cos \lambda_0 \right) \qquad\qquad \textbf{2.5}$$

So the shear strain can be found if the initial orientation of the glide elements (χ_0 and λ_0), and the amount of extension is known. If, however, the final positions of the glide elements are known, then ϵ can be determined thus:

$$\epsilon = \frac{\cos \lambda_1}{\sin \chi_1} - \frac{\cos \lambda_0}{\sin \chi_0} \qquad\qquad \textbf{2.6}$$

2.8   Stress–strain curves of metal crystals

The same arguments used to justify the use of the critical resolved shear stress τ_0 show that the whole stress–strain curve of a crystal is best plotted in terms of the resolved shear stress and the shear or glide strain which has been defined in the previous section. This procedure reduces, but does not eliminate, the scatter of curves obtained from crystals of different orientations. When the stress–strain curves of a number of metals are plotted in this way (Fig. 2.16) certain distinctions can be made between, on one hand, the face-centred cubic metal crystals and on the other, the familiar hexagonal metals cadmium, zinc and magnesium. Figure 2.16 shows that in all cases the resolved shear stress increases as deformation proceeds. This is the phenomenon of *work* or *strain hardening* which is fundamental to the deformation of crystals. The rate of work hardening of the face-centred cubic metal crystals is clearly much greater than that of the hexagonal metals cadmium, zinc and magnesium, although the differences are less pronounced if metals of similar melting point are compared, e.g. aluminium and magnesium. The hexagonal metals are capable of deforming to very large shear strains, but only in the case of suitably oriented crystals. To understand these differences we must first look more closely at the geometrical aspects of slip in face-centred cubic and close-packed hexagonal crystals.

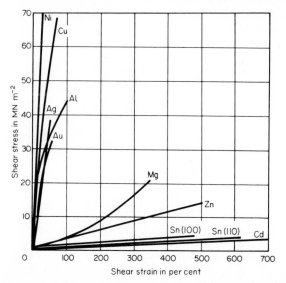

Fig. 2.16   Shear stress/shear strain curves of metal crystals. (After Schmid and Boas, 1935, *Kristallplastizität*, Springer Verlag, Berlin)

2.9   Hexagonal metals—geometrical considerations

In this chapter, the 'ideal' behaviour of hexagonal metals as exemplified by cadmium, zinc and magnesium will be considered. At room temperature crystals of these metals deform essentially by basal slip, so only one slip system operates for a large part of the deformation. The stress–strain curve is very much influenced by the orientation of the basal plane relative to the tension axis prior to deformation, which determines to a large extent the plastic strain the crystal will undergo before fracture.

In Section 2.7 it was shown that in a crystal pulled in tension the slip plane rotates in such a fashion that the slip direction approaches the tension axis. The greater the possible rotation of the slip plane, the greater will be the resulting plastic strain. Clearly a crystal oriented as in A, Fig. 2.17 will give the largest plastic strain, while a crystal of orientation C will be unable to slip on the basal plane at all; orientation B will have the maximum resolved shear stress and undergo moderate shear strains. As we have seen, the ratio of the length of the crystal after deformation (l_1) to the original length (l_0) is defined in terms of the angles χ_0 and χ_1,

$$\frac{l_1}{l_0} = \frac{\sin \chi_0}{\sin \chi_1}$$

from which it can be seen that a crystal of suitable orientation (i.e. large χ_0) can elongate several hundred per cent before χ_1 reaches a limiting low value. The most striking geometrical consequence of large plastic strains on one glide system is that a crystal with an originally circular cross-section gradually becomes elliptical in cross-section (Fig. 2.14), eventually forming at very large strains a thin ribbon. It is not unusual for zinc and cadmium crystals of suitable orientations to sustain elongations of 200–400 per cent at room temperature, and even greater strains at elevated temperatures.

It is implicit in the relationship for the resolved shear stress (eqn. **2.1**), whether at the beginning of plastic deformation (τ_0) or at any other stage (τ), that the applied tensile stress will vary between wide limits for different crystals at the same stage of the deformation. This means that for extreme values of χ, a high tensile stress is necessary to reach the required resolved shear stress in the unfavourably oriented basal plane (e.g. Fig. 2.17A and C). In these circumstances, other slip processes on pyramidal or prismatic

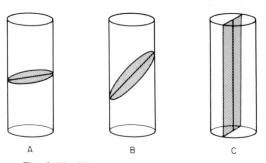

A   B   C

Fig. 2.17   Three orientations of basal plane

planes,† or twinning,† frequently occur, and the behaviour of such crystals can then no longer be related simply to that of crystals deforming solely by basal slip. In general, 'ideal' behaviour can be expected in crystals with χ_0 between 10° and 80°.

A further complication arises from the rotation of the slip plane during plastic deformation. If, for the moment, we ignore the phenomenon of work hardening and assume that a crystal continues to deform at the critical resolved shear stress τ_0, we can write an equation for the stress–strain curve, using eqns. **2.1** and **2.2**.

$$\sigma_t = \frac{\tau_0}{\sin \chi_0} \cdot \frac{1}{\cos \lambda_0} = \frac{\tau_0}{\sin \chi_0} \cdot \frac{1}{\sqrt{1 - \dfrac{\sin^2 \lambda_0}{d^2}}} \quad \text{where} \quad d = \left(\frac{l_1}{l_0} \right) \quad \textbf{2.7}$$

If we make the simplifying assumption that $\lambda_0 = \chi_0$, a series of stress–strain curves can be determined for different values of χ_0, the angle

† See Chapters 4 and 8.

between the slip plane and the tension axis (Fig. 2.18). These curves of course reflect the anisotropy of the tensile yield stress, but also they show that deformation beyond this point occurs often at a reducing stress, the larger χ_0 the greater the drop in stress. The very big stress drops between $\chi_0 = 45°$ and $\chi_0 = 80°$ are readily understood, because a crystal with the basal plane initially nearly normal to the stress axis will require a high tensile stress to reach τ_0 on the slip plane. However, when deformation commences the basal plane will rotate to increasingly favourable positions where the resolved shear stress is higher, so the tensile stress needed to continue deformation will be lower. In practice these stress drops are often hidden by the occurrence of work hardening; however, they are observed in hexagonal crystals of low melting point (low work hardening at room temperature) of the appropriate orientations. The phenomenon is called

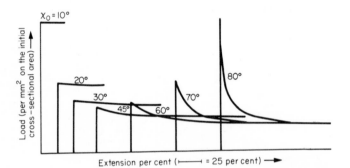

Fig. 2.18   Theoretical stress–strain curves. $\lambda_0 = \chi_0$ is assumed. (After Schmid and Boas, 1935, *Kristallplastizität*, Springer Verlag, Berlin)

'geometrical softening' because it arises from the changing geometry of the crystal, and does not reflect a structural softening of the metal. It is also encountered in alloy single crystals which frequently show low work hardening following a high yield stress.

A large degree of scatter in the stress–strain curves from crystals of different orientations can be eliminated by using resolved shear stress and shear strain, but unlike the critical resolved shear stress τ_0, the resolved shear stress τ at a strain ϵ is not the same for all crystals. This arises primarily from differences in rates of work hardening, which, in turn, depend on structural changes within the metal. However, if we exclude extreme orientations, i.e. very small and very large χ_0, it is possible to approximate to the behaviour of most crystals of a particular metal by a single resolved shear stress/shear strain curve.

The classical work of Schmid and Boas provides much data on the shear stress/shear strain curves of the hexagonal metals, zinc, cadmium and magnesium which illustrate the main features:

1. Large glide strains (for the appropriate orientations).
2. Low linear rates of hardening (at room temperature and above) over a large part of the stress–strain curve.
3. A marked temperature dependence of the stress–strain curves.

Figure 2.19 gives data for cadmium which illustrates these three points. The pronounced effect of the deformation temperature should be particularly noted.

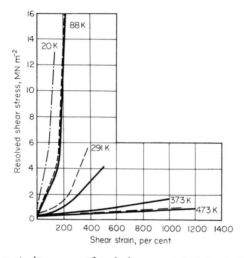

Fig. 2.19   Stress–strain curves of cadmium crystals. The dashed curves were obtained at strain rates 100 times greater than those for the full curves. (After Schmid and Boas, 1935, *Kristallplastizität*, Springer Verlag, Berlin)

2.10   Face-centred cubic crystals—geometrical considerations

The deformation behaviour of face-centred cubic crystals differs from the 'ideal' behaviour of the hexagonal metals zinc and cadmium in so far as there is a much wider choice of slip system, and sooner or later more than one slip system will participate in the deformation.

The system on which deformation commences is called the *primary* system, and this is always that system on which the resolved shear stress is the highest. When the orientation of a crystal is known, the χ_0 and λ_0 values for all the twelve possible slip systems can be measured on a stereographic projection, and thus the Schmid factor ($\sin \chi$. $\cos \lambda$) can be calculated. The operative slip system will have the highest Schmid factor. The system chosen will thus depend on the orientation of the crystal relative to the stress axis.

The relationship between the stress axis and the twelve possible slip systems is best shown on a stereographic projection, where each of the unit triangles defines a region in which a particular slip system operates (see Fig. 2.8). There are four $\langle 111 \rangle$ poles $ABCD$ representing the normals to the octahedral slip planes and six slip directions I to VI. Taking the usual standard triangle $W A$I, we see that the system BIV operates within its boundaries, that is the slip plane of orientation B and the slip direction IV will operate during tensile deformation if the tension axis lies in this triangle. The stereographic projection allows the rotation of the crystal to be followed during deformation. In a tensile test the slip direction tends to rotate towards the tension axis; however, in the projection the tension axis can be considered to rotate towards the slip direction IV. Figure 2.20 shows the sense of the rotations for a range of orientations

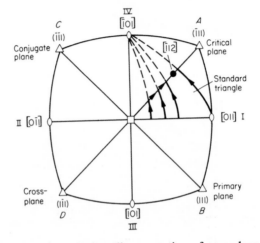

Fig. 2.20   Rotations during slip; operation of secondary systems

within the standard stereographic triangle. However, the rotation sooner or later brings the tension axis to the boundary $W_1 A$ between two stereographic triangles in the second of which an entirely new slip system CI operates (see Fig. 2.8). Normally at this point, deformation proceeds on the two slip systems simultaneously to produce *duplex* or *conjugate* slip. As Figs. 2.8 and 2.20 indicate, the system $(\bar{1}\bar{1}1)$ [011] is the conjugate slip system, while (111) [$\bar{1}$01] is the primary system. The duplex slip causes further movement of the specimen axis along the [001]–[$\bar{1}$11] boundary towards the [$\bar{1}$12] pole which is mid-way between the two operative slip directions [$\bar{1}$01] and [011], and lies on the great circle connecting them. When the stress axis reaches this orientation, it is maintained until localized necking down takes place, followed by fracture. Thus the onset of duplex slip, by interrupting the rotation of the crystal axis towards the slip direction, leads to much smaller extensions in cubic crystals than in hexagonal crystals, where the axis rotates to within a few degrees of the slip direction by slip on one system. A quantitative expression of the difference is easily obtained by use of eqn. **2.3** above. In cubic crystals glide strains rarely exceed 100 per cent, in marked contrast to the behaviour of zinc, cadmium and magnesium crystals.

So far, we have discussed the behaviour of crystals, the axes of which are within the stereographic triangle. Crystals with axes lying on the boundaries of the triangle are in a special category because the critical resolved shear stress will be the same on more than one slip system, so plastic deformation will commence on more than one slip plane. Figure 2.21 summarizes the number of operative slip systems to be expected,

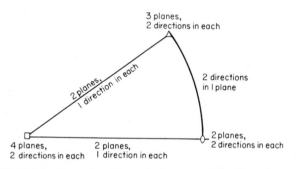

Fig. 2.21   Operative slip systems at special orientations

information which can be deduced also from Fig. 2.8. The most complicated case is the [001] orientation, where no less than four slip systems are equally favoured—actually Fig. 2.8 indicates that eight systems are predicted at the centre of the projection; but this represents four planes each with two slip directions, only one of which can be used at a time. The multiple slip occurring in a cube-oriented crystal of aluminium is shown in Fig. 2.22.

The capacity of cubic crystals to deform on more than one slip system at a time is closely connected with the higher rates of work hardening found in these metals than in metals of hexagonal crystal structure. This important matter is more fully explored in Chapter 4.

2.11   Stress–strain curves of face-centred cubic crystals

The classical work of Taylor and Elam[2] showed that the shear stress/ shear strain curves of face-centred cubic metal crystals, in particular aluminium, were approximately of a parabolic form given by the equation

$$\tau = k\epsilon^{1/2}$$

τ and ϵ being the shear stress and shear strain respectively. The points from many crystals tested in tension fell on the same general curve; furthermore, crystals tested in compression also lay on the same curve, whereas the normal stress varies for the two cases.

In more recent investigations carried out with purer metals, the stress–strain curves have been shown to deviate markedly from a simple parabola, and it is now recognized that there are three well-defined regions of hardening during the deformation of face-centred cubic metal crystals[15]

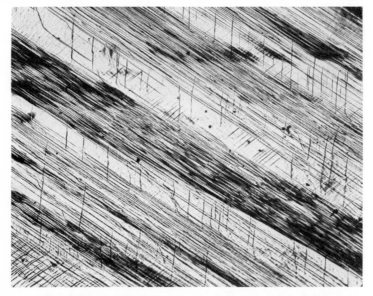

Fig. 2.22   Multiple slip in an aluminium crystal. × 100.

(Fig. 2.23). Stage 1 or '*easy glide*'[16] is a region of low linear hardening which resembles the behaviour of zinc and cadmium up to moderate strains, while Stage 2 is a second linear region with a much greater rate of work hardening, which is frequently interrupted by the early development of Stage 3. This stage represents a period of decreasing rate of hardening, and as it is dominant in aluminium at room temperature, is probably closest to the parabolic hardening curves obtained by Taylor and Elam on less pure aluminium. Work in recent years has shown that with pure metal crystals it is not possible to represent the deformation behaviour by a unique stress–strain curve. For different metals the relative importance of the three stages of hardening changes, and even for a single metal there are important variables which must be considered.

While the different stages of hardening arise from structural changes in the metal during deformation, there is no correlation between the onset for one or other of the stages and the movement of the stress axis to the [001]–[$\bar{1}$11] boundary of the stereographic triangle. When duplex slip commences no discontinuities are obvious in the stress–strain curves. If, however, a 'symmetrical' orientation, e.g. [001], is chosen, from the start the crystal work hardens much more rapidly than crystals with axes towards the middle of the triangle.

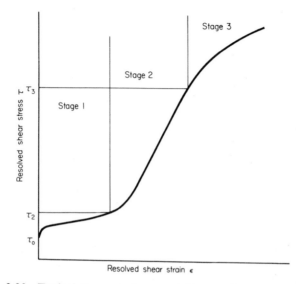

Fig. 2.23   Typical stress–strain curve of a pure f.c.c. metal crystal

General references

1. SCHMID, E. and BOAS, W. (1935). *Kristallplastizität*, Springer Verlag, Berlin, original German edition. English translation published by F. A. Hughes & Co., 1950. *Plasticity of Crystals.*
2. ELAM, C. F. (1936). *The Distortion of Metal Crystals.* Oxford University Press, London.
3. SEEGER, A. (1955). *Handbuch der Physik III.* Springer-Verlag, Berlin. pp. 7, 383.
4. LAWSON, W. D. and NIELSON, T. (1958). *Preparation of Single Crystals.* Butterworth, London.
5. JAOUL, B. (1965). *Étude de la Plasticité et Application aux Métaux.* Dunod, Paris.
6. STANFORD, E. G., PEARSON, J. H. and MCGONNAGLE, W. J. (1964). *Progress in Applied Materials Research.* Vol. 6, article by T. E. MITCHELL on 'Dislocations and Plasticity in Single Crystals of Face-Centred Cubic Metals and Alloys'.

References

7. HONEYCOMBE, R. W. K. (1959). *Metall. Rev.*, **4** (13), 1.
8. PFANN, W. F. (1957). *Metall, Rev.*, **2**, 29.

9. ANDRADE, E. N. DA C. (1937). *Proc. R. Soc.*, **A163**, 16.
10. BURKE, E. C. AND HIBBARD, W. R. (1952). *Trans. AIME*, **194**, 295.
11. JILLSON, D. C. (1950). *Trans. AIME*, **188**, 1129.
12. DIEHL, J. (1956). *Z. Metallk.*, **47**, 331, 411.
13. ROSI, F. D. (1954). *Trans. AIME*, **200**, 1009.
14. GARSTONE, J. and HONEYCOMBE, R. W. K. (1957). *Dislocations and Mechanical Properties of Crystals.* John Wiley, New York and London. p. 391.
15. DIEHL, J., MADER, T. and SEEGER, A. (1956). *Z. Metallk.*, **46**, 650.
16. ANDRADE, E. N. DA C. and HENDERSON, C. (1951–52). *Phil. Trans. R. Soc.*, **244**, 177.

Additional general references

1. GILMAN, J. J. (1969). *Micromechanics of Flow in Solids.* McGraw-Hill, New York.
2. KELLY, A. and GROVES, G. W. (1970). *Crystallography & Crystal Defects.* Longman, London.
3. REID, C. N. (1973). *Deformation Geometry for Materials Scientists.* Pergamon Press, Oxford.
4. PAMPLIN, B. (Ed.) (1980). *Crystal Growth.* International Series in the Science of the Solid State, Vol. 16, Pergamon Press, Oxford.

Chapter 3

Elementary Dislocation Theory

In this chapter the theoretical strength of crystals will be shown to be several orders of magnitude greater than the strengths of real crystals discussed in Chapter 2. The large discrepancy is explained by the presence in most crystals of atomic defects called dislocations, for the existence of which there is now overwhelming evidence.

3.1 The theoretical strength of a crystal

Frenkel first estimated in a simple way the theoretical shear strength of a crystal by taking for a model two rows of atoms subjected to a shear stress (Fig. 3.1), in which the interplanar spacing is a, while the interatomic distance in the slip direction is b. Under a shear stress τ, the atomic rows will move bodily over each other with equilibrium positions at points such as A and B where the shear stress necessary to maintain the configuration is zero. Likewise, it will be zero when the rows are exactly on top of each other at positions C and D. In between these positions the stress will have a finite value which clearly varies in a periodic way throughout the lattice. Now if the displacement is x for a shearing stress τ, the stress will be a periodic function of x with period b. The simplest assumption is a sinusoidal relationship (Fig. 3.1)

$$\tau = k \sin \left(\frac{2\pi x}{b}\right) \qquad \textbf{3.1}$$

For small displacements

$$\tau = k \frac{2\pi x}{b}$$

Applying Hooke's law, another expression for τ is

$$\tau = \frac{Gx}{a}$$

where G is the shear modulus and x/a the shear strain.

So equating,

$$k = \frac{Gb}{2\pi a}$$

and using the value for k in eqn. **3.1**,

$$\tau = \frac{Gb}{2\pi a} \sin\left(\frac{2\pi x}{b}\right) \qquad\qquad \textbf{3.2}$$

The maximum value of τ which represents the stress at which the lattice is rendered unstable is assumed to occur at a displacement of $b/4$, so

$$\tau_{max} = \frac{Gb}{2\pi a} = \tau_0 \qquad\qquad \textbf{3.3}$$

where τ_0 is the critical shear stress.

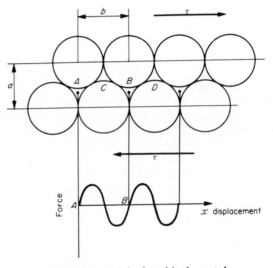

Fig. 3.1   Model of an ideal crystal

It can be assumed that $a \simeq b$ so the theoretical critical shear stress $\simeq G/2\pi$. For copper crystals $G = 45\,110$ MN m^{-2}, so that the theoretical value of τ_0 is ~ 7200 MN m^{-2} compared with ~ 1 MN m^{-2} for actual crystals (see Table 2.2). It is thus clear that the theoretical value is several orders of magnitude too high.

The first reaction to such a discrepancy is that the analysis is wrong, but closer examination shows that although simplifying assumptions yield only an approximate answer, the overall deductions are correct. The calculation has been refined mainly by using a more realistic relationship for the periodic variation of τ; furthermore, the fact has been taken into account that in actual close-packed metal structures stable positions other than A and B can occur, e.g. the twinning configuration. However, even when these factors are taken into consideration, τ_0 can only be reduced to approximately $G/30$, which is still orders of magnitude too high.

The inevitable conclusion from such theoretical work was that the simple model used did not represent the behaviour of real crystals, which must in fact contain defects which reduce the mechanical strength. As early as 1921 Griffith postulated the presence of microscopic cracks to explain the relative weakness of brittle solids, such as glass, where the fracture stress fell well below that predicted theoretically. However, it was not until 1934 that Polanyi,[7] Orowan[8] and Taylor[9] introduced independently the concept of a dislocation in a crystalline solid. The dislocation represents a line defect or discontinuity between part of a crystal which has sheared and a part which has not, so the deformation occurs by passage of such dislocations along the slip plane, and not by unified shear over the whole crystal simultaneously.

3.2   Properties of simple dislocations

3.2.1. *The edge dislocation*

A simple dislocation model can be produced by cutting a slit $ABCD$ in an elastic solid block (Fig. 3.2a) which is made to end in the interior of the block at AB. Material on one side of the slit is then displaced, giving a slip step $CDEF$. The line AB representing one end of the slit also now represents the boundary between deformed and undeformed material. It thus defines the point of emergence of a dislocation line on to the surface of the block. In a simple cubic crystal lattice a dislocation of this type has the structure shown in the model of Fig. 3.3 where the dislocation line is shown at A, having moved half way through the crystal leaving a slip step one interatomic spacing high at B. This type of dislocation is known as an *edge dislocation* which, in the simple cubic lattice, is characterized by an extra half plane of atoms inserted in the crystal. This half plane can either be above the slip plane, in which case the convention is to call it a *positive* edge dislocation, or below the slip plane when the dislocation is a *negative* one. Positive and negative dislocations have to move in opposite directions along the slip plane to produce a shear of the same sign.

To define more precisely the edge dislocation, it is necessary to introduce the Burgers vector, which describes the displacement direction or slip direction characterizing a particular dislocation. It is rigorously determined by means of a Burgers circuit, which is an atomic path involving two lattice directions normal to each other, which is then retraced, using vectors of the same strength but opposite sign. If carried out in a perfect lattice (Fig. 3.4A) the circuit is a closed rectangle, the original point of departure being reached. If, however, the circuit incorporates an edge

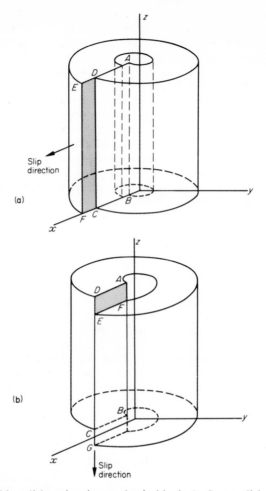

Fig. 3.2 **a,** Edge dislocation in an elastic block. **b,** Screw dislocation in an elastic block

dislocation, there will be a closure failure, the magnitude and direction of which (Fig. 3.4A) defines the Burgers vector *b* of the dislocation.

Returning to the edge dislocation of Fig. 3.3, it is clear that the Burgers vector is parallel to the slip direction and corresponds with the slip vector which is one lattice spacing in this case. Furthermore the dislocation line is *normal* to the Burgers vector, an important characteristic of edge dislocations. In this case the Burgers vector is of unit strength, but it will be seen at a later stage that when the lattice adopts intermediate positions of low energy during slip, the crystallographic configuration changes, with the

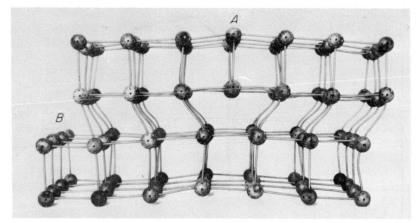

Fig. 3.3   Ball and wire model of an edge dislocation. (After Bilby, 1950, *J. Inst. Metals*, **76**, 613)

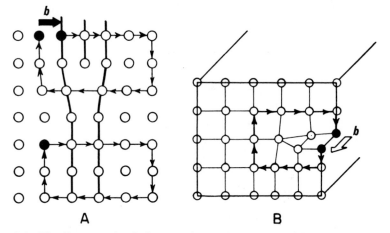

Fig. 3.4   The Burgers circuit in, **A**, edge and, **B**, screw dislocations. (After Friedel, 1964, *Dislocations*, Pergamon Press, Oxford.)

result that partial dislocations are formed with Burgers vectors which are not simple lattice vectors (see Section 3.17).

An important feature of the edge dislocation is that it is the centre of an internal stress field, the nature of which is suggested by the model† in Fig. 3.3. The material each side of the dislocation line is clearly in shear,

† These models are constructed from wooden balls and curtain wire. Many aspects of dislocations are easily grasped by forming and moving dislocations in such models.

while breaking of a bond above the slip plane will demonstrate that a state of compression exists there. Under the dislocation there is a dilation so a state of tension exists. The stress fields of dislocations are extremely important as they influence the interactions between dislocations and thus play a significant role in determining the plastic properties of metals.

3.2.2. *The screw dislocation*

The second basic type of dislocation is the screw dislocation, the nature of which can also be demonstrated macroscopically by shearing an elastic block with a slit *ABCD* in it (Fig. 3.2b) which plays the part of the slip plane. In this case the block is sheared at right angles to the direction in the block demonstrating an edge dislocation, so that the region *ADEF* is the slip step. *AB* represents the line of the screw dislocation from which it can be seen that the slip direction is *parallel* to the dislocation line, i.e. the direction of the dislocation line and the Burgers vector coincide. On the other hand, during deformation the dislocation line moves in a direction at right angles to the slip direction, i.e. parallel to *DA*.

The atomic structure of the screw dislocation is best seen in a ball and wire model of a simple cubic lattice (Fig. 3.5), where a screw dislocation is

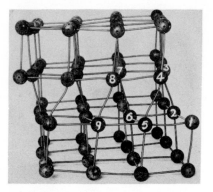

Fig. 3.5   Ball and wire model of a screw dislocation. (After Bilby, 1950, *J. Inst. Metals*, **76**, 613)

shown at the edge of the model, the front of the model having slipped while the rear is still undeformed. The strain pattern around the dislocation is quite different from that of an edge dislocation. There is no extra half plane of atoms, and by introduction of the screw dislocation, the lattice has been changed from a set of discrete planes to a continuous helicoidal surface. This is shown in Fig. 3.5, where if an atomic path is followed around the dislocation 1 to 2 to 3 to 4, etc., a circuit commencing at one surface will eventually emerge on the opposite face of the model. As with edge dislocations, screw dislocations can be of two signs; the screw may be left-handed or right-handed, and the shear resulting from the movement in the same direction of these two types will be in opposite directions. In Fig. 3.4B the Burgers vector of a screw dislocation is defined by the Burgers circuit, which shows that the vector is of the same strength but normal to the Burgers vector of an edge dislocation similarly oriented in a simple cubic material.

A consequence of the absence of the extra half plane of atoms is that the screw dislocation is less restricted in its motion than an edge dislocation. If an edge dislocation wishes to move above or below its present slip plane (Fig. 3.3), then it must do so either by shedding or gaining atoms on the extra half plane. This type of movement is called non-conservative, as distinct from the conservative movement which occurs when a dislocation moves along its slip plane. Non-conservative movement clearly involves diffusion of atoms either to or away from the dislocation, and is thus a process which requires thermal activation. On the other hand, screw dislocations are able to move readily on any cylindrical surface whose axis corresponds with the slip direction, i.e. it can move out of one slip plane into another adjacent one provided it possesses a common slip direction.

3.3   Dislocation loops

So far only straight dislocation lines have been introduced, each end emerging on a crystal face, however it is possible to envisage a limited region of deformation *ABCD* entirely within a crystal. This would be separated from the undeformed material by a dislocation line in the form of a ring or loop (Fig. 3.6). The ring must comprise a mixture of edge and screw dislocation to a varying degree, the line being pure edge at *B* and *D* and pure screw at *A* and *C*. This can be readily demonstrated in the ball model by introducing a rectangular region of slip in the interior whereupon

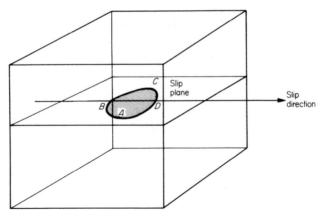

Fig. 3.6   Dislocation ring

it becomes evident that two opposite sides of the rectangle are edge dislocations of opposite sign, and the other two sides are screw dislocations
of opposite sign.

Any arbitrary dislocation line is readily resolved into its edge and screw
components. If AB is a dislocation line which makes an angle θ with its
Burgers vector b, then it can be regarded as the sum of two dislocations
with Burgers vectors b_1 and b_2 (Fig. 3.7)

$$b_1 = b \sin \theta, \quad \text{edge component}$$
$$b_2 = b \cos \theta, \quad \text{screw component}$$

and
$$b = b_1 + b_2$$

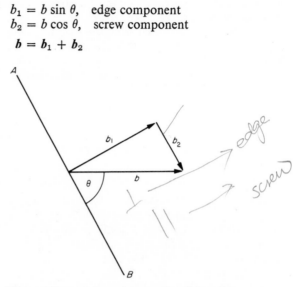

Fig. 3.7   Edge and screw components of a dislocation

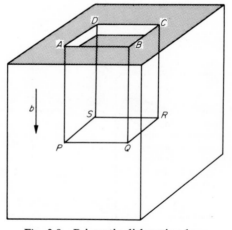

Fig. 3.8   Prismatic dislocation loop

that is, the Burgers vector of the dislocation line AB is the sum of the Burgers vectors of the component dislocations. It is, however, possible to have a dislocation loop which is entirely edge. This is shown in Fig. 3.8, which represents the formation of a prismatic dislocation by indenting a crystal with a prismatic punch at $ABCD$. If the slip direction is AP, the four segments of dislocation PQ, QR, RS, SP, which represent the boundary between sheared and unsheared crystal, are normal to the slip direction, and are thus edge in character. It is clear that in these circumstances an edge dislocation line of various shapes can be obtained by varying the form of the punch, but of course the line will not be confined to one slip plane, and only the slip direction will remain unchanged.

3.4   Force on a dislocation

The result of applying an external stress to a crystal is to cause a force to act on the dislocations within the crystal. This has been determined by Mott and Nabarro[11] as follows.

Consider a rectangular slip plane l_1 wide, l_2 long, to which a shear stress τ is applied. The force on the slip plane is $\tau_1 l_1 l_2$. Now if a dislocation line, l_1 long, with a Burgers vector b, moves from one end of the slip plane to the other, i.e. a distance l_2, a displacement b occurs. So the work done by the slip process is $\tau \cdot l_1 \cdot l_2 \cdot b$.

If the force/unit length on the dislocation line is F, the total force on the dislocation line is $F \cdot l_1$, and the work done by the dislocation in moving a distance l_2 is $F \cdot l_1 \cdot l_2$.

The two expressions for the work done, the one considering the slip plane as a whole, and the other only the dislocation, must be equal so

$$F \cdot l_1 \cdot l_2 = \tau \cdot l_1 \cdot l_2 \cdot b$$

i.e. $$F = \tau \cdot b \qquad\qquad 3.4$$

Thus the force on a dislocation line is the product of the shear stress and the Burgers vector. This force acts along the slip plane normal to the dislocation, whatever its configuration (e.g. complex loop) may be, and the force is directed towards the material through which the dislocation has not yet passed. It is analogous to the pressure exerted by a gas on the walls of a container.

3.5   Stress to move a dislocation

The movement of a dislocation line along a slip plane of a crystal implies that the interatomic forces across the plane are overcome in a series of local movements determined by the periodic stress field of the lattice. This is in marked contrast to a macroscopic shear process where all the bonds are severed simultaneously. Intuitively it is evident that a dislocation will accomplish the overall shear deformation by application of a much smaller external stress, than a process involving simultaneous breaking of all the atomic bonds on the slip plane.

The determination of the magnitude of the shear stress needed to move a dislocation was first calculated by Peierls and Nabarro.[12] They determined the change in energy of misfit across the faces of the slip plane as a dislocation moved from one equilibrium position to the next, assuming that the shear stress acting on the slip plane is a periodic function of the relative displacement of the adjoining planes. They used a sinusoidal approximation from which the stress τ_0 to move a dislocation was shown to be

$$\tau_0 = \frac{2G}{(1 - \nu)} e^{-2\pi a/b(1 - \nu)} \qquad \textbf{3.5}$$

where G = shear modulus; ν = Poisson's ratio; a and b are the lattice constants used in eqn. **3.2**

Substituting a reasonable value for ν, viz. 0·35 and assuming $a = b$, then $\tau_0 = 2 \times 10^{-4}G$ which, although higher than observed values, is a much closer approximation to these than the value from the simple shear model.

The shear stress τ_0 varies exponentially with a/b; with a large a, that is, a large interplanar spacing and consequently a close-packed slip plane, the shear stress is a minimum. In the case of a close-packed plane, the interatomic bonds across the slip plane are weak, with the result that the activation energy and the stress for slip will be much smaller than in the case of planes which are closer together and thus possess looser atomic packing and stronger atomic bonds between adjacent planes. The result is that dislocations tend to move in the closest packed planes, and in the closest packed direction as the Peierls–Nabarro force is smaller for dislocations with short Burgers vectors.

3.6 Multiplication of dislocations—dislocation sources

The dislocation population of annealed metals is very variable but is usually between 10^5 and 10^8 cm^{-2}. These dislocations cannot give rise to the coarse slip steps observed on deformed metal surfaces, so there must be ways in which new dislocations are generated within the crystals. The dimensions of slip bands show that movements of up to several hundred dislocations have to occur on individual slip planes to account for the observed surface steps, and moreover that such numbers of dislocations must be generated at frequent intervals within the crystal.

One of the simplest ways in which dislocations could be continuously generated on a single slip plane was put forward by Frank and Read.[13] They considered a crystal in which a prismatic type dislocation was introduced by slipping part of the crystal from *ABC* to *DEF* which now represents the dislocation line separating the deformed from the undeformed region (Fig. 3.9a). A shear stress is now applied to the slip plane *ABED* and will exert no force on the part of the dislocation line *EF*, which in any case can be assumed to be immovable or *sessile*. However, there will be a force $\tau \cdot b$ on *ED* which will cause it to move forward. As *ED* is anchored at *E* it will thus tend to rotate about this point as the applied stress attempts to expand the area of the slipped region. However, *ED* does not rotate like the hand of a clock for this would mean that the velocity along *ED* would be proportional to the distance from *E*. The velocity is actually constant because the force along *ED* is unchanged, so the outer part of the dislocation line will lag behind the centre, thus producing a curvature which is gradually accentuated.

When *ED* intersects the far end of the crystal block, a displacement equivalent to the Burgers vector of the dislocation progressively occurs (Fig. 3.9b). No displacement takes place on the side walls because they are

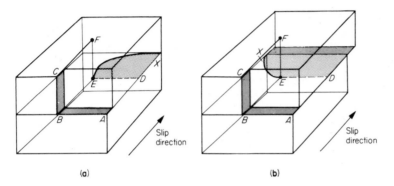

(a)   (b)

Fig. 3.9   Model of a dislocation source

parallel to the slip direction; however, after further rotation the end of the line reaches the front face, and after moving along it the displacement of the whole model by one Burgers vector is complete, and *ED* eventually reaches its original position. The line is now free to commence the process once more under the influence of the applied stress, and this can, in theory, be continued until the shearing of the model is complete. This means that one dislocation of strength b can cause slip displacements of nb, where n can be a large integer, on a single slip plane, and thus produce the shears which are observed along individual slip bands in crystals.

It should be emphasized that while the line *ED* starts off as a length of edge dislocation, as soon as curvature is introduced it acquires a screw component. That part of the line normal to the slip direction will be pure edge, while the pure screw component will be parallel to the slip direction, so the line as a whole is a mixed dislocation.

Another more important mechanism, referred to as the Frank–Read source, involves a dislocation line locked at both ends. A dislocation line CD lies in the slip plane (Fig. 3.10) and is locked at C and D by two immovable dislocations AC and BD. When the shear stress is gradually increased, it will reach a critical value at which the dislocation line will begin to move outwards. This stress τ_0 is defined thus: $\tau_0 = Gb/\Lambda$ where Λ is the length of CD. The applied stress is a maximum for the semi-circular form of the line, but beyond this the dislocation is unstable and

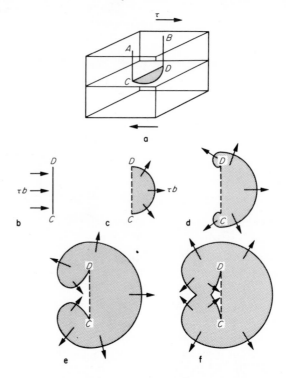

Fig. 3.10   Operation of a Frank–Read source

continually expands (Fig. 3.10b–e). Each point on the line is acted on by the same stress τb, and thus has the same velocity with the result that the line spirals around the two terminal points C and D (Fig. 3.10d). A critical stage is reached in Fig. 3.10e when two parts of the line are approaching towards each other in opposite directions. As they possess the same Burgers vector but now of opposite sign, on contact the disloca-tion will disappear at this point (cf. combination of positive and negative edge dislocations on the same slip plane in a simple cubic model). This results in a closed loop of dislocation which will continue to grow outward under the influence of the applied stress. At the same time, the original dislocation line *AB* has been regenerated and is now free to repeat the whole process. In this way a series of loops will be generated indefinitely, until a back stress is created by some dislocation interaction, which by counteracting the applied stress causes the source to cease operating.

There remains the question of the locking of the line at *C* and *D*. This can be achieved by two dislocations *AC* and *BD* lying on a plane which is not the operative slip plane, or the points *C* and *D* may be nodes representing points where three dislocations meet in a three-dimensional dislocation network. Solute atoms or precipitate particles can also serve to lock parts of a dislocation line.

3.7   Dislocation pile-ups

If an obstacle occurs on a slip plane, the dislocations generated from a Frank–Read source will tend to pile up against it (Fig. 3.11A and B). The later dislocations will exert a force on the earlier ones with the result that those nearest the obstacle will be much more closely spaced than those last emanating from the source and an equilibrium distribution is brought about.

The limiting factor to further movement is the first dislocation in the queue. If it moves a distance δx, all the other dislocations will do likewise.

Work done per unit length of dislocation by such a movement
$$= n \,.\, \delta x \,.\, b \,.\, \tau$$

where n = number of dislocations in pile-up, and τ = applied stress.

Now the leading dislocation does work against the local internal stress τ_l

$$\text{Work done} = \tau_l \,.\, b \,.\, \delta x$$

But in equilibrium these terms are equal, thus

$$n \,.\, \delta x \,.\, b \,.\, \tau = \tau_l \,.\, b \,.\, \delta x,$$

so
$$\tau_l = n\tau \qquad\qquad \textbf{3.6}$$

In other words, the internal stress at the head of a pile-up of n dislocations is n times the applied stress. This stress concentration has important implications in several fields such as work hardening and brittle fracture. The pile-up exerts a back stress τ_b on the dislocation source, which will continue to produce dislocations until the magnitude of the back stress equals that of the applied stress less the stress τ_a necessary to activate the source, i.e.

$$\tau_b = \tau - \tau_a$$

As the head of a pile-up is approached, the dislocations are increasingly closely spaced, as the applied shear stress is tending to push all the

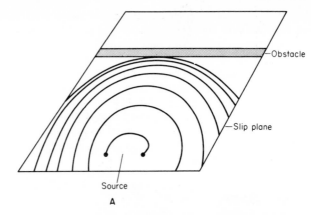

A

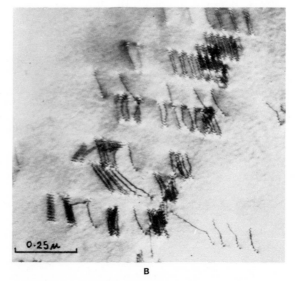

B

Fig. 3.11 **A**, Model of a dislocation pile-up. **B**, Dislocation pile-ups in stainless steel. Thin-foil electron micrograph × 70,000

dislocations from the source towards the obstacle. Eshelby, Frank and Nabarro[48] calculated the distribution of dislocations to be expected in such circumstances. In particular, the number of dislocations which can be piled into length l_0 of a slip plane in which a shear stress τ acts was found to be

$$n = \frac{\pi l_0 \tau k}{Gb} \qquad\qquad 3.7$$

where $k = 1$ for edge and $(1 - \nu)$ for screw dislocations.

The number of dislocations is directly proportional to the stress, and inversely proportional to the shear modulus G and the Burgers vector.

3.8   The experimental detection of dislocations

The theoretical approach briefly described above leaves little doubt that the concept of dislocations is a valid one; however, it is comforting to have in addition experimental evidence for their existence.[4] Fortunately there is now abundant evidence from a variety of different techniques, which will be summarized.

3.8.1. *The bubble model*

One of the most striking early techniques was the use of a crystal analogue composed of a two-dimensional raft of uniformly sized soap bubbles. Bragg and Nye[14] showed that, on deforming these rafts between glass rods, shear took place by the passage of edge dislocations, the characteristic structure of which was clearly revealed in the bubble arrays. Lomer[15] carried out quantitative experiments with bubble rafts, and was able to show that they begin to deform plastically at stress of the order of $10^{-3} \times E$ as predicted by the theoretical work. It was possible also to study simple dislocation interactions, e.g. dislocations of opposite sign on the same slip plane annihilating each other, or combination of dislocations on different planes to form a third dislocation. The bubble model also provided a model for grain boundaries and showed that low angle boundaries could be described in terms of dislocations.

3.8.2. *Growth of crystals*

In 1949 Frank and colleagues[16] showed theoretically that dislocations can play an important part in the growth of crystals from the vapour and liquid states. It was pointed out that in practice many crystals grew much more rapidly than would be expected for perfect crystals and at lower supersaturations. The difficulty with perfect crystals is that while atoms will attach themselves to atomic ledges in growing crystal faces, when a face is made completely smooth by this process the growth slows down until a new nucleus of atoms is created on the smooth face. Frank overcame this difficulty by pointing out that a screw dislocation emerging on a face (Fig. 3.12a) will create a ledge A which is self-perpetuating, because the screw dislocation has converted the crystal into a spiral ramp. A geometrical consequence is that addition of atoms to the step will eventually develop a spiral step (Fig. 3.12b), since atoms added near the emerging dislocation will cause a more rapid advancement of the ledge, than those added in the outer region which will gradually lag behind the core.

At the same time as the dislocation theory of crystal growth was put forward, Griffin[17] reported dislocation spirals in natural beryl, soon to be

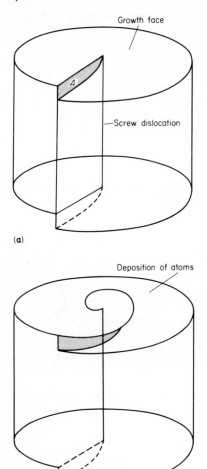

(a)

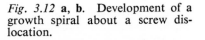

(b)

Fig. 3.12 **a, b.** Development of a growth spiral about a screw dislocation.

followed by many other examples in carborundum, paraffin crystals, cadmium iodide,† etc. The steps were in most cases found to be monomolecular, and were made visible in the optical microscope either by an etching phenomenon or by absorption of impurities to the steps. Forty[18] found similar growth spirals on crystals of cadmium and magnesium grown from the vapour phase.

Interesting geometrical variations occur when two dislocations co-

† The crystallization of cadmium iodide from supersaturated aqueous solution on a microscope slide provides an excellent laboratory demonstration of the dislocation mechanism of crystal growth.

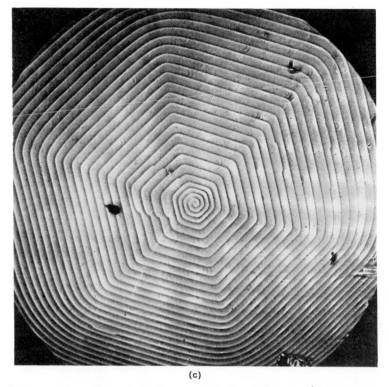

(c)

Fig. 3.12 c.   Growth spiral in silicon carbide. Step height 165 Å. (After Verma, 1953, *Crystal Growth and Dislocations*, Butterworth, London)

operate in the growth process. Two screw dislocations of similar sign combine to produce a growth spiral of double step height, while two dislocations of opposite sign combine to give a series of continuous growth terraces. The spirals and terraces can be almost circular in shape, but often they reflect the symmetry of the crystals by variations in growth rate with orientation (Fig. 3.12c).

3.8.3. *Etch pits*

In one sense etching can be regarded as the reverse of crystal growth, and if the process is sufficiently selective it should be possible to remove atoms from crystals preferentially at atomic imperfections which cause surface steps. In fact spiral etch pits have been observed during chemical etching of germanium[19] while holes have been detected at dislocations during the dissolution of AlB$_2$ crystals.[20]

There is little doubt that dislocations can be preferential sites for chemical attack, primarily because they are small regions of stress with a different electrode potential. In many cases it seems that impurity atoms must be associated with the imperfections, while in others this is apparently not a necessary condition. Wyon and Lacombe[21] demonstrated that in aluminium dislocation arrays could be readily etched provided that impurities had been allowed to migrate to the dislocations. Jacquet[22] made a detailed study of etch pits in α-brass, which revealed the presence of dislocation interactions and pile-ups after small deformations. Dislocations can be readily etched in silicon or germanium, and the method is now used to assess the quality of crystals grown for semi-conductor devices.[23]

Some elegant experiments have been done on lithium fluoride crystals (Fig. 3.13) by Gilman and Johnston,[24] who by careful use of special

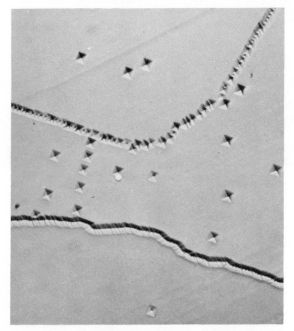

Fig. 3.13   Dislocations revealed by etching in lithium fluoride. Low-angle boundaries and isolated dislocations. × 500 (After Gilman)

etchants were able to distinguish between edge and screw dislocations. By successive removal of layers and etching of a crystal they were able to follow the movements of individual dislocations, to make precise observations on the growth of dislocation loops, and to plot the form of the loops through the crystal. Quantitative measurements were made on dislocation velocities and the densities of dislocations resulting from light deformations. Similar controlled etching experiments have been done on magnesium oxide crystals.[25]

As a result of these and many other similar experiments, the etching technique is a recognized method of detecting dislocations, but care is needed both in the experiments and interpretation of results, which are best confirmed by one of the more recently developed techniques, such as thin-foil electron microscopy.

3.8.4. *Precipitation*

In Chapter 6 the interaction of solute atoms with dislocations is considered and it is shown that these atoms, whether in substitutional or interstitial positions in the metal structure, interact elastically with dislocations with the result that they tend to segregate along dislocations. If the alloy is supersaturated with respect to the solute, then on ageing, precipitation will take place, and as the dislocations are already regions of high solute concentration, they will be preferred sites for nucleation of precipitate particles.

Dislocations have been revealed in many crystalline solids using this technique. For example, Dash[26] diffused copper into silicon crystals, and

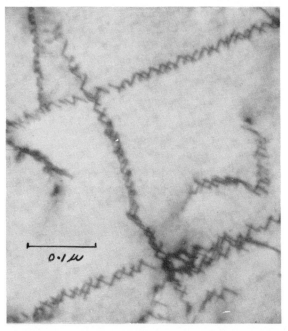

Fig. 3.14   Precipitation of molybdenum carbide on dislocations in a ferritic steel. Thin-foil electron micrograph × 185,000. (After Irani)

by precipitating the copper was able to show up dislocation configurations and in particular Frank–Read sources. Furthermore, he found that etch pits coincided with the emergence of these decorated dislocations on the surface. Similarly dislocations have been revealed in aluminium–copper alloys by precipitation of the θ' precipitate. Dislocations in silver halides are readily revealed by precipitation of colloidal silver[27], while Amelinckx and co-workers[4] have studied dislocation arrangements by similar methods in KCl, CaF_2 and other ionic crystals.

Dislocations in steels[28] are often decorated by precipitates; for example, Fig. 3.14 shows the precipitation of fine needles of molybdenum carbide Mo_2C on dislocations in ferrite. Such reactions are important not only as a means of revealing imperfections, but they are also the basis of a significant strengthening mechanism, particularly in alloys for use at high temperatures (Chapter 13).

3.8.5. *Thin-foil electron microscopy*

Improvement in the resolution of electron microscopes and the development of a method for thinning metals and alloys to such an extent (1000–3000 Å thick) that 100 kV electrons could easily penetrate the specimen, led in 1956[29,30] to the direct observation of dislocations, or more precisely their strain fields, in the electron microscope. The dislocations are revealed as thin dark lines where electrons are preferentially scattered if the diffraction conditions are satisfied. Figure 3.11B shows a typical array of dislocations in a deformed foil of stainless steel. The dislocation lines are normally cut by the top and bottom surfaces of the foil, for it is usually a random section through a specimen; thus it is only possible to view a part of a dislocation line. Consequently, dislocation loops emanating from sources are rarely seen in their entirety, unless the thin section is taken precisely parallel to an active slip plane.

The thin-foil technique[4] has proved to be the most powerful method for the study of dislocations for several reasons. Firstly, the imperfections are revealed directly and optimum contrast can be obtained by suitably tilting the specimen. Secondly, there is no doubt about the nature of the defects, as they can be observed in movement during deformation, and various interactions can thus be directly confirmed and studied. Thirdly, the nature of the dislocation, viz. its Burgers vector and the plane in which it lies, can be determined by crystallographic methods based on electron diffraction and by using the dynamical theory of contrast,[31] which has dealt specifically with dislocations. The facility of selected area electron diffraction makes it possible to carry out detailed crystallographic investigations on the dislocations. There are, however, some difficulties, in particular the process of thinning to 1000–3000 Å is often liable to alter the dislocation configurations and, for example, the more mobile screw components frequently move out of the foil.

Dislocations can be revealed even more directly in terms of atomic planes, if a thin crystal of a substance with a relatively large lattice spacing and large metal atoms is examined by transmission electron microscopy. Using this technique Menter[32] was able to reveal directly lattice planes in platinum and copper phthalocyanines where the lattice spacing is about 12 Å. These planes were usually very regular but occasionally dislocations were observed. A more sensitive method involved the use of overlapping thin crystals at slightly different orientations to form a Moiré pattern, a banded form of contrast which reproduces on a larger scale any defects such as dislocations present in the crystals. In this way dislocation patterns were observed in gold and palladium films.

More recently, the resolution of electron microscopes has improved still further with the result that still smaller lattice spacings have been directly revealed. For example, the {020} planes in Mo_2O_3 ($d = 6.9$ Å) have been photographed, while very recently the {110} planes in copper (1·36 Å) have been successfully observed.

3.8.6. *Field ion microscopy*

It is possible by the use of field ion microscopy[33] to obtain high magnified images of fine points of tungsten and other metals with high melting points which reveal the positions of some of the individual atoms and the crystallography of the specimens.

Consequently dislocations, and also point defects such as vacancies which are beyond the resolution of the electron microscope, can be studied[49]; however, as far as dislocations are concerned, the experimental conditions are rather limiting. This is a very promising direct method for the study of the interaction of vacancies and solute atoms with dislocations.

3.8.7. *X-ray diffraction*

X-rays are scattered in the vicinity of dislocations, so that if a suitable imaging technique is used, sharp images of dislocation lines can be obtained. Lang[34] has shown that dislocation arrays can be revealed in relatively thick sections, for example 100 microns†; however, the method does not possess a very high resolution and is limited to small deformations.

3.9 Stress fields around dislocations

It should be clear from the general account already given and the appearance of dislocation models, that dislocations represent minute centres of internal stress in a material. The internal stress fields of dislocations are of importance both from the point of view of the properties of the individual dislocations, and also when dislocations interact with each other.

† 1 micron (μm) = 0·001 mm = 10^4 Å = 10^3 nm.

Elasticity theory is used to describe the strain fields of dislocations.[1,2] At the core of a dislocation the atomic misfit is so bad that this approach is only an approximation, but as the distance from the centre is increased, elasticity theory gives the strain with greater accuracy. To simplify the problem, the dislocation is assumed to be a straight line imperfection in an elastically isotropic crystal (Fig. 3.2a and b). The stress is described in terms of rectangular co-ordinates x, y and z, or cylindrical co-ordinates $r\theta z$, where in each case the z axis corresponds with the dislocation line. Now, comparing the relative positions of equivalent points in a perfect and in a dislocated crystal, u v and w are taken to be the x, y and z components of the displacement as a result of the dislocation.

3.9.1. Stress field of an edge dislocation

The x axis is taken along the Burgers vector and thus xz is the slip plane. The relations for the displacements u, v and w in the x, y and z directions are as follows.

$$u = \frac{b}{2\pi}\left[\tan^{-1}\frac{y}{x} + \frac{1}{2(1-\nu)}\cdot\frac{xy}{x^2+y^2}\right]$$

$$= \frac{b}{2\pi}\left[\theta + \frac{\sin 2\theta}{4(1-\nu)}\right] \quad \nu = \text{Poisson's ratio} \qquad \textbf{3.8}$$

$$v = \frac{-b}{8\pi(1-\nu)}\left[(1-2\nu)\ln(x^2+y^2) + \frac{x^2-y^2}{x^2+y^2}\right]$$

$$= \frac{-b}{2\pi}\left[\frac{1-2\nu}{2(1-\nu)}\ln r + \frac{\cos 2\theta}{4(1-\nu)}\right] \qquad \textbf{3.9}$$

$$w = 0 \qquad \textbf{3.10}$$

Thus, the displacement parallel to the dislocation is zero and we are considering a plane deformation. Consequently, the only stresses of significance are the normal stresses along the x and y axes, σ_{xx} and σ_{yy}, and the shear stress τ_{xy} acting along the y axis on planes perpendicular to the x axis. The other shear stresses are zero. The stresses which are thus confined to the xy plane are as follows:

$$\sigma_{xx} = -\frac{Gb}{2\pi(1-\nu)}\cdot\frac{y(3x^2+y^2)}{(x^2+y^2)^2}$$

$$= -\frac{Gb}{2\pi(1-\nu)}\cdot\frac{b}{r}\cdot\sin\theta(2+\cos 2\theta) \qquad \textbf{3.11}$$

$$\sigma_{yy} = \frac{Gb}{2\pi(1-\nu)}\cdot\frac{y(x^2-y^2)}{(x^2+y^2)^2}$$

$$= \frac{Gb}{2\pi(1-\nu)}\cdot\frac{\sin\theta\cos 2\theta}{r} \qquad \textbf{3.12}$$

$$\tau_{xy} = \frac{Gb}{2\pi(1 - \nu)} \cdot \frac{x(x^2 - y^2)}{(x^2 + y^2)^2}$$

$$= \frac{Gb}{2\pi(1 - \nu)} \cdot \frac{\cos\theta \cos 2\theta}{r} \qquad \textbf{3.13}$$

The largest normal stress σ_{xx} is along the x axis, and is compressive above the slip plane and tensile beneath. This can be appreciated by studying the distortion around an edge dislocation in a wire-ball model (Fig. 3.3).

3.9.2. Stress field of a screw dislocation

The configuration of atoms in the screw dislocation suggests that the displacement is confined to the direction of the dislocation line, i.e. the z component (see Fig. 3.5). The displacements u and v in the x and y directions are both zero while

$$w = \frac{b}{2\pi} \tan^{-1}\frac{y}{x}$$

$$= \frac{b}{2\pi} \cdot \theta \qquad \textbf{3.14}$$

Analysis of the stress field shows that it is entirely of a shear character with no dilatation. The strain components are the derivatives of u, v and w, from which are obtained the stress components, of which there are only two when expressed in rectangular co-ordinates

$$\tau_{xz} = -\frac{Gb}{2\pi} \cdot \frac{y}{x^2 + y^2} \qquad \textbf{3.15}$$

$$\tau_{yz} = \frac{Gb}{2\pi} \cdot \frac{x}{x^2 + y^2} \qquad \textbf{3.16}$$

and only one when cylindrical co-ordinates are used

$$\tau_{\theta z} = \frac{Gb}{2\pi r} \qquad \textbf{3.17}$$

The stress field thus has radial symmetry, being independent of θ (contrast eqns. **3.11–3.13** for the edge dislocation). This can be grasped also from an inspection of the models (Figs. 3.3 and 3.5), where the edge dislocation with its extra half plane of atoms clearly has an asymmetric field, while that of the screw dislocation is symmetrical about any plane containing the dislocation line.

3.10   The stored energy associated with a dislocation

The presence of the localized stress fields referred to above will mean that dislocations represent regions of stored elastic energy. The energy will vary markedly from point to point in the region of the dislocation, but if the local stress and strain are known then the elastic energy is the area under the elastic region of the stress–strain curve, that is the linear part where Hooke's law applies.

$$U = \tfrac{1}{2}\sigma\epsilon = \tfrac{1}{2}G\epsilon^2 \qquad\qquad \textbf{3.18}$$

where U = strain energy and ϵ = strain.

The total energy is then obtained by integration of the elastic energy over the whole volume between r_0, near the core of the dislocation, and r, the radius of the crystal (Cottrell[1]).

We shall refer to Fig. 3.2b and calculate the strain energy for a screw dislocation. The distance $DE = b$, $AD = r$ and $CD = l$, while r_0 is the radius of the central core. The shear strain ϵ is easily defined if we unroll the surface of the cylinder giving a sheared rectangle $CDEG$ of sides $l\,(CD)$ and $2\pi r\,(DE)$, thus

$$\epsilon = \frac{b}{2\pi r}$$

and the shear stress

$$\tau = \frac{Gb}{2\pi r}$$

Now   The strain energy U (or work done per unit volume) $= \tfrac{1}{2}G\epsilon^2$

$$= \tfrac{1}{2}G\left(\frac{b}{2\pi r}\right)^2 \qquad\qquad \textbf{3.19}$$

Now we define the unit volume δv in terms of a section of the cylinder δr in thickness

$$\delta v = 2\pi r \,.\, \delta r \,.\, l$$

thus the total strain energy is now obtained by integrating the energy between the core radius r_0 and the maximum radius r

$$\text{Total strain energy} = \int_{r_0}^{r} \tfrac{1}{2}G\left(\frac{b}{2\pi r}\right)^2 2\pi r \,.\, l\,dr$$

$$= \frac{Gb^2 l}{4\pi} \int_{r_0}^{r} \frac{dr}{r}$$

$$= \frac{Gb^2 l}{4\pi} \ln\left(\frac{r}{r_0}\right)$$

$$\therefore \; \begin{array}{c}\text{Energy/unit length}\\ \text{(screw)}\end{array} = \frac{Gb^2}{4\pi}\ln\left(\frac{r}{r_0}\right) \qquad\qquad \textbf{3.20}$$

A similar expression is obtained for an edge dislocation:

$$\text{Energy/unit length (edge)} = \frac{Gb^2}{4\pi(1 - \nu)} \ln\left(\frac{r}{r_0}\right) \qquad \textbf{3.21}$$

from which it can be seen that the energy of a screw dislocation is somewhat less than that of an edge in the same material. Substituting reasonable values for r_0 and r, G and b, energies of the order of $10^{-6}\,\text{mJ m}^{-1}$ or several electron volts/atomic spacing are obtained, e.g. an edge dislocation in copper has a strain energy of about $5 \times 10^{-6}\,\text{mJ m}^{-1}$.

3.11   Line tensions of dislocations

A dislocation has elastic strain energy and, as it is a line defect, it possesses strain energy per unit length. Consequently it will always attempt to reduce its length to attain a condition of minimum energy, and can be said to possess a line tension analogous to the surface tension of films. While the dislocations seen in crystals are rarely straight, continuous observation of moving dislocations has confirmed that they attempt to iron out irregularities along their length during movement under stress. The line tension of a dislocation per unit length has been shown by Nabarro[35] to be approximately

$$T = Gb^2$$

The existence of a line tension has a large influence on the behaviour of dislocation lines in crystals. For example, in the case of the bowing dislocation of the Frank–Read source (Fig. 3.10) the line tension T exerts a restoring force F which tends to eliminate the curvature, and must be overcome by the applied stress. The magnitude of F is T/r where r is the radius of curvature of the dislocation. As the dislocation line expands r decreases until the line is a semicircle (Fig. 3.10c), at which point r has a minimum value. Instability of the line occurs at this stage:

$$F \simeq \frac{T}{r_{min}} = \frac{Gb^2}{r_{min}} = \tau b \qquad \textbf{3.22}$$

so the critical stress for the growth of the loop is

$$\tau \simeq \frac{Gb}{r_{min}} \simeq \frac{\alpha Gb}{\Lambda} \qquad \textbf{3.23}$$

where Λ is the distance between pinning points on the dislocation, $\alpha =$ constant. Bowing of dislocations occurs in many other circumstances where ends of the dislocation lines are locked, e.g. on the two surfaces of a thin foil, or when a dislocation line has to move through a dispersion of particles (Chapter 7). Another consequence is that when several dislocations meet at a point, the angles will be such that the line tensions are in equilibrium (cf. grain boundaries).

3.12   Networks of dislocations

Frank[36] first pointed out that the equilibrium distribution of dislocations in a crystal should be a three-dimensional network similar to the poly-hedral cellular networks obtainable in some foams. In such a network, dislocations meet at nodes (Fig. 3.15a) which satisfies the line tension requirements, and also the criterion that the resultant Burgers vector should be zero. In actual crystals both two- and three-dimensional net-works are obtained, where the dislocation lines between the nodes contract until they are as short as possible. A typical network observed in iron is shown in Fig. 3.15a and b. Part of the symmetry of the networks arises from the ease with which dislocations align themselves in certain crystallo-graphic directions for energetic reasons.

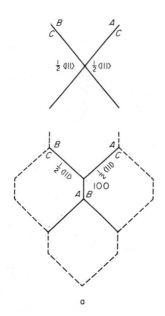

Fig. 3.15 **a**. Equilibrium network of dislocations expected for iron. (After Carrington, Hale and McLean, 1960, *Proc. R. Soc.*, **A259**, 203)

Particularly good two-dimensional dislocation networks are observed at low angle grain boundaries and sub-boundaries which are frequently formed during the recovery of cold worked metals. Such sub-boundaries can be demonstrated in the bubble model which shows the increasing con-centration of dislocations in a boundary as the angle of disorientation across the boundary increases. The simplest low angle boundary is com-posed of a series of edge dislocations (Fig. 3.16) indicated by the extra half planes of atoms in the boundary.[3]

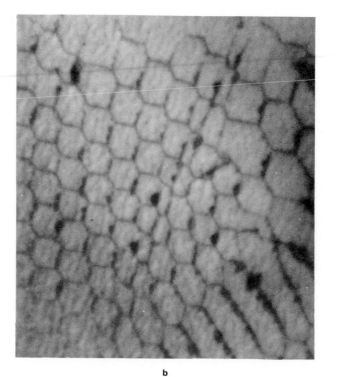

b

Fig. 3.15 **b.**   Dislocation network in normalized iron. Thin-foil electron micrograph × 50,000. (After McLean (Crown Copyright reserved))

If the Burgers vector of the dislocation is b, it is obvious that the disorientation θ is related to the spacing D (Fig. 3.16B), thus

$$D = \frac{b}{2 \sin (\theta/2)} \simeq \frac{b}{\theta} \quad \text{for small angles} \qquad \textbf{3.24}$$

As θ increases, the spacing of the dislocations becomes smaller, until at
high angles the boundary can no longer be realistically described in terms
of dislocations, but is rather a region of local disorder. Boundaries composed entirely of edge dislocations are referred to as *tilt* boundaries,
because the disorientation can be described by tilting about the plane containing the dislocations; the axis of tilt is normal to the plane of the
diagram in Fig. 3.16, and the boundary is thus a symmetrical tilt boundary.
    If a series of screw dislocations is introduced instead of edge dislocations,
the disorientation occurs by rotation about an axis in the plane of the

diagram and a *twist* boundary is formed. It is clearly possible to have a wide range of disorientations by combining networks of edge and screw type dislocations. In favourable circumstances, low-angle boundaries can be etched up, and the individual dislocations revealed by etch pits (Fig. 3.13), the spacing of which can be readily measured. It has been shown that the disorientation calculated using eqn. **3.24** and the experimentally determined value of *D*, is the same as that found using an X-ray diffraction method, which measures the disorientation directly.[3]

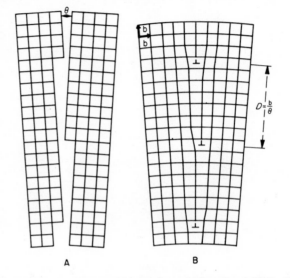

A   B

Fig. 3.16   Structure of a low angle boundary. (After Read 1953, *Dislocations in Crystals*. McGraw Hill, New York and Maidenhead)

3.13   Non-conservative movement of dislocations

In normal glide, an edge dislocation moves on the slip plane in the direction of its Burgers vector, and this involves only small localized displacements of atomic rows in this plane. The situation is however very different if one tries to move such a dislocation line perpendicular to its Burgers vector into a parallel adjacent plane (Fig. 3.17A and B). This movement involves the removal of the row of atoms forming the end of the extra half plane typical of the edge dislocation. This can only be done by diffusion of the atoms away from the dislocation, either by the atoms becoming interstitial and diffusing, or by diffusion of atomic vacancies to the sites. As the base of the extra half plane is free to move above or below the slip plane the moving dislocation can be either a source or a sink for vacancies. The process is usually referred to as *climb* and it is clearly dependent on energy being provided, usually in the form of thermal activation.

In the case of the screw dislocation where the Burgers vector is parallel to the axis of the dislocation there is no extra half plane of atoms. The screw dislocation is thus free to slip on any plane which contains the dislocation line and the Burgers vector. In this case no mass transfer of atoms is needed, although a higher stress or activation energy may be needed to cause the screw dislocation to move on to another plane, where the resolved shear stress is not as high as on the original slip plane.

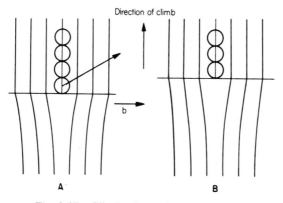

Fig. 3.17 Climb of an edge dislocation

3.14 Dislocation jogs

While the simplest form of plastic deformation involves the movement of dislocations on one parallel set of slip planes, more commonly dislocations are generated and move on more than one slip system simultaneously, e.g. in suitably oriented single crystals or in polycrystalline aggregates. In these circumstances, the dislocations in the different systems must cut across each other with predictable changes in the geometry. In general the intersection will cause *jogs* in one or both dislocations equivalent to the Burgers vectors, so that the lengths of the dislocation lines are increased and consequently also their energy. So the intersection of dislocations involves the expenditure of additional energy. A second consequence is that the jogged dislocations will move less readily through the crystal than previously.

There are many possible combinations but one of the simplest is that of the intersection of two edge dislocations moving on planes normal to each other. In Fig. 3.18 the dislocation lines and their Burgers vectors are at right angles to each other. The edge dislocation AB is moving over its slip plane and on intersecting the edge dislocation CD, it introduces a jog EF in the dislocation CD (Fig. 3.18a), the jog having the same Burgers vector as the rest of the dislocation so it will move freely with it. The length of the jog EF is clearly equal to b_{AB}, the Burgers vector of the moving dislocation

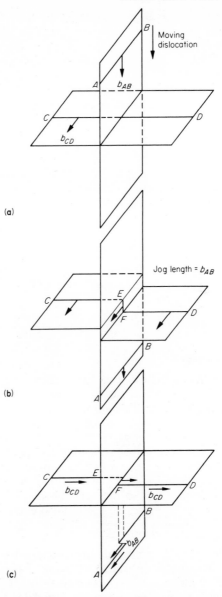

Fig. 3.18   **a** and **b,** Intersection of two edge dislocations; **c,** intersection of two
             screw dislocations

AB. Since the line tension of a dislocation is approximately Gb^2 and the
jog is of length b, the energy of the jog is roughly Gb^3.

In many cases jogs can be serious obstructions to dislocation movement, for example if two screw dislocations intersect (Fig. 3.18c). As in the previous case, a jog *EF* is formed when the moving screw dislocation *AB* cuts the stationary screw dislocation *CD*, but the Burgers vector of the dislocation line *CEFD* is now *normal* to the jog *EF*, which is thus a small section of edge dislocation. The direction of easy movement of an edge dislocation is in a plane containing the dislocation line and the Burgers vector, i.e. in the case of the jog *EF* along the axis of the screw dislocation *CD*. But the dislocation *CD* moves in a direction normal to its screw axis, so the jog *EF* is forced to accompany the other segments *CE* and *FD* of the dislocation. This it can only do by non-conservative movement, i.e. climb, so that a series of vacant lattice sites will be generated during the forward movement of the jog, or alternatively a row of interstitial atoms is formed depending on the direction of the slip vector. The extra half plane of the edge dislocation must either move by absorption of vacancies or interstitial atoms. Such processes require additional energy, with the result that the jog cannot move as fast as the rest of the dislocation, and so tends to become elongated. In similar fashion, an edge-type jog is formed on the moving dislocation *AB*.

These geometrical conclusions suggest that in conditions where intersection of dislocations is occurring, screw dislocations will tend to be less mobile than edge dislocations. On the other hand, screw dislocations can move more readily from one plane to another if they possess a common slip direction. There is now experimental evidence for the formation of jogs on dislocations which are frequently seen during electron microscopic examination of thin foils (Chapter 4). Furthermore, a number of different experiments indicate that point defects, in particular vacant lattice sites, are created during plastic deformation (Chapter 10).

3.15 Dipoles

Recent observations by thin-foil electron microscopy[37] have indicated that one of the important features of the deformation of metal crystals is the formation of fairly stable parallel pairs of positive and negative dislocations which are referred to as dipoles (Fig. 3.19). A number of mechanisms have been put forward to account for these arrays, several of which are shown schematically in Fig. 3.20. The first mechanism involves a large jog formed in a screw dislocation by coalescence of a number of small jogs (Fig. 3.20A). If the jog is of edge character it will trail behind gradually forming two long edge dislocations of opposite sign on two slip planes the distance of the jog apart. Alternatively, two adjacent jogs can form in a dislocation, for example by cross-slip (see Chapter 4) (Fig. 3.20B), which are connected by a piece of dislocation which subsequently bows out to form a dipole in a single slip plane under the action of the applied stress[38] (Fig. 3.20B). The third mechanism illustrated (Fig. 3.20C) involves dislocations of opposite sign on parallel slip planes, which become aligned at least over part of their lengths. Cross-slip at one end of the section leads to the formation of a dipole and a jogged dislocation.[39] This mechanism might well be expected to operate where the slip line spacing is close (~ 300 Å), which is approximately the observed average width of dipoles. One difficulty about mechanisms involving cross-slip is that dipoles can form at 42 K when in the absence of thermal activation cross-slip would only be expected at very high stresses.

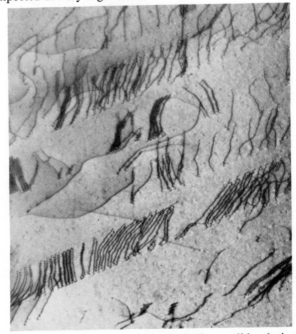

Fig. 3.19   Dislocation dipoles in a copper–indium solid solution. Thin-foil electron micrograph. $\times 40{,}000$ (Corderoy)

While the above mechanisms can account for isolated dipoles, the queues that are observed in thin foils are probably nucleated by one such event. The first dipole creates an obstacle to other dislocations, which avoid it by cross-slip and in doing so generate other dipoles in turn.

3.16   Forces between dislocations

As dislocations possess their own internal stress fields, it will be expected that when two dislocations approach close to each other, some interaction will take place. The simplest cases to consider are those of two edge or two screw dislocations on parallel, close slip planes (Fig. 3.21A and B).

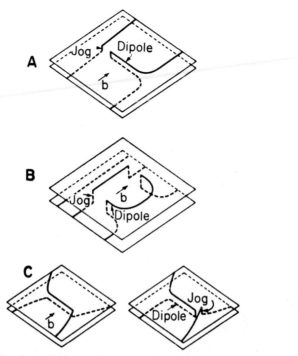

Fig. 3.20   Mechanisms for formation of dipoles. (After Mitchell, 1964, *Progress in Applied Materials Research*, **6**)

3.16.1. *Parallel edge dislocations on parallel planes*

The orientation of one dislocation is assumed to be the same as that for which stress fields were described, viz. the z axis is the dislocation line and xz the slip plane (Fig. 3.21A). The force exerted on the other edge

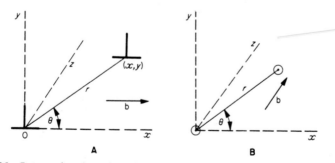

Fig. 3.21   Interaction between, **A**, parallel edge dislocations and **B**, parallel screw dislocations

dislocation at (x, y) parallel to the first, and in a parallel slip plane, has two components F_x and F_y, one in the x direction and the other in the y.

$$F_x = b \cdot \tau_{xy}$$
$$F_y = b \cdot \sigma_{xx}$$

where τ_{xy} is the shear stress at $(x \cdot y)$ caused by the first dislocation.

Now the values for τ_{xy} and σ_{xx} have been given by eqns. **3.13** and **3.11**, so expressions can be obtained for F_x and F_y to give the force on a dislocation at (xy) exerted by the dislocation at (00):

$$F_x = \frac{Gb^2}{2\pi(1 - \nu)} \cdot \frac{x(x^2 - y^2)}{(x^2 + y^2)^2} \qquad \textbf{3.25}$$

$$F_y = -\frac{Gb^2}{2\pi(1 - \nu)} \cdot \frac{y(3x^2 + y^2)}{(x^2 + y^2)^2} \qquad \textbf{3.26}$$

Alternatively the forces can be expressed in terms of polar co-ordinates.

$$F_r = \frac{Gb^2}{2\pi(1 - \nu)} \cdot \frac{1}{r} \quad \text{(radial force)} \qquad \textbf{3.27}$$

$$F_\theta = \frac{Gb^2}{2\pi(1 - \nu)} \cdot \frac{\sin 2\theta}{r} \quad \text{(tangential force)} \qquad \textbf{3.28}$$

From these relations we see that the force between the dislocations is inversely proportional to the distance r between them. There will be a repulsive force along a line joining two like dislocations, while there will be an attractive force between unlike dislocations.

3.16.2. *Parallel screw dislocations*

The two Burgers vectors are assumed to be parallel to the z axis (Fig. 3.21B). Because of the radial symmetry of the stress field, only F_r the radial component exists. It can be shown that $F_r = b \cdot \tau_{\theta z}$

but
$$\tau_{\theta z} = \frac{Gb}{2\pi r} \quad \text{(eqn. } \textbf{3.17}\text{)}$$

so
$$F_r = \frac{Gb^2}{2\pi r} \qquad \textbf{3.29}$$

In cartesian co-ordinates there are two forces F_x and F_y

$$F_x = \frac{Gb^2 x}{2\pi(x^2 + y^2)} \qquad \textbf{3.30}$$

$$F_y = \frac{Gb^2 y}{2\pi(x^2 + y^2)} \qquad \textbf{3.31}$$

As in the case of the edge dislocations, the force is attractive with unlike dislocations and repulsive between dislocations of the same sign.

3.17   Dissociation of dislocations in close-packed structures

In the simple cubic models discussed so far, movement of one Burgers vector results in an identity translation, that is, the atoms form the same configurations as before movement, and so the dislocation is said to be *perfect*. If, however, the movement results in a new atomic configuration, the dislocation is said to be *imperfect* or *partial*. The particular crystal structure concerned will in part determine which type of dislocation is prevalent, but there is also an energy criterion to be taken into account. For example, a perfect dislocation may dissociate into two or more imperfect dislocations if, in doing so, the energy of the system is reduced.

Whether dissociation occurs or not can be determined by application of Frank's rule, which states that the strain energy of a dislocation is proportional to the square of its Burgers vector (see eqns. **3.20** and **3.21**). If the sum of the squares of the Burgers vectors of the component dislocations is less than the square of that of the original dislocation, the dissociation is energetically favoured.

The simplest case is a dislocation with a Burgers vector of several lattice spacings, e.g. with a Burgers vector of $2b$. This can obviously dissociate into two separate dislocations, each with Burgers vector b.

$$2b \rightarrow b + b$$

Applying Frank's criterion, $(4b^2) \rightarrow (b^2) + (b^2)$, the dissociation is clearly favourable from the energetic point of view. If we consider the simple cubic lattice with atoms only at the corners, the basic lattice vectors are the

$$a[100], \quad a[110] \quad \text{and} \quad a[111]$$

the magnitudes of which are a, $a\sqrt{2}$ and $a\sqrt{3}$ respectively. It is clear that $a[100]$ has the lowest energy and also the closest atomic packing. It will be recalled that the Peierls–Nabarro force needed to move a dislocation is least in close-packed directions, and in actual crystals the direction of closest packing is almost always the observed slip direction.

In the simple cubic lattice, the dislocations already referred to do not dissociate into imperfect ones because the crystal cannot exist in any alternative structure; however, in the close-packed face-centred cubic structure such translations are possible.

It has already been pointed out that slip in face-centred cubic metal crystals is almost always on {111} planes in ⟨110⟩ directions, but there is a multiplicity of such systems, so this leads to dislocation reactions which have an important effect on the behaviour during plastic deformation. Figure 3.22A shows the close-packed array of {111} octahedral planes in the face-centred structure. The shortest lattice vector is $\frac{1}{2}a\langle 110\rangle$, represented by the line between an atom at a cube corner and one in the centre of a cube face, Fig. 3.22B, which is frequently the observed slip direction.

This is shown in Fig. 3.22A as $b_1 = \frac{1}{2}a[10\bar{1}]$, the unit translation being from one B position to the next; however, energetically this can be more readily achieved via C, a fact which can be demonstrated qualitatively on a close-packed ball model. Heidenreich and Shockley[40] first suggested that

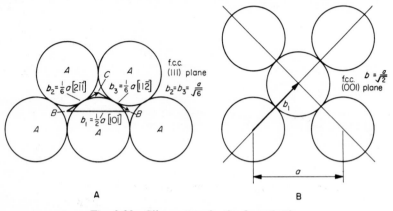

Fig. 3.22   Slip vectors in the f.c.c. lattice

it would be energetically favourable for the primary dislocation $\frac{1}{2}a[10\bar{1}]$ to dissociate thus into two imperfect or partial dislocations

$$\tfrac{1}{2}a[10\bar{1}] \to \tfrac{1}{6}a[2\bar{1}\bar{1}] + \tfrac{1}{6}a[11\bar{2}]$$

It is helpful to remember an interplanar spacing d in a crystal can be expressed in terms of the lattice constants and the Miller indices of the particular plane (hkl). In the cubic system,

$$d_{hkl} = \frac{a}{\sqrt{h^2 + k^2 + l^2}}$$

so it is a simple matter to determine the magnitude of the slip vectors in any given direction $\langle hkl \rangle$. Thus the slip vectors in the above dislocation reaction are as follows:

$$\frac{a}{\sqrt{2}} \to \frac{a}{\sqrt{6}} + \frac{a}{\sqrt{6}}$$

Applying Frank's rule,

$$\frac{a^2}{2} > \frac{a^2}{6} + \frac{a^2}{6}$$

The Burgers vector b of the undissociated dislocation is in the $[10\bar{1}]$ direction so

$$b = d_{[10\bar{1}]} = \frac{a}{\sqrt{2}}$$

and so the squares of the vectors can also be expressed as

$$b^2 > \frac{b^2}{3} + \frac{b^2}{3}$$

Either way, the sum of the squares of the resultant slip vectors is less than the square of the vector of the undissociated dislocation, so the formation of the partial dislocation is energetically favoured. The resultant partial dislocations move successively through the crystal, and together cause the same displacement which movement of the original undissociated dislocation would have achieved. They are referred to as *glissile* partial dislocations.

A useful way of visualizing the possible dislocation reactions is provided by the Thompson tetrahedron,[41] which is a regular tetrahedron $ABCD$ with vertices at the points $\frac{1}{2}a(011)$, $\frac{1}{2}a(101)$, $\frac{1}{2}a(110)$ and (000) as shown in Fig. 3.23. The mid-points of the faces are referred to as α, β, γ

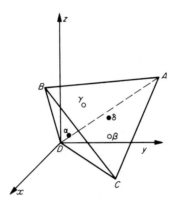

Fig. 3.23   The Thompson tetrahedron

and δ. The four possible slip planes {111} are represented by the faces of the tetrahedron a, b, c and d, while the twelve slip directions $\langle 110 \rangle$ are represented by the edges of the figure, AB, BC, etc. So from the tetrahedron we can represent

1. Normal slip dislocations

$$b = \frac{1}{2}a\langle 110 \rangle = AB, \ BC, \text{ etc.}$$

2. Formation of glissile partial dislocations (Shockley type)

$$\frac{1}{2}a[110] \rightarrow \frac{1}{6}a[211] + \frac{1}{6}a[112]$$

i.e. $$AB \rightarrow A\gamma + \gamma B$$

Alternative vector equations are

$$AB \rightarrow A\delta + \delta B$$
$$AC \rightarrow A\delta + \delta C, \text{ etc.}$$

The tetrahedron can be also used to demonstrate the formation of *sessile* dislocations (section 3.20).

3.18 Dislocations and stacking faults

The packing of the close-packed {111} planes in the face-centred cubic lattice is *ABCABCABC* so that the fourth layer is in the same relative position as the first. If, however, the third layer is placed in the same relative position as the first, then the sequence *ABABAB* is obtained, which is the arrangement of the close-packed basal planes (0001) in the hexagonal structure. It is possible to change from one type of close packing to another within the one crystal. For example, the following sequence in the face-centred cubic array

ABCACABCA

contains a sequence *CACA* which is close-packed hexagonal in structure. Such a local change in arrangement of the planes is known as a *stacking fault*. Equally a stacking fault of face-centred cubic type can occur in a hexagonal structure thus

ABABCABAB

Turning again to Fig. 3.22A, if slip takes place along vector b_1 then atoms in '*B*' positions go to '*B*' positions and the correct order of packing is preserved. However, if slip occurs along the vector b_2, then atoms in '*B*' positions go to '*C*' positions, thus introducing locally a change in packing of the type *ABCACABC*. So if the primary dislocation $\frac{1}{2}a[10\bar{1}]$ dissociates in the Heidenreich–Shockley way into two partial dislocations, $\frac{1}{6}a[2\bar{1}\bar{1}]$ and $\frac{1}{6}a[11\bar{2}]$, as the primary dislocation divides, a region of stacking fault is formed between the partial dislocations.

This is best shown in Fig. 3.24A and B, where the structure of the [110] dislocation is shown for a face-centred cubic lattice represented by a series of (110) planes, normal to the slip plane, which have the packing *ABABAB*. To preserve this packing it is shown that the edge dislocation in this structure must have *two* extra half planes of (110) orientation. The arrangement of atoms in these (110) planes is shown in Fig. 3.25a where the packing of the ($\bar{1}$11) planes *normal* to the plane of the paper are shown as *ABCABC*. This is an atomic view of the vertical end planes of Fig. 3.24A and B. If, however, the (110) dislocation dissociates (Fig. 3.24B) the two half planes separate by a distance d_0, and in the shaded region between, a stacking fault forms. This is seen when the arrangement of atoms is studied in a (110) plane cutting through the faulted area, when '*a*' half

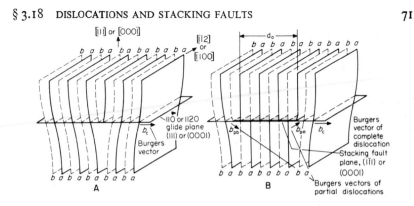

Fig. 3.24 Edge dislocation in an f.c.c. lattice, **A**, undissociated, **B**, dissociated. (After Seeger, 1957, *Dislocation and Mechanical Properties of Crystals*, John Wiley, New York and London)

planes above the fault have to match up with 'b' half planes beneath. Figure 3.25b shows that the packing in the unfaulted lattice is replaced by the sequence *ABCACAB* which includes a stacking fault.

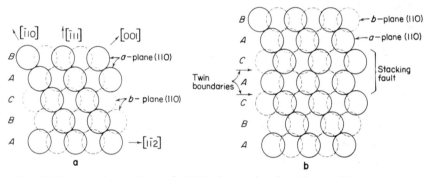

Fig. 3.25 Atomic packing of (110) planes showing the stacking order of (111) planes viewed edge-on: **a**, normal; **b**, across a stacking fault. (After Seeger, 1957, *Dislocation and Mechanical Properties of Crystals*, John Wiley, New York and London)

The stacking fault has an energy γ per unit area which can be regarded as a surface tension which opposes the repulsive force between the two partial dislocations of like sign, so that the equilibrium separation distance d_0 is reached.

Seeger[42] has shown that the equilibrium separation d_0 of the partial dislocations is determined by a dimensionless parameter $\gamma c/Gb^2$ where γ is the stacking fault energy, G is the shear modulus in the glide plane, and c the separation of neighbouring glide planes. If this parameter is greater than 10^{-2}, d_0 is almost equal to b, and the material has a high stacking fault energy, e.g. aluminium where $\gamma \simeq 200$ mJ m^{-2}. On the other hand, in the case of copper, $\gamma c/Gb^2 = 4 \times 10^{-3}$, the separation distance is of the order of $12b$ and γ is about 40 mJ m^{-2}. It is thus clear that dislocations in copper will normally possess different characteristics to those in aluminium, which lead to differences in the behaviour of these metals during work hardening and annealing.

The close-packed hexagonal structure can be considered in a similar way to the face-centred cubic, and in Fig. 3.24 the vertical planes can also represent the $(11\bar{2}0)$ planes of the hexagonal structure. When dissociation of the primary dislocation occurs there is again a faulted region formed in the sequence $ABABCAB$.

3.19 Evidence for the dissociation of dislocations

While the above ideas about the dissociation of dislocations were developed theoretically, there is now substantial experimental evidence to support the analysis. The use of thin-foil electron microscopy methods has demonstrated that stacking faults occur in close-packed metallic structures, for example they are frequently seen in evaporated thin films. Moreover, it has been possible to observe the stacking faults produced by the dissociation of perfect dislocations in face-centred cubic metals. These were first observed by Whelan et al.[43] in stainless steel foils as a characteristic banded contrast (Fig. 3.26a) between partial dislocations. Hirsch and colleagues[31] have analysed such problems in terms of the dynamical theory of electron diffraction contrast, and shown that stacking faults give such patterns under particular diffracting conditions. If the diffraction conditions are modified it is easy to reveal (Fig. 3.26b) the two partial dislocations at the ends of the stacking fault. Such structures are never observed in pure aluminium where the dislocations are always represented as sharp

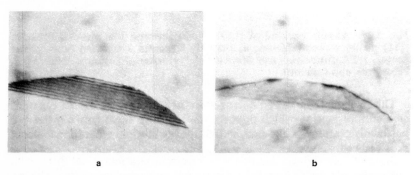

a   b

Fig. 3.26 Stacking fault in a thin foil of stainless steel. **a**, Characteristic banded contrast; **b**, contrast condition showing the terminal partial dislocations.

black lines; later on other evidence will be presented which confirms that dislocations in aluminium are dissociated to a very small degree, and thus the metal has a high stacking-fault energy. On the other hand, stacking faults have been seen in thin films of zinc, copper, gold, silver, cobalt and silicon and numerous alloys.

The stacking fault energy influences the nature of the dislocation networks frequently observed in deformed metals. Dislocation networks provide a quantitative means of determining the stacking fault energy over a limited range. In metals with fairly low γ, the dislocation networks develop nodes in which stacking fault regions are apparent. These arise from the interaction of extended dislocations, which on combination, give series of extended and contracted nodes (Fig. 3.27a and b) where the dislocations are alternately dissociated and associated. The geometry of these interactions has been examined by Whelan,[44] who has shown that the equilibrium radius of curvature R of an extended node can be defined (Fig. 3.27a)

$$R = \frac{Gb^2}{2\gamma} \qquad\qquad 3.32$$

The stacking fault energy γ can thus be determined by measurement of node radii in thin-foil electron micrographs. The method has been modified by Howie and Swann[45] who used the expression

$$\gamma = \frac{1}{R}\left[\frac{Gb^2}{4\pi k}\right]\ln\frac{R}{r_0} \qquad\qquad 3.33$$

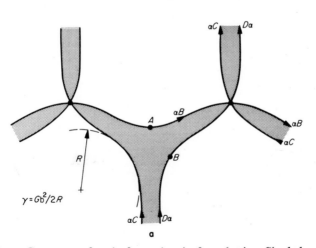

Fig. 3.27 a.   Geometry of node formation in f.c.c. lattice. Shaded areas are stacking fault. The nomenclature of the partial dislocations is the same as that in Fig 3.23 (After Whelan, 1959, *Proc. R. Soc.*, **A249**, 114)

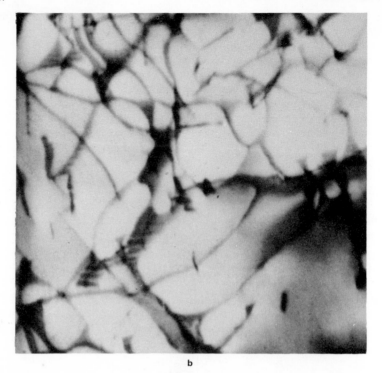

b

Fig. 3.27 **b.**   Dislocation nodes in austenitic stainless steel. Thin-foil electron micrograph. (After Liebert)

for the determination of stacking fault energies of a number of face-centred cubic metals and alloys where r_0 = radius of dislocation core, $\simeq b$; k = constant = 1 for screw, = $(1 - \nu)$ for edge dislocation, where ν = Poisson's ratio.† The method has also been applied to a number of non-

Table 3.1   Stacking fault energies determined by node method

Material	$mJ\,m^{-2}$
Ag	25
Ag–25 atomic per cent Zn	3
Cu	40
Cu–25 atomic per cent Zn	7
Co	20
Graphite	0·5
AlN	4

† This method has been refined by L. M. Brown (*Phil Mag 1964*, **10**, 441) and by P. J. C. Gallagher (*Met Trans. 1970*, **1**, 2429) and as a result, values of γ for f.c.c. metals and alloys have been modified. The latter paper is a useful detailed review of this subject.

metallic crystalline solids which have low stacking fault energies and exhibit very clear extended nodes in their crystals,[4] e.g. graphite, aluminium nitride, molybdenum disulphide. Some typical values of stacking fault energies obtained by this method are given in Table 3.1.

3.20   Sessile dislocations

The dislocations which have been so far considered are dislocations which can move readily along the slip plane, i.e. glissile dislocations. There is, however, another important group of dislocations which for one reason or another are unable to glide readily—these are sessile dislocations. This type is important because they are natural obstacles to glissile dislocations, and thus play an important part in phenomena where dislocations move with increasing difficulty, e.g. in work-hardened metals.

3.20.1. *Frank dislocations*

A Frank partial dislocation is formed in a face-centred cubic array of close-packed planes (Fig. 3.28) by the removal of part of one of the close-

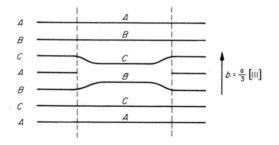

Fig. 3.28   The Frank sessile dislocation

packed planes or by insertion of an extra part-plane. The former can be achieved by collecting a disc of vacant lattice sites which then collapses to form a *negative* Frank partial dislocation which, when viewed normal to the close-packed planes, is in the form of a continuous loop. The dislocation is of edge character with a Burgers vector of $\frac{1}{3}a[111]$. Alternatively, a *positive* Frank dislocation is obtained by insertion of an extra part-plane. In each of these cases the partial dislocation cannot glide on the slip plane (i.e. parallel to the $ABC\ldots$ planes)—it can only glide normally to them in the [111] direction, but is restrained because this would result in a disruption of the close-packing array $ABCABC\ldots$. The only way left for the dislocation to move is by non-conservative motion on the close-packed plane either contracting or extending the dislocation ring, i.e. by transport of atoms or vacancies to or from the dislocation ring.

One important result of forming a Frank dislocation of either type is that within the dislocation ring there is an error in stacking and thus a stacking fault exists in this region. The negative Frank dislocation gives rise to the following sequence across the region surrounded by the dislocation ring, by removal of an A plane

$$ABCBCA$$

This is an *intrinsic* stacking fault. If, on the other hand, a B plane is added as in the positive Frank dislocation, an *extrinsic* fault is obtained

$$ABCBABC$$

The possible Frank partial dislocations can be represented on the Thompson tetrahedron (Fig. 3.27) by the vectors $A\alpha$, $B\beta$, $C\gamma$ and $D\delta$. They can be visualized by condensing a disc of vacancies on any face of the tetrahedron, e.g. BCD, and then allowing them to collapse to form the sessile dislocation ring. It has already been pointed out that the Burgers vector will not lie in the slip plane BCD but normal to it, i.e. the vector A. The dislocation is thus sessile.

3.20.2. *Formation of Lomer–Cottrell dislocations*

Lomer[46] first showed that a sessile dislocation could be produced by interaction of slip dislocations on two different {111} planes. The resultant dislocation cannot move freely in either plane. One form of the reaction uses the primary and conjugate slip dislocations thus

$$\tfrac{1}{2}a[101] + \tfrac{1}{2}a[\bar{1}\bar{1}0] \rightarrow \tfrac{1}{2}a[0\bar{1}1]$$

This reaction can be best understood by reference to the Thompson tetrahedron (Fig. 3.23) where it is

$$(AB)_d + (BC)_a \rightarrow (AC)$$

The two interacting dislocations are on planes d and a (ABC and DBC), and in two directions $120°$ apart, which is a general criterion for this type of interaction. The reaction occurs along the line of intersection of the two planes, i.e. BC, and the resulting dislocation line and its Burgers vector both lie in (001). The dislocation cannot move easily in this plane as it is less close-packed, and thus not a usual slip system.

Cottrell[47] pointed out that a further reaction occurs if the primary dislocations are considered to dissociate in the normal Shockley manner. This is illustrated by another pair of dislocations which give the basic reaction

$$\tfrac{1}{2}a[0\bar{1}1] + \tfrac{1}{2}a[10\bar{1}] \rightarrow \tfrac{1}{2}a[1\bar{1}0]$$
$$(AD)_b + (DB)_a \rightarrow AB$$

but if each primary dislocation dissociates into two Shockley partials

$$\tfrac{1}{2}a[0\bar{1}1] \to \tfrac{1}{6}a[1\bar{1}2]^* + \tfrac{1}{6}a[\bar{1}\bar{2}1]$$
$$\tfrac{1}{2}a[10\bar{1}] \to \tfrac{1}{6}a[21\bar{1}] + \tfrac{1}{6}a[1\bar{1}\bar{2}]^*$$

In the Thompson notation (see Fig. 3.23)

$$(AD)_b \to A\beta + \beta D$$
$$(DB)_a \to D\alpha + \alpha B$$

then one partial of each reaction combines thus

$$\tfrac{1}{6}a[\bar{1}\bar{2}1] + \tfrac{1}{6}a[21\bar{1}] \to \tfrac{1}{6}a[1\bar{1}0]^*$$
$$\beta D + D\alpha \to \beta\alpha$$

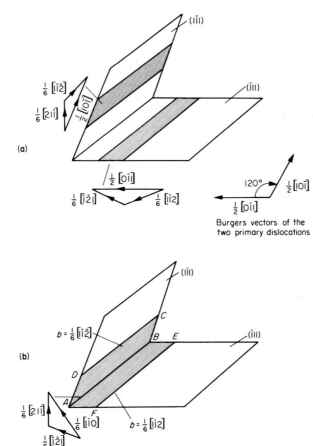

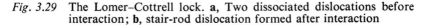

Fig. 3.29 The Lomer–Cottrell lock. **a**, Two dissociated dislocations before interaction; **b**, stair-rod dislocation formed after interaction

The dislocations asterisked are thus the total product of the interaction of the two original dislocations, and the dislocation $\frac{1}{6}a[1\bar{1}0]$ or $\beta\alpha$ is the Lomer–Cottrell dislocation. Figure 3.29A and B show the reaction in diagrammatic form. The reaction leads to a 'stair-rod' dislocation AB separated from the other two dislocations of the type $\frac{1}{6}a\langle 112\rangle$, by two regions of stacking fault $ABCD$ and $ABEF$ which form a wedge shaped obstacle to further slip, bounded by partial dislocations.

It can be seen that this reaction is a direct consequence of the basic structure of the glide dislocations on {111} slip planes in a face-centred cubic structure, and that it depends on the fact that the Shockley dissociations take place first, before two of the resultant partial dislocations can combine to form a dislocation with a Burgers vector lying in *neither* of the slip planes.

So in metals with dislocations which dissociate readily, i.e. metals of low stacking-fault energy, the Lomer–Cottrell reaction will lead to substantial wedge-shaped bands of stacking fault which, together with the sessile dislocation at the junction, will be a formidable barrier to slip on *either* of the two original slip planes. On the other hand, such barriers would not be expected to be as effective in metals of high stacking-fault energy, because the distance between the partial dislocations would be small.

General references

1. COTTRELL, A. H. (1953). *Dislocations and Plastic Flow in Crystals*. Oxford University Press, London.
2. FRIEDEL, J. (1956). *Les Dislocations*. Gauthier-Villars, Paris. English edition (enlarged) (1964), Pergamon Press, Oxford.
3. READ, W. T. (1953). *Dislocations in Crystals*. McGraw-Hill, New York and Maidenhead.
4. AMELINCKX, S. (1964). 'The Direct Observation of Dislocations', *Solid State Physics*, Supplement 6. Academic Press, New York and London.
5. WEERTMAN , J. and WEERTMAN, J. R. (1964). *Elementary Dislocation Theory*. MacMillan, New York.
6. HULL, D. (1975). *Introduction to Dislocations*. Second Edition. Pergamon Press, Oxford.

References

7. POLANYI, M. (1934). *Z. Phys.*, **89**, 660.
8. OROWAN, E. (1934). *Z. Phys.*, **89**, 605, 614, 634.
9. TAYLOR, G. I. (1934). *Proc. R. Soc.*, **A145**, 362.
10. BILBY, B. A. (1950). *J. Inst. Metals*. **76**, 613.
11. MOTT, N. F. and NABARRO, F. R. N. (1948). *Report on Strength of Solids*. Physical Society, London.
12. PEIERLS, R. (1940). *Proc. phys. Soc.*, **52**, 34.
    NABARRO, F. R. N. (1947). *Proc. phys. Soc.*, **59**, 256.
13. FRANK, F. C. and READ, W. T. (1950). *Phys. Rev.*, **79**, 722.

14. BRAGG, W. L. and NYE, J. F. (1947). *Proc. R. Soc.*, **A190**, 474.
15. LOMER, W. M. (1949). *Proc. R. Soc.*, **A196**, 182.
16. BURTON, W. K., CABRERA, N. and FRANK, F. C. (1949). *Phil. Trans. R. Soc.*, **A243**, 299.
17. GRIFFIN, L. J. (1950). *Phil. Mag.*, **41**, 196.
18. FORTY, A. J. (1952). *Phil. Mag.*, **43**, 949.
19. ELLIS, J. (1955). *J. appl. Phys.*, **26**, 1140.
20. HORNE, F. H. (1952). *Nature*, **169**, 927.
21. WYON, G. and LACOMBE, P. (1955). Bristol Conference on Defects in Crystalline Solids, p. 187.
22. JACQUET, P. A. (1954). *Acta metall.*, **2**, 752, 770.
23. CRESSWELL, J. G. and POWELL, J. A. (1957). *Progress in Semi-conductors.* Ed. by A. F. GIBSON *et al.* Vol. 2, 139. Heywood.
24. GILMAN, J. J. and JOHNSTON, W. G. (1956). *J. appl. Phys.*, **27**, 3.
25. STOKES, R. J., JOHNSON, T. L. and LI, C. K. (1959). *Phil. Mag.*, **3**, 718.
26. DASH, W. C. (1956). *J. appl. Phys.*, **27**, 1193.
27. HEDGES, J. M. and MITCHELL, J. W. (1953). *Phil. Mag.*, **44**, 233.
28. HONEYCOMBE, R. W. K. (1964). Sorby Centenary Conference, Iron and Steel Institute Special Report No. 80.
29. BOLLMANN, W. (1956). *Phys. Rev.*, **103**, 1588.
30. HIRSCH, P. B., HORNE, R. W. and WHELAN, M. J. (1956). *Phil. Mag.*, **1**, 677.
31. HIRSCH, P. B., HOWIE, A. and WHELAN, M. J. (1960). *Phil. Trans. R. Soc.*, **A252**, 499.
32. MENTER, J. W. (1956). *Proc. R. Soc.*, **A236**, 119.
33. MULLER, E. W. (1960). *Advances Electron. Phys.*, **13**, 83.
34. LANG, A. R. (1958). *Acta metall.*, **5**, 358.
35. NABARRO, F. R. N. (1952). *Adv. Phys.*, **1**, 269.
36. FRANK, F. C. (1950). Pittsburgh Conference on Plastic Deformation of Crystals. Carnegie Institute of Technology. p. 100.
37. KUHLMANN-WILSDORF, D. and WILSDORF, K. G. F. (1962). *Electron Microscopy and the Strength of Crystals.* Edited by G. Thomas and J. Washburn. Interscience, London. p. 575.
38. GILMAN, J. J. and JOHNSTON, W. G. (1962). *Solid St. Physics*, **13**, 148.
39. TETELMAN, A. T. (1962). *Acta metall.*, **10**, 813.
40. HEIDENREICH, R. D. and SHOCKLEY, W. (1948). Report on Strength of Solids. Physical Society, London. p. 57.
41. THOMPSON, N. (1953). *Proc. Phys. Soc. Lond.*, **366**, 481.
42. SEEGER, A. (1957). *Dislocations and Mechanical Properties of Crystals.* Edited by FISHER *et al.* John Wiley, New York and London.
43. WHELAN, M. J., HIRSCH, P. B., HORNE, R. K. and BOLLMANN, W. (1957). *Proc. phys. Soc.*, **A240**, 524.
44. WHELAN, M. J. (1959). *Proc. R. Soc.*, **A249**, 114.
45. HOWIE, A. and SWANN, P. R. (1961). *Phil. Mag.*, **8** (6), 1215.
46. LOMER, W. M. (1951). *Phil. Mag.*, **42**, 1327.
47. COTTRELL, A. H. (1952). *Phil. Mag.*, **43**, 645.
48. ESHELBY, J. D., FRANK, F. C. and NABARRO, F. R. N. (1951). *Phil. Mag.*, **42**. 351.
49. FORTES, M. A. and RALPH, B., (1967). *Acta metall.*, **15**, 707.

Additional general references

1. NABARRO, F. R. N. (1967). *The Theory of Crystal Dislocations*. Clarendon Press, Oxford.
2. GILMAN, J. J. (1969). *Micromechanics of Flow in Solids*. McGraw-Hill, New York.
3. CAHN, R. W. and HAASEN P. (Eds.) (1970). *Physical Metallurgy* (3rd ed.) Chapter on Dislocations by J. P. HIRTH, p. 1223, North Holland, Amsterdam.
4. HIRSCH, P. B. (Ed.) (1975). *The Physics of Metals*, Vol. 2, Defects, Cambridge University Press.
5. NABARRO, F. R. N. (Ed.) (1979). *Dislocation in Solids*, Vol. 1, The Elastic Theory, North Holland, Amsterdam.

Chapter 4

Deformation of Metal Crystals

4.1 Deformation of face-centred cubic metal crystals

It is now realized that the shapes of the shear stress/shear strain curves of face-centred cubic metal crystals are more complex than the parabolic curves found in earlier work. The use of purer materials and more rigorous testing techniques has shown that the stress–strain curves possess three distinctive stages, 1, 2 and 3, which were described briefly in Chapter 2 (Fig. 2.23).

It should be emphasized that the three stages are not always present; one or more may disappear if the testing conditions are varied, e.g. if the temperature of deformation is substantially raised Stage 3 tends to predominate. It is thus impossible to characterize the stress–strain curves of single crystals by one typical curve or work-hardening parameter, and it is important to explore the variables which alter the hardening behaviour. The most important of these variables are:

1. the metal,
2. purity,
3. orientation of crystal,
4. temperature of deformation,
5. crystal size and shape,
6. surface condition.

Each stage of the stress–strain curve will be considered in turn, to determine the effect of the variables on the length of the stage and the rate of hardening in it. Subsequently, the structural changes as revealed by slip markings and dislocation arrangements in the crystal will be described.

4.1.1. *Stage 1 hardening*

This is a stage of low linear hardening which may be absent, or account for as much as 40 per cent shear strain. It was first described as a general phenomenon in gold and silver crystals by Andrade and Henderson,[9] who referred to it as 'easy glide', but earlier workers had occasionally noted

initial low rates of work hardening in single crystals of pure metals and
solid solutions.

(a) THE METAL AND ITS PURITY

Stage 1 hardening is now well established in crystals of aluminium,
copper, gold, silver and nickel; however, under similar testing conditions
these metals do not show Stage 1 hardening of the same duration. For
example, Stage 1 hardening is at most 4–5 per cent shear strain in aluminium crystals at room temperature, whereas in copper crystals the strain in
Stage 1 can be as high as 20 per cent. The differences cannot be attributed
to impurity levels.

Impurities can, however, influence in a pronounced way the extent of
Stage 1, but the mode of dispersion of the impurity is important. In
general, impurities which form a dispersion of a second phase, even when
present in low concentration, tend to reduce and finally eliminate Stage 1
hardening. The less pure aluminium used in the earlier single crystal experiments did not exhibit Stage 1 because the impurities present (primarily
silicon and iron) formed a fine dispersion of other phases. These small
inclusions encourage localized slip on other than the primary slip plane,
and this eliminates Stage 1 hardening. This effect is illustrated in Fig. 4.1
where stress–strain curves for two crystals of different purity aluminium,
but of similar orientation, are plotted. In the curve from the purer crystal
some Stage 1 hardening is visible, but it is completely absent in the less
pure metal which exhibits a parabolic stress–strain curve markedly different from that of the purer metal. On the other hand, impurities which go

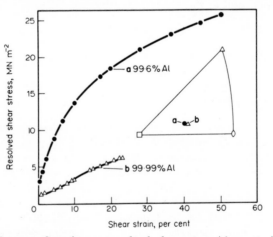

Fig. 4.1 Influence of purity on resolved shear stress/shear strain curves of
single crystals. **a**, 99·6 per cent Al; **b**, 99·99 per cent Al. (After Lücke and
Lange, 1952, *Z. Metallk.*, **43**, 55)

into solid solution tend to *increase* the extent of Stage 1, this is shown in Fig. 4.2 for three crystals of silver of different impurity level. As the impurity is mainly copper, the resulting crystal is a very dilute solid solution. This effect of extending the Stage 1 region of the stress–strain curve is even more pronounced in single crystals of more concentrated solid solutions, where this mode of hardening can dominate the stress–strain curve (Chapter 6).

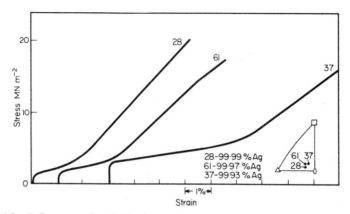

Fig. 4.2 Influence of soluble impurities on stress–strain curves of silver crystals. (After Rosi, 1954, *Trans. AIME*, **200**, 1009)

(b) CRYSTAL ORIENTATION

Crystals of different orientations exhibit a wide variation in both the extent of Stage 1 hardening and the rate of hardening in this stage. The extremes of behaviour are shown by 'soft' and 'hard' orientations. The former crystals possess orientations in the shaded area of the stereographic triangle (Fig. 4.3) which is remote from the sides of the triangle, particularly the [001]–[$\bar{1}$11] side which is approached by the tension axis during deformation. Soft crystals are thus those least likely to exhibit slip on the conjugate system, for the resolved shear stress on this system in the soft region of the triangle is low. On the other hand, hard orientations lie near or on the boundaries of the triangle where two or more slip systems have the same resolved shear stress, and will thus tend to operate; the steepest stress–strain curves are from crystals at [001] and [$\bar{1}$11], where 4 and 3 slip systems respectively operate. The distinction is thus clearly based on the probability of slip taking place to a limited degree on systems other than the primary system, even although the crystals are not precisely oriented for multiple slip to occur.

Figure 4.4. shows stress–strain curves from copper crystals covering a wide range of orientations. In the hardest orientation, Stage 1 hardening is absent and the crystal is exhibiting primarily Stage 2 hardening. On the

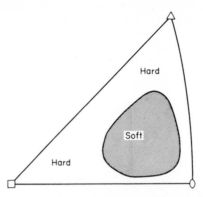

Fig. 4.3 Stereographic triangle showing areas of soft and hard orientations

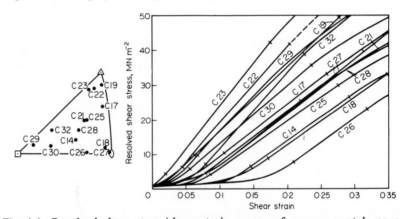

Fig. 4.4 Resolved shear stress/shear strain curves of copper crystals as a function of orientation. (After Diehl, 1956, *Z. Metallk.*, **47**, 331)

other hand, the softest crystal gives about 15 per cent shear strain in Stage 1. This behaviour has been confirmed for aluminium[10] and for silver.[11] The rate of hardening during Stage 1 as a function of orientation has been particularly considered by Diehl[12] for copper crystals. He found that θ_1 (dτ/dϵ in Stage 1) increased in the same way as the length of Stage 1 decreased, with maximum values near [$\bar{1}$11] and [001] and minimum values towards [011] (Fig. 4.5).

The extent of Stage 1 hardening is clearly defined by the degree of rotation of the crystal axes possible during (tensile) testing. However, the end of Stage 1 does *not* correspond with the tensile axis coinciding with the [001]–[$\bar{1}$11] symmetry boundary of the stereographic triangle because, prior to this point, limited slip on other systems occurs, which is sufficient to alter the rate of hardening substantially. If the rotation of the specimen axis is prevented during slip, much more extensive Stage 1 hardening should result. This has been achieved in two ways.

1. Experiments in shear† with no tensile or compressive stress components. Shear tests have been carried out on copper[13] which gave strains in Stage 1 over 20 per cent, much greater than observed in similar crystals deformed in tension.
2. Deformation in alternating tension and compression (high strain fatigue). Patterson[14] showed that such tests on copper crystals did not result in a rotation of crystal axes, with the result that shear strains of over 60 per cent were obtained in Stage 1 at room temperature. As in normal tensile tests, crystals with near symmetrical orientations showed no Stage 1 hardening.

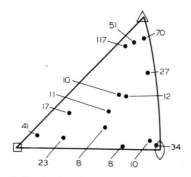

Fig. 4.5   Rates of work hardening (MN m⁻²) of 99·98 per cent copper crystals in Stage 1, as a function of orientation. (Diehl, 1956, *Z. Metallk.*, **47**, 331)

(c) TEMPERATURE

In general as the temperature of deformation is lowered the extent of Stage 1 hardening is increased.[15] This change is associated with a rise in the critical shear stress for slip, and thus acts in the same sense as that of increasing the amount of impurity in solid solution. When the critical shear stress τ_0 is raised on the primary slip system, it is raised also on the various secondary systems. Thus, assuming that $d\tau/d\epsilon$ is unchanged, larger strains are necessary before these other critical shear stress values are approached, and the Stage 1 hardening ended.

In Fig. 4.6 the stress–strain curves of identically oriented copper crystals (near [011]) are plotted for a wide range of deformation temperatures.[16] At 623 K only a short Stage 1 is apparent, but this stage becomes more prolonged as the temperature is dropped to 93 K. In these tests there was little change in θ_1, particularly at the lower temperatures, a fact which has

† See Section 4.3.5.

been confirmed by other experiments on copper.[17] Similar results have been obtained on aluminium crystals.[18] However, recent experiments[19] indicate that θ_1, corrected for the change in shear modulus with temperature, increases with increasing temperature. The effect is greater with crystals nearer the symmetrical orientations, i.e. in crystals exhibiting a small easy-glide range. This increase in work-hardening rate can be accounted for by the lowering of the critical shear stress for slip in secondary systems as the temperature is raised, which makes slip on these systems more likely to take place.

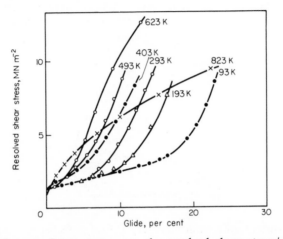

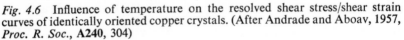

Fig. 4.6 Influence of temperature on the resolved shear stress/shear strain curves of identically oriented copper crystals. (After Andrade and Aboav, 1957, *Proc. R. Soc.*, **A240**, 304)

(d) CRYSTAL SIZE AND SHAPE

Many results on crystals of one metal of similar orientation and purity show distinct differences, which can only be accounted for by changes in size and shape of the specimen. The effect of crystal size has been exhaustively studied by Suzuki *et al.*[20] on copper crystals (99·98 per cent) of very similar orientation but varying in radius from 0·1 to 0·89 mm.

Typical results are shown in Fig. 4.7 where the easy-glide strain varies from about 16 per cent for the 0·89 mm radius crystal to over 40 per cent for a crystal of 0·105 mm radius. In general, θ_1 was unchanged as the crystal size was altered. Garstone *et al.*[18] found similar results for identically oriented copper crystals in the size range 2·4–4·8 mm diameter. These results suggest that low rates of hardening associated with Stage 1 are a result of providing short glide paths for dislocations, so that they mostly reach the surface of the crystal and disappear, rather than become trapped inside the crystal.

Such a view is confirmed by experiments which demonstrate that crystal shape can be a significant influence on the deformation process. For example, in rectangular aluminium crystals, the first slip system to operate is not necessarily the one with the highest resolved shear stress, but that which involves the smallest glide path through the crystal.[21] More recently McKinnon† has shown that in the case of two identically oriented aluminium crystals of different section, one with a short glide path and the other with a long path on the primary slip plane, the crystal with the short glide path gave a longer easy glide region.

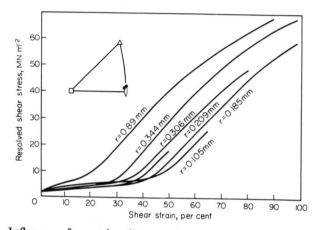

Fig. 4.7 Influence of crystal radius on the extent of Stage 1 hardening in copper crystals. (After Suzuki *et al.*, 1956, *J. phys. Soc. Japan*, **11**, 382)

(e) SURFACE CONDITION

Andrade and Henderson found that silver crystals showed much less easy glide when the surface was covered with a thin oxide film. Many experiments illustrating a similar role of an oxide film have been done on the hexagonal metals zinc and cadmium. Stage 1 hardening is also reduced or eliminated by plating the surface of copper crystals with a thin metal layer.[18] Similar experiments on copper crystals by Rosi[22] demonstrated that an electro-deposited silver layer, 4×10^{-4} cm thick, reduced Stage 1 in the case of soft crystals. On subsequent annealing to diffuse the silver into the copper, the easy-glide range increased very substantially in accord with the results already described for solid solutions.

The effects of crystal size and shape, also the role of surface layers, give some further clues as to the real nature of the deformation process in Stage 1 hardening. The fact that the length of slip path or the state of surface can be influential indicates that the dislocations are moving

† Quoted in Ref. 6.

relatively large distances in the crystals without meeting obstructions. A high density of internal obstructions (e.g. slip on other systems) would lead to a higher work-hardening rate as, for example, in the case of symmetrically oriented crystals. On the other hand, a low density of obstructions will mean that the surface assumes a significant role as a macroscopic obstacle, particularly if an oxide film prevents dislocations from moving completely out of the crystal.

4.1.2. *Microstructure during Stage* 1

There are three ways in which the deformation process can be studied directly, using techniques of microscopy. Firstly, the surface features, slip lines and bands can be examined in the optical and electron microscopes to give information about the dislocations which have left the crystal during the deformation. Secondly, the dislocation population remaining in the crystal after a given deformation can be studied indirectly by use of selective etching techniques, or in some cases by using a precipitate to mark the dislocations. Thirdly, the dislocation arrays can be studied directly using thin-foil electron metallography but, as the process of thinning is likely to disturb and alter the arrays, it is preferable to examine thin foils prepared parallel to the primary slip plane. The latter is a specialized technique which has only been used occasionally (e.g. Refs. 25, 94).

In a typical virgin crystal of copper a dislocation density of about $10^{10}\,m^{-2}$ is usual, but this can be reduced to $10^8\,m^{-2}$ by annealing using temperature cycling.[7] As the stress approaches the macroscopic yield stress, up to 75 per cent of the pre-existing dislocations have already moved to produce microflow,[23] that is, small plastic strains prior to the macroscopic yield stress, and by the time Stage 1 deformation commences the dislocation density of all crystals is not less than $10^{10}\,m^{-2}$.

During Stage 1, long fine slip lines are produced which cover the crystals uniformly (Fig. 4.8), but towards the end of the easy-glide range traces of slip on other systems can be detected with some difficulty. McKinnon[24] has been able to show that secondary slip is more common in the *interior* of a deformed crystal, by revealing the slip traces by a precipitation reaction. As many as three systems were detected in the interior when only the primary system could be seen on the surface. Other evidence for slip on secondary systems has come from the study of inhomogeneities which develop even during Stage 1 deformation, for example kink bands and bands of secondary slip,† both types of band being preferred regions for slip on secondary systems.[18]

Etch pit studies[7,23] have been made of dislocation distributions during easy glide, which have revealed piling up of dislocations on the primary planes at sub-boundaries. Also there is a substantial and early increase of

† See Chapter 8.

Fig. 4.8   Slip lines in Stage 1 deformation of copper crystals. Replica. (After Diehl, 1955, *Z. Metallk.*, **46**, 650)

Fig. 4.9   Etch pits on the surface of a copper crystal deformed 1·5 per cent in tension. × 140. (After Livingston, 1962, *Acta metall.*, **10**, 229)

dislocations on the cross slip plane. A typical etch pit distribution is shown in Fig. 4.9 for a copper crystal after 1·5 per cent elongation. At this stage the average slip distance L, calculated from $\epsilon = NbL$, where N = number of dislocations/cm², gives a value of 1–2 mm. This is the macroscopic distance which the other indirect evidence supports (e.g. the size effect).

Recently thin-foil electron microscopy studies[8, 25, 94] have been made on specimens cut parallel to the primary slip plane. These confirm the etch pit studies in so far as the dislocation distributions are very irregular, and tangles (corresponding to etch pit sub-boundaries) occur periodically. In addition, there is much evidence for the existence of dislocation dipoles in between the sub-boundaries. These probably start out as very long dipoles which are continuously cut during subsequent deformation. A typical arrangement in a copper crystal during Stage 1 is shown in Fig. 4.10. The ends of dipoles in the primary plane tend to lie in the [10$\bar{1}$] direction (for example, at S), which is the trace of the direction of intersection of the cross-slip plane with the primary. This strongly suggests that dislocations

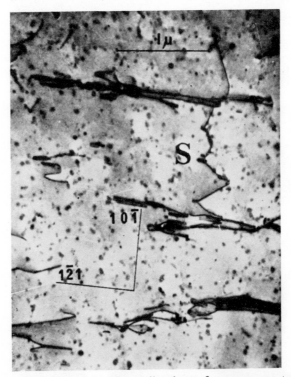

Fig. 4.10 Section along the primary slip plane of a copper crystal deformed in Stage 1. Thin-foil electron micrograph. (After Hirsch and Steeds, 1964, *Phil. Mag.* Advances in Physics, **13**, 50)

have moved on the cross-slip plane, and thus is confirmation of the earlier observations about secondary slip in Stage 1.

4.1.3. Dislocation densities

All methods of detecting dislocations show that the dislocation density in a metal increases with increasing strain. Mitchell[5] has collected data from several sources on the dislocation densities determined from etch pits in copper crystals deformed in Stage 1, and found the following relationship was valid.

$$\tau = \tfrac{1}{2}Gb\rho^{1/2} \qquad\qquad \textbf{4.1}$$

where ρ = dislocation density, τ = flow stress, and G = shear modulus.

In Fig. 4.11 τ/G is plotted against $b\rho^{1/2}$, and the data can be fairly

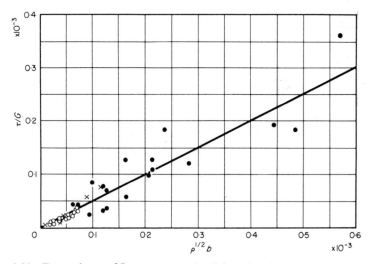

Fig. 4.11   Dependence of flow stress on the dislocation density. (After Mitchell, 1964, *Progress in Applied Materials Research*, 6)

accurately represented by a straight line of slope 0·5. Young[26] has found a direct proportionality between ρ and the shear strain ϵ.

$$\rho = 2\cdot8 \times 10^{8}\epsilon \qquad\qquad \textbf{4.2}$$

Further work on dislocation densities has been done on silver, copper,[27] copper alloys,[28] nickel and nickel alloys,[29] using thin-foil electron microscopy which has enabled higher densities to be determined. These results, which have been correlated by Mitchell,[5] fit the relationship **4.1** to much higher stress levels, well within Stage 2 hardening.

4.1.4. *Stage 2 hardening*

The work-hardening coefficient θ_{II} in this stage is approximately ten times as large as θ_I, and in general it is Stage 2 hardening which accounts for the much greater work hardening exhibited by face-centred cubic metal crystals in contrast to crystals of the hexagonal metals such as zinc and cadmium. At low temperatures Stage 2 hardening dominates the stress–strain curve, and is thus a phenomenon of considerable importance. It should be emphasized that Stage 2 usually commences when the crystal tension axes are still well within the stereographic triangle: the onset of Stage 2 is not correlated with the commencement of duplex slip on reaching the $[001]$–$[\bar{1}11]$ boundary.

(a) THE METAL

Aluminium does not exhibit a well-defined Stage 2 hardening at room temperature, Stage 1 merging into Stage 3 (Fig. 4.12). On the other hand,

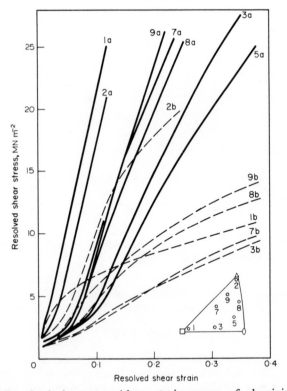

Fig. 4.12 Resolved shear stress/shear strain curves of aluminium crystals at **a,** 77 K and **b,** room temperature. (After Staubwasser, 1957, *Dislocations and the Mechanical Properties of Crystals*, John Wiley, New York and London)

copper crystals possess a well-defined Stage 2 at room temperature (Fig. 4.4), because in contrast to aluminium the transition to Stage 3 occurs only at high stress levels. In aluminium Stage 2 hardening is best studied at 77 K when Stage 3 is largely suppressed (Fig. 4.12).

(b) ORIENTATION

There is a definitely orientation dependence of the rate of Stage 2 hardening but it is not nearly as pronounced as for Stage 1. However, the orientation dependence follows the same general trend in so far as crystals with tension axes towards the [011] corner of the stereogram have a lower rate of work hardening in Stage 2 than crystals oriented near the [001]–[$\bar{1}$11] symmetry boundary. This is well illustrated for copper in Fig. 4.4 at room temperature, but this behaviour is also typical of other metals such as silver[22], nickel[30], and at the appropriate temperature (e.g. 77 K) for aluminium.[31] Absolute rates of hardening for Stage 2 cannot be given, because these values are sensitive not only to purity and nature of the metal but also to the size and shape of the specimen. Figure 4.13 gives

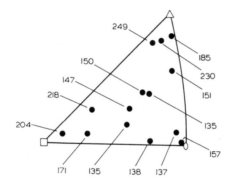

Fig. 4.13   Rates of work hardening (MN m^{-2}) of copper crystals in Stage 2 as a function of orientation. (After Diehl, 1956, z. Metallk. **47**, 331.)

some values of θ_{II} for copper crystals at room temperature, where the rate of hardening is shown to vary from 135 to over 200 MN m^{-2}. Tests at lower temperatures on copper confirm that this is also the general pattern of behaviour at 93 K and as low as 4·2 K.

(c) TEMPERATURE

Temperature has a significant effect on the duration of Stage 2 hardening but the rate of hardening, other factors being constant, is very insensitive to temperature (Fig. 4.6). The transition from Stage 2 to Stage 3 is raised to higher stresses the lower the temperature, until at very low temperatures Stage 2 predominates over most of the stress–strain curve. Conversely, as the temperature is raised, Stage 3 becomes more pronounced until Stage 2 is entirely eliminated and the stress–strain curve is essentially composed of Stage 3. Such results were obtained by Andrade and Aboav[16] for copper over the temperature range 93–673 K, while Blewitt, Coltman and Redman[32] have obtained similar results for 99·999 per cent copper in the range 4·2–300 K. They found that at 300K Stage 2 ended at about 20 per cent shear strain ($\tau \simeq 30$ MN m^{-2}), whereas at 4·2 K, Stage 2 comprised over 70 per cent glide and a resolved shear stress of 120 MN m^{-2} was reached without deviation from linearity.

(d) CRYSTAL SIZE, SHAPE AND CONDITION

Some influence of size and shape of the crystal on the stress–strain curves during Stage 2 has been demonstrated, but the effects are small relative to those observed during Stage 1. The work of Suzuki et al.[20] on copper revealed that with 'soft' crystals, i.e. oriented near [011] or the centre of the triangle, there was no size effect. However, near [001] there was a definite size effect, small crystals hardening more rapidly than large crystals. McKinnon,† on the other hand, found that aluminium crystals with a long glide path hardened more readily than those with a short glide path.

The role of surface films or electro-deposited layers is also much less significant in Stage 2 than in Stage 1 hardening; in tests on copper crystals θ_{II} was uninfluenced by electro-deposited layers of silver, but after annealing to allow diffusion, θ_{II} was measurably reduced.

(e) MICROSTRUCTURE DURING STAGE 2

During Stage 2 the slip line patterns are very dependent on orientation,[6,33] particularly if the crystal tension axis is near the symmetry boundaries of the stereographic triangle, when the occurrence of slip on one or more of the secondary planes in addition to the primary slip plane is commonly observed. Such crystals always exhibit a higher work hardening rate than 'softer' crystals with axes towards the [011] corner of the triangle. Garstone and co-workers[18] found that copper crystals with 'soft' orientations showed slip on the cross-slip plane in limited regions soon after the end of easy glide. As the stress in Stage 2 increased, slip on several {111} secondary systems was observed. Similar observations were made by McKinnon[24] on aluminium crystals. In these cases the amount of secondary slip was small, and it should be emphasized that the rotation of the axes during deformation had not yet brought the crystals near the [001]–[$\bar{1}$11] boundary.

† Quoted in ref. 6.

More recently it has been shown by electron microscopy that the length of slip lines is related inversely to the shear strain in Stage 2.

$$L_s = \frac{\Lambda}{(\epsilon - \epsilon')}$$ **4.3**

where L_s is the slip line length on crystal surfaces where *edge* dislocations *emerge*, and is thus related to the slip distance of *screw* dislocations, Λ is a constant, ϵ is the total strain and ϵ' the strain at the end of Stage 1 hardening. Measurements of slip lines in copper, nickel and nickel–cobalt alloys have been made by Mader,[34] Kronmuller[35] and Pfaff[36] and summarized by Mitchell[5] as shown in Table 4.1.

Table 4.1   Slip line measurements on f.c.c. metals (Stage 2 hardening)

Metal	Temperature K	Λ 10^{-4} cm	Slip line height (Burgers vectors)
Copper	293	4·0	20
Nickel	90	5·9	31
Nickel–20 per cent cobalt	293	6·2	32
–40 per cent cobalt	293	6·5	25
–50 per cent cobalt	333	6·0	15

These results show that the slip line height which has been calculated from the strain and L_s is fairly constant, and is independent of strain.

Etch pit techniques have shown that the dislocation patterns typical of Stage 1 are still evident at least in the early part of Stage 2. There is a coarse substructure of dislocations developing which can be primarily related to the presence of inhomogeneities such as deformation bands, but on a finer scale there are more complex tangles of dislocations lying roughly parallel to the principal slip plane.

Thin-foil studies by Hirsch and Steeds[8,94] reveal that the dislocations are continuing to form cell boundaries during Stage 2. While individual micrographs tend to vary a great deal, most sections along the active slip plane reveal regions clear of dislocations surrounded by dislocation tangles which tend to be aligned along ⟨110⟩ directions (Fig. 4.14). It is possible to distinguish between dislocations on the primary and secondary systems by choice of the diffraction image in the electron microscope; this technique has revealed that dipoles and edge dislocations of the primary slip system are a prominent feature, especially early in Stage 2, and they help build the dislocation walls (Fig. 4.15a). However, on adjusting the diffracting conditions so that the primary dislocations are invisible, the walls are seen also to contain dislocations on other systems (Fig. 4.15b) (forest dislocations). Seeger and colleagues[8] have also carried out thin-foil studies on sections parallel and normal to the primary slip plane in nickel–cobalt alloy crystals and on copper crystals. One of the main features

Fig. 4.14 Copper crystals deformed into Stage 2. Thin-foil section along slip plane. (After Steeds, 1966, *Proc. R. Soc.*, **A292**, 343)

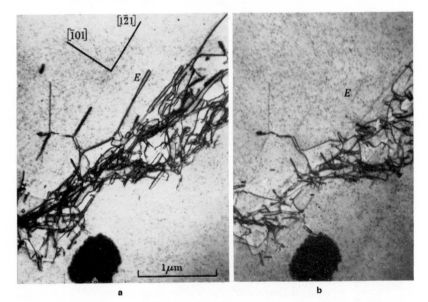

a   b

Fig. 4.15 Copper crystals in Stage 2. Slip plane section. Two diffraction conditions. (After Steeds, 1966, *Proc. R. Soc.*, **A292**, 343)

observed was linear arrays of dislocations, referred to as dislocation 'braids' as they are essentially one-dimensional, in contrast to planar arrays found in sub-boundaries. On close examination the braids are seen to contain many small dislocation loops and small dipoles. Typical dimensions of the braids are 2–4 microns diameter and 15 microns long. The density and length of the braids was comparable with the slip lines observed on the surface, and it may be that they are a significant feature of the work-hardening process. However, recent work[94] suggests that the braids are really sections through larger structures such as incipient cell walls.

4.1.5. *Role of secondary slip in Stages* 1 *and* 2

The experimental results show that the stress–strain curves are very dependent on orientation in Stages 1 and 2. Seeger and co-workers have suggested[1] that the soft orientations which give not only the longest easy glide but also the lowest value of θ_I are deforming entirely on the primary system in Stage 1. These crystals in Stage 1 are thus very close in their behaviour to that of suitably oriented crystals of cadmium or zinc. Soft crystals oriented in the region of the stereographic triangle towards [011] are the least likely to exhibit occasional secondary slip, and are thus least likely to form Lomer–Cottrell locks, which result from the combination of primary dislocations with certain suitably oriented secondary dislocations (Chapter 3). Seeger *et al.* assumed that the higher the resolved shear stress on a secondary system the more likely localized slip would be on this system. Their analysis shows that in general the occurrence of Lomer–Cottrell locks is more probable as the orientation moves away from the area near [110] towards the [001]–[Ī11] boundary, with a maximum probability at the [Ī11] and [001] poles. This is the same trend as for the slope θ_I of the stress–strain curves in Stage 1, which reaches a maximum at these orientations.

Clarebrough and Hargreaves[6] have taken this approach a stage further by adopting the suggestion of Friedel[37] that to block a primary dislocation source it is necessary to have sessile dislocations in at least two directions in the primary slip plane, i.e. any two of the ⟨110⟩ slip directions.

The pairs of slip systems on which dislocations will combine to give Lomer–Cottrell locks in these three directions are shown in Table 4.2; the symbols used correspond to those in Fig. 2.8.

Table 4.2 Slip system combinations giving Lomer–Cottrell locks in the primary slip plane[6]

Direction of sessile dislocation	Pairs of slip systems
[011]	BIV (11Ī) [101]—CI (1Ī1) [Ī10]
	BII (11Ī) [1Ī0]—CIII (1Ī1) [Ī01]
[1Ī0]	BIV (11Ī) [101]—AVI (111) [01Ī]
	BV (11Ī) [011]—AIII (111) [10Ī]
[101]	BV (11Ī) [011]—DI (1ĪĪ) [ĪĪ0]
	BII (11Ī) [1Ī0]—DVI (1ĪĪ) [01Ī]

It is necessary to know the values for the critical resolved shear stress on the secondary systems for various orientations in the stereographic triangle. These have been calculated[38] and curves of constant shear stress plotted in the stereographic triangle (Fig. 4.16), the stresses being expressed

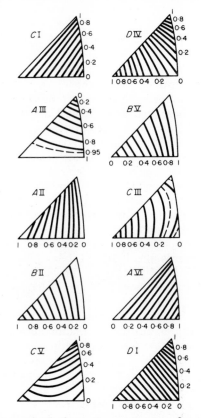

Fig. 4.16   Constant resolved shear stress contours for the ten most likely secondary slip systems. (After Diehl *et al.*, 1954, *Z. Metallk.*, **45**, 489)

as a fraction of the resolved shear stress on the primary system, *B*IV, in the standard nomenclature (Fig. 2.8). Use of these diagrams together with the information in Table 4.2 about Lomer–Cottrell locks gives a fairly satisfactory qualitative explanation of the orientation dependence of the stress–strain curves. The argument has been applied both to Stage 1 and Stage 2 hardening because these stages exhibit the same general trends as far as dependence on orientation is concerned. There is experimental evidence for secondary slip even in Stage 1, so it is reasonable to suggest that sessile dislocations are already forming; however, in Stage 2 the axes of the crystals have rotated towards the regions of the triangle where secondary slip is more likely, and consequently the incidence of sessile dislocations is much higher.

Thin-foil electron microscopy has recently provided some convincing evidence for the occurrence of Lomer–Cottrell locks during deformation.[4] However, it seems likely that other types of barrier exist, which also arise as a result of interaction of dislocations on secondary planes with those on primary planes. The electron micrographs reveal the presence of dipoles which could arise by the chance meeting of dislocation loops slipping in adjacent planes. It is possible that these configurations lead to some hardening, but the main feature in Stage 2 is the formation of dislocation networks, complex in nature, which are difficult to treat theoretically in a simple manner.

4.1.6. *Stage* 3 *hardening*

This stage is the parabolic part of the stress–strain curve following Stage 2, and is characterized by a gradually decreasing rate of hardening. It is much less sensitive to the variables discussed in connection with Stages 1 and 2 with the exception of temperature, which provides the clue to the mechanisms characteristic of this stage.

TEMPERATURE

Stage 3 hardening becomes more pronounced the higher the temperature and eventually dominates the whole stress–strain curve at high temperatures (Fig. 4.6). Aluminium crystals even at room temperature show very

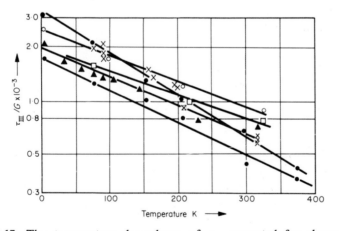

Fig. 4.17 The temperature dependence of τ_{III} corrected for changes in the shear modulus. (After Mitchell, 1964, *Progress in Applied Materials Research*, **6**)

limited Stage 2 hardening while Stage 3 is very extensive. At temperatures below 100 K, Stage 3 almost disappears (see Fig. 4.12) and the high work hardening characteristic of low temperatures is achieved by the predominance of Stage 2. It thus seems likely that the mechanisms in Stage 3 are thermally activated. It is further obvious from the data already presented that aluminium requires a lower temperature to enter into Stage 3 than does copper or, alternatively, at a given temperature, aluminium passes into Stage 3 hardening at a much lower stress than does copper.

The stress τ_{III} for the commencement of Stage 3 hardening decreases rapidly with increasing temperature according to a logarithmic law.

$$\ln\left(\frac{\tau_{III}}{G}\right) = \ln\left[\frac{\tau_{III}(o)}{G(o)}\right] - BT \qquad \textbf{4.4}$$

where $\tau_{III}(o)$ and $G(o)$ are the stress and shear modulus at 0 K. This temperature dependence has been confirmed for crystals of copper, silver, gold, nickel, aluminium and lead as well as for several copper-base alloys. The results of a number of workers for copper are plotted in Fig. 4.17.

4.1.7. *Microstructure during Stage 3*

At the start of Stage 3, the slip lines undergo a marked change and the fine lines are replaced by well-defined coarse slip bands. These broad bands, which are easily visible in an optical microscope, were formerly thought to be typical of single crystal deformation, largely because they are found on aluminium crystals even after small strains at room temperature. However, this is now explained by the fact that aluminium at this temperature is essentially deforming in Stage 3.

Seeger and co-workers have established that the development of coarse slip in copper and aluminium is associated with the onset of Stage 3 and have studied the detailed structure by examination of surface replicas in the electron microscope. This work shows at this stage of the deformation, the slip lines cluster together to form the bands which are comparatively short (Fig. 4.18) and connected by slip on other systems. The connecting slip (for example, at A) is called 'cross-slip', which has been known as a familiar feature of slip in aluminium for many years.[39] The cross-slip plane is also {111} and the slip direction is identical with that in the primary slip plane. Figure 4.19a shows a model of the process where the slip plane, because of frequent cross-slip, has become quite corrugated. Slip bands viewed on a face normal to the slip direction show the cross-slip clearly (Fig. 4.19b), but the bands are quite straight when viewed on faces which contain the slip direction.

The slip bands in Stage 3 are comparable in length to the slip lines in the latter part of Stage 2, so it is assumed that the dislocations are blocked periodically by substantial obstacles, and that this process of obstructing slip dislocations has continued to a stage when some of the blocked dislocations have been able to sidestep the obstacles by localized slip on the

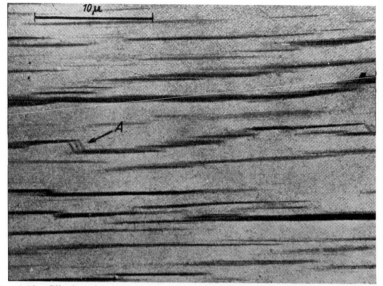

Fig. 4.18   Slip bands on a copper crystal in Stage 3 (replica). (After Mader)

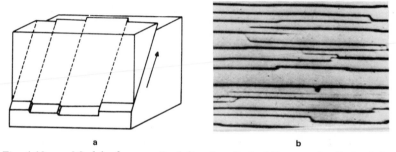

a   b

Fig. 4.19   **a,** Model of cross-slip (after Jaoul); **b,** Micrograph of aluminium corresponding to top face of model. × 300. (After Cahn, 1950, *J. Inst. Metals,* **79,** 129)

cross-slip plane. An edge dislocation on the primary slip plane would find movement on an alternative octahedral plane very difficult, because this would involve non-conservative movement, i.e. the dislocation would have to climb. On the other hand, a screw dislocation, or the screw-component of a dislocation loop, would not have this difficulty *provided* the alternative plane had a common slip direction so the dislocation would possess the *same* Burgers vector.

To understand the marked difference in behaviour of copper and aluminium, in so far as Stage 3 occurs much more readily in the latter, it is necessary to discuss differences in the nature of the dislocations in the two metals. Copper has a lower stacking fault energy (40 mJ m^{-2}) than aluminium (200 mJ m^{-2}), so that the dislocations in copper will be more widely dissociated into partial dissociations separated by a stacking fault layer (Chapter 3). Seeger[1] gives a value of $12b$ for the separation distance in copper, but the exact value is very sensitive to the actual value of γ.

The model which Seeger advances for late Stage 2 hardening envisages pile-ups of dislocations at obstacles; these could be Lomer–Cottrell locks, but recent studies suggest that they are more likely to be dislocation tangles. A sessile dislocation model is shown in Fig. 4.20, where for simplicity there

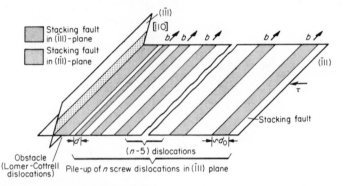

Fig. 4.20 Pile-up of screw dislocations against a Lomer–Cottrell lock. (After Seeger, 1957, *Dislocations and Mechanical Properties of Crystals*, John Wiley, New York and London)

is a series of extended screw dislocations blocked by a Lomer–Cottrell lock. The magnified stress at the head of the pile-up of dislocations has caused a reduction in the dissociation distance d_0 of the leading dislocations. However, if these dislocations are to by-pass the obstacle by cross-slip in Stage 3, they must first associate and remove the stacking fault. Otherwise a difficult geometrical situation would be created which would be extremely unfavourable energetically. The method requiring the least energy involves the formation of a constriction in the leading dislocation of length, say, $2b_0$ which, being screw in character, is able to cross-slip on an alternative slip plane $(1\bar{1}1)$ containing the same slip direction [110] as the primary slip plane $(\bar{1}11)$. As soon as the loop of constricted screw dislocation moves on to the new cross-slip plane, it again is free to dissociate, which it does forming a double loop bounded by partial dislocations and containing a layer of stacking fault (Fig. 4.21). When sufficiently far away from the original obstacle this loop is free, under the applied stress, to reassociate and to move back into the original slip system, but on a plane some distance away from the original slip plane.

For the process of constriction and movement on to the cross-slip plane, an activation energy is necessary and this in turn will depend on the size of the constriction. Schoeck and Seeger[40] have calculated a value of about 1·0 eV for aluminium, which indicates that cross-slip should be frequent in this metal at a stress of around 1 MN m⁻² just above room temperature. However, a much higher energy is needed for copper because of the much larger equilibrium separation of partial dislocations, so cross-slip would need either much higher stresses or temperatures as the experimental results have shown. Seeger derived the following relationship for the shear

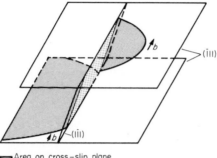

Area on cross–slip plane swept out by dislocation
Area on primary plane swept out by dislocation

Fig. 4.21   Cross-slip of part of an extended screw dislocation by formation of a constriction. (After Seeger, 1957, *Dislocations and Mechanical Properties of Crystals*, John Wiley, New York and London)

stress necessary to re-associate two partial dislocations in the pile-up model.

$$\tau = \frac{1}{n}\left(\frac{\sqrt{2}}{4\pi}\,G - \frac{2\gamma}{b}\right)$$ 4.5

n = number of dislocations in the pile-ups from which the dependence of τ on γ the stacking fault energy can be seen.

This approach thus explains why the shear stress/shear strain curves of aluminium and copper crystals are so different, for example at room temperature. It is possible to account for differences in the dominance of Stages 2 and 3 of the work hardening in terms of the very different stacking fault energies these two metals possess. This in turn influences important dislocation movements such as cross-slip, which is one of the ways in which obstacles to slip can be surmounted.

On the other hand, if instead of comparing the deformation behaviour of different metals at the same *strain*, the comparison is made at the same stage of *hardening*, it is seen that the face-centred cubic metals fall into a regular pattern of behaviour. Each metal displays the same phenomena in the same order, and only differ in the stresses and/or temperatures needed to bring about the dislocation reactions which control the work-hardening behaviour.

4.1.8. *The temperature dependence of the flow stress*

The detailed results described in the previous section are concerned primarily with the stress–strain curve as a whole. To examine the effect of temperature, crystals of identical orientation were tested separately at the different temperatures. However, this approach suffers from the fault that similar strains at different temperatures even in identical crystals involve different dislocation distributions, and in this way irreversible effects of temperature can take place. To eliminate this difficulty a method was developed by Cottrell and Stokes[41] which enables the flow stress at two different temperatures to be compared in the same crystal for the same dislocation distribution. A crystal is deformed to a given strain at a temperature T_1 and the flow stress τ_{T_1} measured (Fig. 4.22). The crystal is then

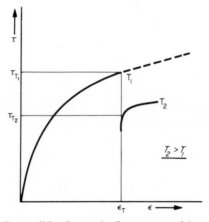

Fig. 4.22   Reversible change in flow stress with temperature

unloaded, the temperature changed to T_2 and the new flow stress τ_{T_2} measured at the same strain ϵ_T. It is normally better to carry out the first measurement at the lower temperature T_1 in case the higher temperature T_2 causes structural changes after the flow stress is measured. The change in flow stress measured is referred to as the *reversible* change in flow stress to distinguish it from irreversible changes which arise from different dislocation distributions obtained when crystals are deformed at different temperatures.

It was found for aluminium crystals that the ratio τ_{T_2}/τ_{T_1} was a constant, independent of strain, if the very early stages of the stress–strain curve were neglected. This is usually referred to as the Cottrell–Stokes law. Subsequently, the law has been confirmed for copper,[42] silver[43] and nickel crystals.[44] The law holds reasonably well in Stages 2 and 3 of the stress–strain curve, but it breaks down in Stage 1.[45]

If an aluminium crystal is deformed well into Stage 2 at a low temperature, e.g. 90 K, on restraining at a much higher temperature, e.g. 293 K, Cottrell and Stokes have shown that the low temperature hardening is partly removed in a catastrophic way with the development of a sharp yield point. They called this phenomenon *work softening*; it is really a way of causing a sharp transition from Stage 2 to Stage 3 hardening by an upwards temperature change during deformation. Similar results have been obtained with copper crystals,[42] but a somewhat higher temperature is needed to obtain a comparable effect. It has been confirmed[46] that work softening occurs in aluminium crystals because dislocations blocked by obstacles in the primary plane are able to avoid them by cross-slip. Microscopic examination of work-softened crystals reveals short coarse slip bands with regions of cross-slip.

While the flow stress ratio is constant with strain, it is very dependent on the temperature. This has been examined by referring flow stress ratios to 0 K by extrapolation, and plotting the ratio against temperature (T). To achieve greater precision, the ratios in Fig. 4.23 have been corrected

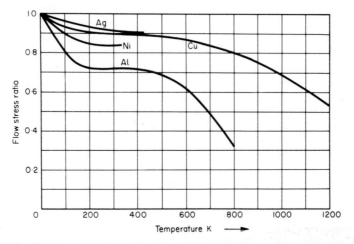

Fig. 4.23 Temperature dependence of the flow stress ratio for several f.c.c. metals. (After Mitchell, 1964, *Progress in Applied Materials Research*, **6**)

for the variation of shear modulus with temperature so that $\left(\dfrac{\tau_T/G_T}{\tau_0/G_0}\right)$ is plotted against T. The experiments reveal that the ratio is fairly constant for Ag, Cu, Ni and Al in the range 200–400 K, but rises steeply below 200 K and falls off rapidly above 500 K, in the case of aluminium and copper.

Changes in strain rate during deformation can cause similar effects to changes in temperature. An increase in strain rate gives less time for thermally activated events: consequently, it is equivalent to a lowering of the temperature of deformation. In practice, an increase in strain rate by, say, a factor 10 will increase the flow stress by $\Delta\tau$. Experiments on copper crystals[47] have shown that $\Delta\tau$ is proportional to τ beyond the early stages of plastic deformation. Basinski[43] has used an expression for the strain rate sensitivity of the flow stress which is temperature compensated,

$$\frac{1}{T}\left(\frac{\partial \ln \tau}{\partial \ln \dot{\epsilon}}\right)_T$$

which varies with temperature in a similar way to the flow stress ratio discussed above.

4.2   The deformation of body-centred cubic metal crystals

4.2.1.   *Crystallographic observations*

It is not surprising that the first body-centred cubic metal crystals to be studied during plastic deformation were those of α-iron. The careful classical work of Taylor and Elam[48,49] showed that the slip always occurred in the closest packed directions, viz. $\langle 111 \rangle$, on a number of different planes which belonged to the $\langle 111 \rangle$ zone, and proposed the name 'pencil' glide. If a bundle of hexagonal pencils is taken as an analogy, then the pencils can slide relative to one another in the slip direction on several different planes defined by the sides of the pencils. Slip viewed parallel to the slip direction could then be straight, but, in any other direction and particularly normal to the slip direction, would appear wavy (Fig. 2.6).

Fahrenhorst and Schmid[50] found that a $\{123\}$ slip plane and a $\langle 111 \rangle$ slip direction best suited their observations. Barrett, Ansel and Mehl,[51] after a detailed investigation of α-iron and silicon iron crystals in the range 77 K to 770 K, found that all slip traces could be attributed to the $\{110\}$, $\{112\}$ and $\{123\}$ planes, but in all cases the $\langle 111 \rangle$ slip direction remained unchanged. Opinsky and Smoluchowski[52] concluded that the choice of plane was determined by the maximum resolved shear stress criterion, and divided the basic stereographic triangle into areas indicating the orientations where planes of the type $\{110\}$, $\{112\}$ and $\{123\}$ would be preferred.

It should, however, be emphasized that in these earlier experiments the iron used was not of a high degree of purity, particularly with respect to interstitial atoms such as carbon, nitrogen and oxygen. There is also the complicating factor that at low temperatures deformation also occurs by twinning; narrow twins, often called Neumann bands, accompany slip bands (Chapter 8). Recent work indicates that at low temperatures and high rates of strain, slip in iron is confined to $\{110\}$ planes but, at room temperature and above, pencil glide is predominant.[53,54]

Andrade and co-workers[55] investigated single crystals of the body-centred cubic alkali metals sodium and potassium, also iron and molybdenum, and found that the precise glide plane operating depended on the temperature of deformation. At low temperatures the {112} plane predominated, while at intermediate temperatures the {110} plane operated, and at high temperatures the {123} plane was found to participate in the deformation, but in all cases the ⟨111⟩ direction was unchanged.

In recent years attention has also turned to the high melting body-centred cubic metals niobium, tantalum, tungsten and molybdenum, which present a wide temperature range for investigation, but their behaviour, like that of iron, is sensitive to interstitial impurities. Maddin and Chen[56] found that molybdenum and niobium crystals slipped normally on the system {110} ⟨111⟩; this behaviour has been confirmed by Mitchell and co-workers[57] using niobium crystals grown by electron beam zone melting. A similar method was used to prepare tantalum crystals, which when deformed in compression slipped both on {110} ⟨111⟩ and {112} ⟨111⟩ systems.[58]

Summing up, in general body-centred cubic metal crystals tend to slip on the {110} ⟨111⟩ system at low and moderate temperatures, but at higher temperatures planes of higher indices can operate, and pencil glide conditions are more likely. The behaviour is probably more complicated when the mode of deformation is not simple, for example in polycrystalline aggregates.

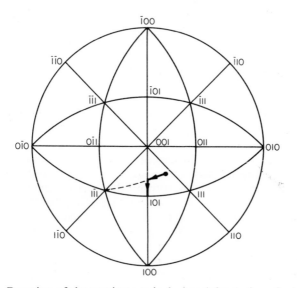

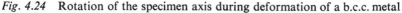

Fig. 4.24 Rotation of the specimen axis during deformation of a b.c.c. metal

4.2.2. *Geometrical aspects of glide in body-centred cubic metals*

To describe rotation of axes during glide, we shall make the simplifying assumption that slip takes place entirely on {110} planes in the ⟨111⟩ directions. Figure 4.24 shows the rotation which occurs when a crystal is deformed in tension and slips on the primary system (011) [1$\bar{1}$1]. The tensile axis moves along a great circle towards the slip direction [1$\bar{1}$1] until the [001]–[101] boundary is reached, when the conjugate system (0$\bar{1}$1) [111] will operate together with the primary system, and the tension axis moves along the symmetry boundary towards [101].

4.2.3. *The flow stress*

In contrast to face-centred cubic metal crystals the yield stress or critical resolved shear stress of body-centred cubic single crystals is markedly temperature dependent, in particular at low temperatures. Figure 4.25

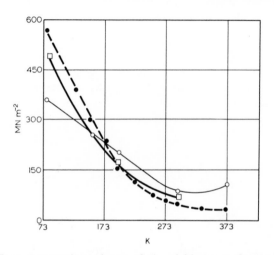

Fig. 4.25 Temperature dependence of the yield stress of pure iron crystals (3 batches). (After Allen *et al.*, 1956, *Proc. R. Soc.*, **A234**, 221)

shows the temperature dependence for the yield stress of three batches of iron crystals.[59] The temperature sensitivity of the yield stress of body-centred cubic crystals has been attributed to the presence of interstitial impurities on one hand, and to a temperature-dependent Peierls–Nabarro force on the other. Because of the difficulty of reducing interstitial impurities to levels where they would be only sparsely distributed along dislocations, it is not possible to distinguish with any certainty between these two possibilities. However, in recent work with highly purified iron containing 10^{-3} parts per million of carbon,[60] the yield strength is very low when the impurities reach this level, but there is still a substantial temperature dependent contribution to the flow stress, which must arise from the Peierls–Nabarro force.[96]

The temperature dependence of the flow stress of molybdenum crystals has been studied as a function of the number of purifying passes during zone refining.[61] The results (Fig. 4.26) show that the temperature dependence decreases with increasing purity, but at interstitial levels of 10–20 p.p.m. the temperature dependence is still substantial.

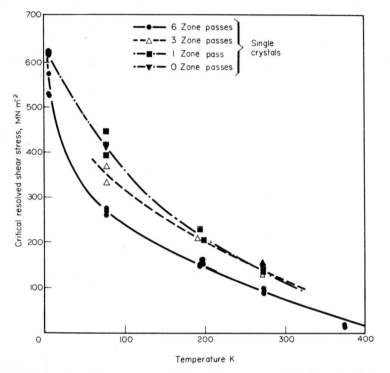

Fig. 4.26 Effect of temperature and purity on the critical resolved shear stress of molybdenum crystals. (After Lawley *et al.*, 1962–3, *J. Inst. Metals*, **91**, 23)

Conrad[8] has examined a number of possible mechanisms involving dislocations which could account for the strong temperature dependence of the yield stress.

1. Overcoming the Peierls–Nabarro force.
2. Escape of dislocations from interstitial atmospheres.
3. Breaking away from fine precipitates.
4. Non-conservative movement of dislocation jogs.
5. Cross-slip.

Using detailed thermodynamic arguments he has concluded that the Peierls–Nabarro force is the effective rate controlling mechanism which leads to the strong temperature dependence. However, further experiments are needed on body-centred cubic crystals of extremely low interstitial content before this matter is satisfactorily resolved.

4.2.4. *Stress–strain curves of body-centred cubic crystals*

The early part of the stress–strain curve of a body-centred cubic crystal is frequently interrupted by the presence of a sharp yield point due to interstitial impurities such as carbon, nitrogen or oxygen. The subsequent form of the stress–strain curve is also sensitive to such impurities as shown by recent work on niobium[57] crystals of identical orientation, but subject to different numbers of zone-purifying passes during their preparation (Fig. 4.27). The purest crystal shows no yield point and has the lowest stress–strain curve.

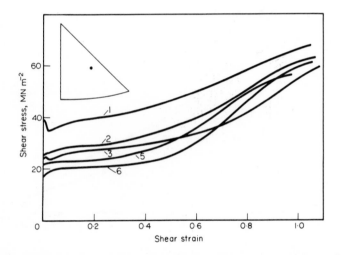

Fig. 4.27 Stress–strain curves of niobium crystals of increasing purity at 295K. (Figures refer to the number of zone passes.) (After Mitchell *et al.*, 1963, *Phil. Mag.*, **8**, 1895)

The most general stress–strain behaviour of body-centred cubic metal crystals seems to be represented by a parabolic curve of the form

$$\sigma = \sigma_0 + \alpha\epsilon^{1/2} \qquad\qquad \textbf{4.6}$$

as exemplified by work on molybdenum,[56] tantalum[58] and iron,[62] but deviations from this behaviour appear to be frequent. Typical shear stress/shear strain curves for zone-refined niobium are reproduced in Fig. 4.28 for the temperature range 77 to 513 K, from which it can be seen that the work hardening rate is low compared to Stage 2 in face-centred cubic crystals, and that it *decreases* with decreasing temperature over the range studied. The deformation process at all temperatures was slip; twinning was not observed.

Niobium crystals have been deformed over larger strains than most other body-centred cubic crystals, and exhibit three stages of hardening (Fig. 4.27), which at first sight resemble those in face-centred cubic crystals. However, it was found that the onset of the steeper Stage 2 corresponded fairly closely with the onset of duplex slip unlike Stage 2 in face-centred

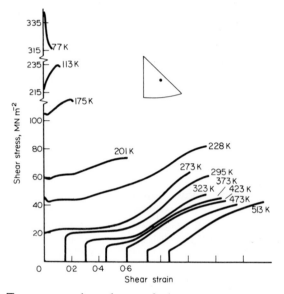

Fig. 4.28 Temperature dependence of the stress–strain curves of zone-refined niobium. (Strain rate $= 4.5 \times 10^{-5}$ s^{-1}) (After Mitchell *et al.*, 1963, *Phil. Mag.*, **8**, 1895)

cubic metals. This observation is coupled with the fact that Stage 1 hardening increases with the distance from the [001]–[101] symmetry boundary, and is at least superficially equivalent to the orientation dependence of easy glide in face-centred cubic crystals. There does not appear to be a well-defined broad slip band formation on the primary system which is familiar in Stage 3 deformation of face-centred cubic crystals.

The stress–strain curves are also sensitive to crystal orientation. Jaoul and Gonzalez,[63] for example, found three stages of hardening only in iron crystals of 'soft' orientations, an observation confirmed by Keh[64] who obtained three stage curves when slip occurred on a single system, but a parabolic stress–strain curve from crystals deforming on more than one system.

4.2.5. *Dislocations in the body-centred cubic lattice*

The slip direction in the body-centred cubic crystal is always the closest packed direction $\langle 111 \rangle$ and the shortest lattice vector is $\frac{1}{2}a\langle 111 \rangle$. The most important dislocation reaction appears to be

$$\frac{1}{2}a[111] + \frac{1}{2}a[1\bar{1}1] = a[100]$$

from the evidence that $a\langle 100 \rangle$ type dislocations are frequently observed in dislocation networks.[65] Another reaction which Crussard[66] has suggested is

$$\frac{1}{2}a[111] \rightarrow \frac{1}{3}a[111] + \frac{1}{6}a[111]$$

The movement of dislocations in body-centred cubic crystals has been studied indirectly by etch pits, and directly by electron microscopy. In iron (with 3 per cent Si) the edge dislocation components move much further than the screws as the slip observations suggest, so the screw components are long and are usually aligned along $\langle 111 \rangle$ directions. However, the screw components cross-slip very readily and so the slip line broadens. This behaviour has also been seen in niobium. The simple dislocation patterns visible after 1 or 2 per cent strain quickly degenerate into a cellular structure, where the cell walls contain complex tangled arrays of dislocations, the density increasing up to strains around 20 per cent. In iron at very low temperatures, the cell structure is developed with greater difficulty, i.e. at much higher strains.

Recently an extensive study of dislocations in α-iron has been carried out, using a special dark field electron microscope technique.[95] The Burgers vectors of a large number of dislocations were determined in this way, and the anticipated $\frac{1}{2}a\langle 111 \rangle$ type of dislocation found (60 per cent). However, there were also large proportions of both $a\langle 100 \rangle$ (20 per cent), and $a\langle 110 \rangle$ (20 per cent) types of dislocation, consequently the reaction between $\frac{1}{2}a\langle 111 \rangle$ dislocations to form $a\langle 100 \rangle$ dislocations cannot be the only interaction. The proportions of different dislocations seem to be relatively insensitive to temperature of deformation, rate and amount of strain, and alloy content.

4.3 The deformation of hexagonal metal crystals

A large number of metals possess the hexagonal crystal structure; however, detailed single crystal investigations on plastic deformation have been mainly restricted to zinc, cadmium and magnesium. While the behaviour of these metals has helped to frame the basic laws underlying crystal deformation, it should not be assumed that they are typical of all hexagonal metals. The c/a ratio of the hexagonal structure can vary markedly from 1·633, the value for an ideally close-packed lattice of hard spheres; zinc and cadmium represent one extreme in so far as the axial ratio is 1·856 and 1·886 respectively, while in beryllium the value is as low as 1·567. These changes in crystal geometry cause changes in the relative close packing of various crystal planes, which in turn influence the slip behaviour during plastic deformation.[99]

4.3.1. *Axial ratios in the hexagonal lattice*

The important crystallographic planes in the hexagonal structures are shown in Fig. 4.29. In the ideal close-packed hexagonal structure formed by the packing of uniformly sized hard spheres, the (0001) basal plane is the closest packed, and the lattice is built up by stacking the basal planes in the order $ABABAB$... where the third layer is directly above the first layer.

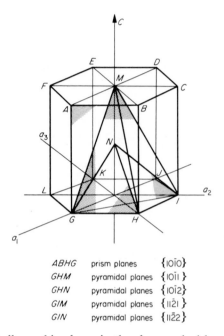

$ABHG$	prism planes	$\{10\bar{1}0\}$
GHM	pyramidal planes	$\{10\bar{1}1\}$
GHN	pyramidal planes	$\{10\bar{1}2\}$
GIM	pyramidal planes	$\{11\bar{2}1\}$
GIN	pyramidal planes	$\{11\bar{2}2\}$

Fig. 4.29   Crystallographic planes in the close-packed hexagonal structure

The c parameter is the distance between the first and third layer while the a parameter is the interatomic distance on the basal plane in any of the three close-packed directions $\langle 11\bar{2}0 \rangle$. Clearly as the c/a ratio is altered from the ideal value, the relative close packing of the crystal planes will vary. For example, as c/a increases, the distance between adjacent basal planes which are normal to the c axis, increases, so they will appear relatively more close-packed than alternative planes, for example the prism planes $\{10\bar{1}0\}$. A survey of hexagonal metal deformation reveals that, while basal slip is very common, both prismatic slip $\{10\bar{1}0\}$ and pyramidal slip of the type $\{10\bar{1}1\}$ are frequently found in a number of metals. These alternative slip planes are shown in Fig. 4.29, from which it can be seen that a reduction in the c/a ratio will increase the relative atomic packing in these planes, and make them more likely slip planes.[67] Table 4.3 tabulates the relative packing densities of the basal, prismatic and pyramidal planes, assuming that of the basal plane is unity, for several hexagonal metals of different c/a ratios. The observed slip in order of ease of operation confirms that basal slip is predominant only in metals with a high c/a ratio.

Table 4.3   Packing densities in planes of hexagonal metals[67]

Metal	c/a	Packing densities			Observed slip planes in
		(0001)	{10$\bar{1}$0}	{10$\bar{1}$1}	order of ease of operation
Cadmium	1·886	1·000	0·918	0·816	(0001) (1$\bar{1}$00) (10$\bar{1}$1)
Zinc	1·856	1·000	0·933	0·846	(0001) (1$\bar{1}$00) ($\bar{1}\bar{1}$22)
Magnesium	1·624	1·000	1·066	0·940	(0001) (10$\bar{1}$1) (1$\bar{1}$00)
Titanium	1·587	1·000	1·092	0·959	(1$\bar{1}$00) (0001) ($\bar{1}$011)

There is a substantial correlation between the ease with which slip occurs on non-basal planes and the c/a ratio of the metal.

4.3.2. *Crystallography of slip in hexagonal metals*[100]

1. ZINC AND CADMIUM

As we have seen in Chapter 2, these two metals in the form of suitably oriented single crystals normally deform almost exclusively on the basal plane in one of the close-packed directions $\langle 11\bar{2}0\rangle$. However, if the basal plane is unfavourably oriented to the stress axis, non-basal slip systems will operate. Zinc crystals with their basal planes parallel to the tension axis exhibit slip on a pyramidal plane $\{\bar{1}\bar{1}22\}\langle11\bar{2}3\rangle$ at room temperature[68]; however, the critical shear stress to initiate slip on this system is between 10 and 16 MN m^{-2} compared with 0·35 MN m^{-2} for basal slip in zinc of similar purity. Pyramidal slip on $\{10\bar{1}1\}$ has been observed in cadmium[69] at room temperature. Prismatic slip on $\{10\bar{1}0\}\langle11\bar{2}0\rangle$[70] has been found in zinc and cadmium and seems to be limited to elevated temperatures, whereas pyramidal slip is not usually observed in these circumstances.

2. MAGNESIUM

The axial ratio of magnesium is close to the ideal value 1·633. It is thus much more prone to slip on non-basal planes than zinc or cadmium, although at room temperature, basal slip predominates. The early work of Schmid[71] showed that above 500 K pyramidal slip occurred on the system $\{10\bar{1}1\}\langle11\bar{2}0\rangle$, but later work suggests that such a slip system will operate only in crystals unfavourably oriented for basal slip. For example, Burke and Hibbard[72] detected pyramidal slip at 298 K when the stress axis was 6° from the basal plane, while Reed-Hill and Robertson,[73] using crystals oriented to suppress basal slip, have found prismatic slip $\{10\bar{1}0\} \langle 11\bar{2}0 \rangle$ at 298 K and 77 K, but pyramidal slip on $\{10\bar{1}1\}$ at 423 K and 559 K.

3. TITANIUM

The crystallography of slip in titanium is very dependent on the concentration of interstitial impurity atoms, particularly oxygen and nitrogen.[74] Crystals with interstitial content of 0·01 wt per cent deform predominantly at room temperature on prism $\{10\bar{1}0\}$ planes, but some basal slip occurs. This behaviour would be anticipated as the c/a ratio is 1·587. However, with less pure metal (0·1 per cent O and N) slip was also observed on the pyramidal planes $\{10\bar{1}1\}$. In the more pure metal, the critical resolved shear stress for prismatic slip was well below that for basal slip (Table 4.4) but

Table 4.4   Slip in titanium crystals

Slip plane	τ_0, MN m^{-2}	
	0·01 per cent O and N	0·1 per cent O and N
$\{1\bar{1}00\}$	13·7	90
$\{0001\}$	62	107
$\{10\bar{1}1\}$	—	97

in the less pure metal this marked difference did not persist. This has been explained in terms of the more marked effect of interstitial oxygen atoms on the movement of dislocations along $\{10\bar{1}0\}$ planes, than on dislocations moving on $\{10\bar{1}1\}$ planes.[74]

4. BERYLLIUM

Beryllium possesses the lowest c/a ratio of the more common hexagonal metals, and so might be expected to be least likely to exhibit basal slip. Nevertheless, at room temperature, basal slip predominates[75] and prismatic slip on $\{10\bar{1}0\}$ planes occurs only when the basal plane is unfavourably oriented.[76] More recent work[77] has confirmed this behaviour and shown that the critical resolved shear stress for both the $\{0001\} \langle 11\bar{2}0 \rangle$ and $\{10\bar{1}0\} \langle 11\bar{2}0 \rangle$ systems decreases with increasing purity. Basal slip becomes even more favoured with increase in purity, so that beryllium behaves in an anomalous way. Nevertheless, the results with both titanium and beryllium indicate that interstitial impurities can drastically alter the deformation behaviour of these metals.

Because of the lack of reliable experimental results, the other hexagonal metals will not be discussed in detail; however, the information available on operative slip systems is summarized in Table 4.5. It should be mentioned that many intermetallic phases have hexagonal structures and in

Table 4.5   Slip systems in hexagonal crystals

Metal or alloy	Axial ratio c/a	Slip systems (in order of importance)	Remarks	Reference
Cadmium	1·886	(0001) [11$\bar{2}$0]		Price[2]
		(1$\bar{1}$00) [11$\bar{2}$0]	High temperatures	
		(1$\bar{1}$01) [11$\bar{2}$0]	Room temperature	
Zinc	1·856	(0001) [11$\bar{2}$0]		Price[2]
		(1$\bar{1}$00) [11$\bar{2}$0]	High temperatures	Price[2]
		(11$\bar{2}\bar{2}$) [11$\bar{2}$3]		Bell and Cahn[68]
Magnesium	1·624	(0001) [11$\bar{2}$0]		
		(1$\bar{1}$01) [11$\bar{2}$0]	Mainly high temperatures	{Burke and Hibbard[72]
		(1$\bar{1}$00) [11$\bar{2}$0]		{Reed-Hill and Robertson[73]
Cobalt	1·621	(0001) [11$\bar{2}$0]		
		(11$\bar{2}$2) [11$\bar{2}$3]		
Titanium	1·587	(1$\bar{1}$00) [11$\bar{2}$0]		Churchman[74]
		(0001) [11$\bar{2}$0]	More common in impure metal	Rosi et al.[67]
		(10$\bar{1}$1) [11$\bar{2}$0]	Only in impure metal	
Zirconium	1·593	(1$\bar{1}$00) [11$\bar{2}$0]		
		(0001) [11$\bar{2}$0]		
Beryllium	1·567	(0001) [11$\bar{2}$0]		
		(1$\bar{1}$00) [11$\bar{2}$0]		Levine et al.[77]
		(1$\bar{1}$01) [11$\bar{2}$0]		Tuer and Kaufmann[75]
εAg–Zn solid solution	1·557 to 1·571	(0001) [11$\bar{2}$0]	Non-basal slip increases with decreasing c/a	Stoloff and Davies[97]
		(1$\bar{1}$00) [11$\bar{2}$0]		
		(10$\bar{1}$1) [11$\bar{2}$0]		
γAg$_2$Al	1·588	(0001) [11$\bar{2}$0]		Mote et al.[98]
		(1$\bar{1}$00) [11$\bar{2}$0]		

some cases their deformation behaviour has been studied. Several examples are listed but there are many more referred to in the literature.[78] Solid solutions and intermetallic phases possess an experimental advantage that the effect of *c/a* ratio can be often investigated by changing the concentration of the solution. For example, the addition of lithium to form a solid solution in magnesium suppresses basal slip, and encourages the operation of prismatic slip,[79] which has a marked beneficial effect on ductility.

4.3.3. *Dislocations in hexagonal crystals*[93, 99, 100]

The close-packed hexagonal lattice is closely related to the face-centred cubic lattice in so far as they can both be constructed by simple variations in packing of the closest packed planes.

$$ABCABCABC \qquad \text{face-centred cubic}$$
$$ABABABAB \qquad \text{close-packed hexagonal}$$

We have seen that perfect dislocations in the face-centred cubic lattice find certain dissociations into imperfect or partial dislocations energetically favourable, and the possible reactions have been demonstrated by use of the Thompson tetrahedron.

In the close-packed hexagonal case, a similar representation is possible except that it is necessary to use a double tetrahedron (Fig. 4.30).

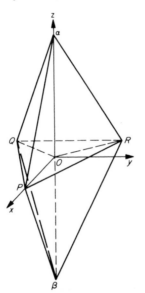

Fig. 4.30   Geometry of dislocations in the c.p.h. structure. (After Berghezan *et al.*, 1961, *Acta metall.*, **9**, 464)

This figure illustrates several different types of dislocation; vectors joining letters of the same type (except O) are perfect dislocations, whereas vectors joining dissimilar letters are partial dislocations.

1. Perfect dislocations in the basal plane *PQR* with their Burgers vectors lying along the close-packed directions, *PQ*, *QR* and *RP*.

2. Perfect dislocations with vectors perpendicular to the basal plane and represented by $\alpha\beta$, etc.

3. Perfect dislocations with a Burgers vector the sum of $\alpha\beta$ and PQ, which is usually referred to as $c + a$ and is of the type $\frac{1}{3}[11\bar{2}3]$. The strength of the vector is equal to twice the distance between the mid-points of $R\alpha$ and $Q\beta$.

4. Imperfect dislocations in the basal plane (PO, QO and RO). These are of the Shockley type, and as in the face-centred cubic lattice are formed by dissociation of perfect dislocations in the close packed plane.

$$PQ = PO + OQ$$

Such a reaction, as in the face-centred cubic case, leads to a region of stacking fault separating the two partial dislocations PO and OQ. This has been illustrated in Fig. 3.24, which serves to illustrate this case with appropriate changes in indices. In terms of Miller indices the reaction is

$$a[11\bar{2}0] = \frac{a}{\sqrt{3}}\,[01\bar{1}0] + \frac{a}{\sqrt{3}}\,[10\bar{1}0]$$

Applying the Frank rule it is clear that this reaction results in a reduction in energy.

5. Imperfect dislocations perpendicular to the basal plane $O\alpha$, $O\beta$, and those of opposite sign.

6. Imperfect dislocations which combine the imperfect dislocations types 4 and 5, viz. $P\alpha$, $Q\alpha$ and $R\alpha$, etc. This leads to other possible dislocation reactions such as

$$Q\alpha = QO + O\alpha$$

Using the rule that the energy of a dislocation is proportional to the square of its Burgers vector, and assuming ideal close packing ($c/a = 1\cdot633$), we can tabulate the relative energies of the possible dislocations (Table 4.6). The Burgers vectors can be deduced from the geometry of Fig. 4.30.

Table 4.6   Dislocations in the hexagonal structure

Dislocation type		Energy factor
Perfect $\begin{cases} PQ \\ \alpha\beta \\ \alpha\beta + PQ \end{cases}$ = a = c = $c + a$		a^2 $c^2 = \frac{8}{3}a^2$ $c^2 + a^2 = \frac{11}{3}a^2$
PO	= p	$p^2 = \frac{1}{3}a^2$
αO	= $c/2$	$c^2/4 = \frac{2}{3}a^2$
$P\alpha$	= $c/2 + p$	a^2

4.3.4. *Stress–strain curves of hexagonal metals—work hardening*

In Chapter 2 we have briefly referred to stress–strain curves of zinc, cadmium and magnesium crystals which exhibit very low rates of work hardening at room temperature up to shear strains of about 200 per cent. This type of deformation is the result of slipping on the basal plane when it is favourably oriented to the tension axis. The most fundamental studies of basal slip have been carried out on zinc crystals deformed in simple shear (Fig. 4.31) in a specially designed apparatus[80] which permitted the

Fig. 4.31 Zinc single crystals chemically machined to shear specimens. **a**, etched spherical crystal; **b**, cleaved to reveal basal plane; **c**, acid-machined shear specimen; **d**, after a shear test (100 per cent strain). (After Parker and Washburn, 1953, *Modern Research Techniques in Physical Metallurgy*, American Society of Metals)

specimen to be sheared along the basal plane in one of the close-packed directions. Such experiments gave almost linear stress–strain curves (Fig. 4.32) up to 50–100 per cent shear strain with very low rates of work hardening ($\sim 10^{-4}\,G$), which are typical of metals deforming exclusively on one slip system, for example face-centred cubic crystals deforming during easy glide. The effect of introducing two slip systems is readily observed by rotating the shear specimen so that the direction of shear bisects the angle between two close-packed directions. The slope of the resulting stress–strain curve is substantially greater than that of the specimen deforming by single slip (Fig. 4.32). Basal slip does not cause serious

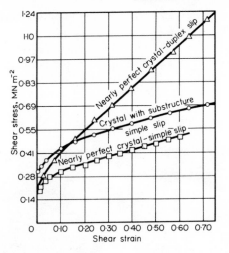

Fig. 4.32   Shear stress/shear strain curves from sheared zinc crystals. (After Parker and Washburn, 1953, *Modern Research Techniques in Physical Metallurgy*, American Society of Metals)

lattice distortions, and the X-ray Laue spots from a deformed crystal are still sharp after 100 per cent elongation,[81] moreover, on annealing the crystal will recover completely without recrystallization. In the shear experiment it was found possible to shear a crystal to 50 per cent shear

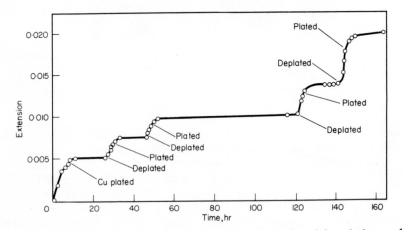

Fig. 4.33   Changes in creep rate of a zinc crystal produced by plating and deplating during test. (After Pickus and Parker, 1951, *Trans. AIME*, **191**, 792)

strain, carry out a recovery anneal and then retrace the original stress–strain curve as many as nine times in succession.[80]

Such experiments indicate that the dislocations must be moving very readily through the crystals meeting few obstacles. Only a very small percentage of the dislocations generated remain in the crystal after deformation and these are readily removed by a recovery anneal. That the average distance moved by the dislocations is large is supported by experiments on the effect of surface films such as oxides or plated layers on the deformation of crystals. For example, the deformation rate of a zinc crystal under constant stress increases instantly if the surface oxide film is suddenly removed by hydrochloric acid.[82] In another series of experiments the rate of deformation was rapidly decreased when a thin surface layer of copper was deposited and increased again when the copper was removed[83] (Fig. 4.33). Barrett[84] twisted oxide-coated zinc crystals and found that on removal of

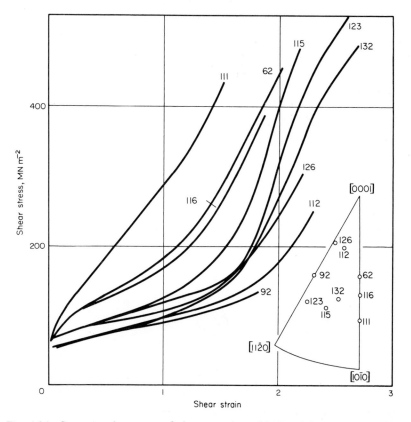

Fig. 4.34   Stress–strain curves of zinc crystals at 294 K. (After Lücke, Masing and Schröder, 1955, *Z. Metallk.*, **46**, 792)

the stress some untwisting usually occurred; however, if the oxide layer was removed prior to unloading the crystal, further twisting occurred in the same sense as the original deformation. All these experiments support the view that during deformation in Stage 1, dislocations tend to move out of the crystals in large numbers. Surface layers impede this process and lead to dislocation pile-ups near the metal–oxide interface, which are relieved when the surface layer is removed.

In general, Stage 1 hardening extends to shear strains between 100 and 200 per cent depending on the orientation of the crystal and the nature of the metal. The subsequent change in rate of work hardening is well illustrated by work on zinc crystals[85] deformed at 294 K where the Stage 1 hardening ends after 100–150 per cent shear strain (Fig. 4.34) and is replaced by a region of higher work hardening. At 90 K, the stress-strain curves are linear, but fracture usually occurs after about 50 per cent glide strain and the work-hardening coefficient is much higher than at room temperature. On the other hand, cadmium[86] is more ductile at 77 K but, like zinc at room temperature, exhibits a transition from a low to a high rate of work hardening between 100 and 200 per cent shear strain (Fig. 4.35). More recent work on zinc crystals[87] has confirmed the existence of Stages 1 and 2 hardening and added a third parabolic stage which is prevalent at room temperature. Moreover, Stage 2 hardening was found to be greater at room temperature than at 243 K, a result which has been attributed to the formation of vacancies during basal glide, which cluster

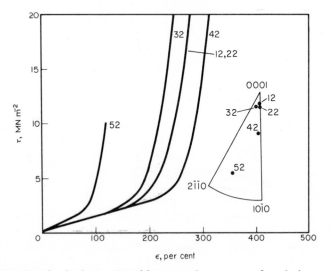

Fig. 4.35 Resolved shear stress/shear strain curves of cadmium crystals, 99·9999 per cent purity at 77 K. (Reproduced by permission of the National Research Council of Canada, from Davies, 1963, *Con. J. Phys.*, **41**, 1456)

and condense to form dislocation loops. These loops then act as obstacles to further basal slip.

Similarly, experiments with magnesium single crystals[88] gave linear stress–strain curves with low strain-hardening coefficients at 364 K up to shear strains of 70 per cent, but a marked increase in the work-hardening coefficient occurred in the range 300–200 K. At 78 K deviations from linearity occurred in the stress–strain curves after only small strains. Figure 4.36 shows the strain-hardening coefficients plotted against tem-

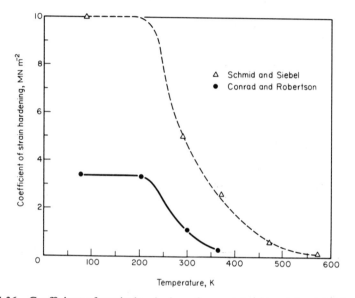

Fig. 4.36   Coefficient of strain hardening of magnesium crystals as a function of temperature. (After Conrad and Robertson, 1957, *Trans. AIME,* **209,** 503)

perature for two sets of results both of which show the same general trend. The coefficient is approximately constant at low temperatures but rapidly falls to nearly zero at higher temperatures. The difference in magnitude of the coefficients from the two sets of experiments is due to the fact that in the earlier work, Schmid and Siebel[89] used less pure magnesium. The absence of strain hardening at elevated temperatures is a result of the occurrence of recovery during the tensile test. More recent[90] work has confirmed that Stage 2 hardening occurs at room temperature but disappears above 473 K. Hirsch and Lally[91] found that Stage 1 hardening is parabolic at low strains and associated with long slip lines. In Stage 2 twinning was observed together with dislocation networks. The stress–strain curves of zinc crystals (Fig. 4.34) also illustrate the point that curves for differently oriented crystals do not overlap when plotted on a shear stress/shear strain basis. Crystals with orientations towards the centre of the stereographic triangle, and on the [11$\bar{2}$0]–[0001] boundary show normal behaviour, while crystals with axes on the [0001]–[10$\bar{1}$0] boundary show a much higher degree of work hardening.

Finally, the rate of straining can have an important influence on the work-hardening coefficient. The data included in Fig. 2.19 for cadmium crystals include curves obtained at a testing speed 100 times greater. At low temperatures (88 K) little difference is evident but, at room temperature, the strain-hardening coefficient is increased appreciably. However, at an elevated temperature (473 K) when recovery is taking place fairly rapidly there is again little influence of the higher rate of straining. Lücke et al.[85] examined the effect of strain rate on the work hardening of zinc crystals and found a three-fold variation in the coefficient when the rate was changed by a factor of 100 (Table 4.7).

Table 4.7   Influence of strain rate on hardening of zinc crystals at room temperature[85]

Shear strain rate, s^{-1}	Rate of hardening, MN m^{-2}	
	1st linear stage	2nd linear stage
2 × 10^{-4}	0·52	4·31
8 × 10^{-4}	0·77	6·85
2 × 10^{-2}	1·50	8·83

Little systematic work has been done on stress–strain curves of pure hexagonal metals of higher melting points. In titanium the shape of the stress–strain curves is very dependent on the oxygen and nitrogen contents. Crystals of 99·7 per cent titanium possess a work-hardening coefficient of 48 MN m^{-2},[67] which is much higher than those of zinc and cadmium. Rhenium[92, 101] crystals exhibit particularly high work-hardening rates at room temperature and 90 K. Higher rates of work hardening would be anticipated as basal slip is replaced by slip on more than one alternative system, but little work has been done on this aspect of the problem.

4.3.5. Observations on dislocations in hexagonal metals

Examination of the surface of deformed crystals of zinc, cadmium and magnesium provides ample evidence of the predominant basal slip system.[44] The slip lines are usually very long, closely spaced and typical of Stage 1 hardening in both hexagonal and face-centred cubic metals. Thin-foil electron microscopy has provided recently much more information about dislocation arrays in zinc and cadmium, but the temperature of examination, viz. room temperature, is relatively high for these metals, so recovery processes have often occurred before observation.

Dissociated dislocations have been observed in the basal plane in both zinc and cadmium[93] showing pronounced formation of stacking faults, in contrast to magnesium where no dissociation is detectable. Assuming analogous behaviour to face-centred cubic metals, this would put the stacking fault of zinc and cadmium in the range 30-50 mJ m^{-2} and magnesium much higher.

A striking feature of deformed cadmium, zinc and magnesium crystals[93, 69] is the formation of sessile dislocation loops, which has been followed directly in the electron microscope. Many loops have been shown to be created from jogs, the formation of which has been discussed in Chapter 3. Elongated loops are pinched off the jogs and gradually become more circular in form (Fig. 4.37). The process was observed to occur very

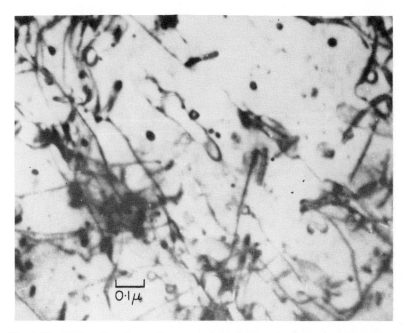

Fig. 4.37   Dislocation loops in magnesium deformed 5 per cent in tension. (After Lally)

rapidly even at 153 K, so it seems unlikely that the loops are in any way formed by diffusion of point defects. Hirsch and Lally[91] found that dislocations in Stage 1 hardening of magnesium crystals comprised mainly bands of edge dipoles formed by edge dislocations on parallel slip bands trapping each other. There is a correlation between dipole density and slip line lengths.

General references

1. FISHER, J. C., JOHNSTON, W. G., THOMSON, R. and VREELAND, T. (Eds.) (1957). *Dislocations and Mechanical Properties of Crystals* (Lake Placid Conference). John Wiley, New York and London.
2. THOMAS, G. and WASHBURN, J. (Eds.) (1963). *Electron Microscopy and the Strength of Crystals*. Interscience, New York and London.
3. NABARRO, F. R. N., BASINSKI, Z. S. and HOLT, D. B. (1964). 'Plasticity of Pure Single Crystals.' *Adv. Phys.*, **13** (50).
4. *Dislocations in Solids*. Discussions of the Faraday Society, No. 38, 1964.
5. MITCHELL, T. E. (1964). 'The Deformation of Metal Crystals.' *Progress in Applied Materials Research*, Vol. 6, 119.
6. CLAREBROUGH, L. M. and HARGREAVES, M. E. (1959). 'Work Hardening of Metals.' *Progress in Metal Physics*, **8**, 1. Pergamon Press, Oxford.
7. NEWKIRK, J. B. and WERNICK, J. H. (Eds.) (1962). *Direct Observation of Dislocations in Crystals*. Interscience, New York and London.
8. *The Relation between the Structure and Mechanical Properties of Metals*. (1963). National Physical Laboratory, Symposium No. 15. H.M.S.O.

References

9. ANDRADE, E. N. DA C. and HENDERSON, C. (1951). *Phil. Trans. R. Soc.*, **A244**, 177.
10. LUCKE, K. and LANGE, H. (1952). *Z. Metallk.*, **43**, 55.
11. ROSI, F. D. (1954). *Trans. AIME*, **200**, 1009.
12. DIEHL, J. (1956). *Z. Metallk.*, **47**, 331.
13. EDWARDS, E. H., WASHBURN, J. and PARKER, E. R. (1953). *Trans. AIME*, **197**, 1525.
14. PATTERSON, M. S. (1955). *Acta metall.*, **3**, 491.
15. HONEYCOMBE, R. W. K. (1961). *Progress in Materials Science*, **9**, 93.
16. ANDRADE, E. N. DA C. and ABOAV, D. A. (1957). *Proc. phys. Soc.*, **A240**, 304.
17. ADAMS, M. A. and COTTRELL, A. H. (1955). *Phil. Mag.*, **46**, 1187.
18. GARSTONE, J., HONEYCOMBE, R. W. K. and GREETHAM, G. (1956). *Acta metall.*, **4**, 485.
19. MITCHELL, T. E. and THORNTON, P. R. (1963). *Phil. Mag.*, **8**, 1127.
20. SUZUKI, H., IKEDA, T. and TAKEUCHI, T. (1956). *J, phys. Soc. Japan.*, **11**, 382.
21. WU, T. W. and SMOLUCHOWSKI, R. (1950). *Phys. Rev.*, **78**, 468.
22. ROSI, F. D. (1957). *Acta metall.*, **5**, 348.
23. YOUNG, F. W. (1963). *J. phys. Soc., Japan*, **18**, Supplement 1.
24. MCKINNON, N. A. (1955). *Phil. Mag.*, **46**, 1150.
25. FOURIE, J. T. and MURPHY, R. J. (1962). *Phil. Mag.*, **7**, 1617.
26. YOUNG, F. W. (1962). *J. appl. Phys.*, **33**, 963.
27. BAILEY, J. E. (1963). *Phil. Mag.*, **8**, 223.
28. VENABLES, J. A. (1962). *Phil. Mag.*, **7**, 1969.
29. MADER, S., SEEGER, A. and KIERINGER, H. M. (1963). *J. appl. Phys.*, **34**, 3376.
30. HAASEN, P. (1958). *Phil. Mag.*, **3**, 384.
31. NOGGLE, T. S. and KOEHLER, J. S. (1957). *J. appl. Phys.*, **28**, 53.
32. BLEWITT, T. H., COLTMAN, R. R. and REDMAN, J. K. (1955). *Defects in Crystalline Solids*. Physical Society, London. p. 369.

33. SAWKILL, J. and HONEYCOMBE, R. W. K. (1954). *Acta metall.*, **2**, 854.
34. MADER, S. (1957). *Z. Phys.*, **149**, 73.
35. KRONMULLER, H. (1959). *Z. Phys.*, **154**, 574.
36. PFAFF, F. (1962). *Z. Metallk.*, **53**, 411, 466.
37. FRIEDEL, J. (1955). *Phil. Mag.*, **46**, 1169.
38. DIEHL, J., KRAUSE, M., OPPENHAUSER, W. and STAUBWASSER, W. (1954). *Z. Metallk.*, **45**, 489.
39. CAHN, R. W. (1951). *J. Inst. Metals*, **79**, 129.
40. SCHOECK, G. and SEEGER, A. (1955). Report of a Conference on Defects in Crystalline Solids. Physical Society, London. p. 340.
41. COTTRELL, A. H. and STOKES, R. J. (1955). *Proc. R. Soc.*, **A233**, 17.
42. ADAMS, M. A. and COTTRELL, A. H. (1955). *Phil. Mag.*, **46**, 1187.
43. BASINSKI, Z. S. (1959). *Phil. Mag.*, **4**, 393.
44. SEEGER, A. and TRAUBLE, H. (1960). *Z. Metall.* **51**, 435.
45. DIEHL, J. and BERNER, R. (1960). *Z. Metallk.*, **51**, 522.
46. KELLY, A. (1956). *Phil. Mag.*, **1**, 835.
47. THORNTON, P. R., MITCHELL, T. E. and HIRSCH, P. B. (1962). *Phil. Mag.*, **7**, 337.
48. TAYLOR, G. I. and ELAM, C. F. (1926). *Proc. R. Soc.*, **A112**, 337.
49. TAYLOR, G. I. and ELAM, C. F. (1936). *Proc. R. Soc.*, **A153**, 273.
50. FAHRENHORST, W. and SCHMID, E. (1932). *Z. Physik.*, **78**, 383.
51. BARRETT, C. S., ANSEL, G. and MEHL, R. F. (1937). *Trans. ASM*, **25**, 702.
52. OPINSKY, A. J. and SMOLUCHOWSKI, R. (1951). *J. Applied Phys.*, **22**, 1488.
53. HULL, D. (1961). *Acta metall.*, **9**, 909.
54. LESTAK, B. and LIBOVICKY, T. (1963). *Acta metall.*, **11**, 1190.
55. ANDRADE, E. N. DA C. and CHOW, Y. S. (1940). *Proc. R. Soc.*, **A175**, 290.
56. MADDIN, R. and CHEN, N. K. (1951). *Trans. AIME*, **191**, 937.
57. MITCHELL, T. E., FOXALL, R. A. and HIRSCH, P. B. (1963). *Phil. Mag.*, **8**, 1895.
58. MORDIKE, B. L. (1962). *Z. Metallk.*, **53**, 586.
59. ALLEN, N. P., HOPKINS, B. E. and MCLENNAN, J. E. (1956). *Proc. R. Soc.*, **A234**, 221.
60. STEIN, D. F., LOW, J. R. and LEYBOLT, A. V. (1963). *Acta metall.*, **11**, 1253.
61. LAWLEY, A., VAN DEN SYPE, J. and MADDIN, R. (1962–63). *J. Inst. Metals*, **91**, 23.
62. MORDIKE, B. L. and HAASEN, P. (1962). *Phil. Mag.*, **7**, 459.
63. JAOUL, B. and GONZALEZ, D. (1964). *J. Mech. Phys. Solids*, **9**, 16.
64. KEH, A. T. (1965). *Phil. Mag.*, **12**, 9.
65. CARRINGTON, W., HALE, K. F. and MCLEAN, D. (1960). *Proc. R. Soc.*, **A259**, 203.
66. CRUSSARD, C. (1961). *Compt. Rend. Acad. Sci.*, **252**, 273
67. ROSI, F. D., DUBE, C. A. and ALEXANDER, B. H. (1953). *AIME, J. Metals*, **5**, 257.
68. BELL, R. L. and CAHN, R. W. (1957). *Proc. R. Soc.*, **A239**, 494.
69. PRICE, P. B., Reference 2, p. 41.
70. GILMAN, J. J. (1956). *Trans. AIME*, **206**, 998; (1961), *Ibid.*, **241**, 456.
71. SCHMID, E. (1931). *Z. Elektrochem.*, **37**, 347.
72. BURKE, E. C. and HIBBARD, W. R. (1952). *Trans. AIME*, **194**, 295.
73. REED-HILL, R. E. and ROBERTSON, W. D. (1957). *Trans. AIME*, **209**, 496.
74. CHURCHMAN, A. T. (1954). *Proc. R. Soc.*, **A226**, 216.

75. TUER, G. L. and KAUFMANN, A. R. (1955). *The Metal Beryllium*. American Society for Metals. p. 372.
76. LEE, H. T. and BRICK, R. M. (1956). *Trans. ASM*, **48**, 1003.
77. LEVINE, E. D., KAUFMAN, D. F. and ARONIN, L. R. (1964). *Trans. AIME*, **230**, 260.
78. WESTBROOK, J. H. (1960). *Mechanical Properties of Intermetallic Compounds*. John Wiley, New York and London.
79. HAUSNER, F. E., LANDON, P. R. and DORN, J. E. (1958). *Trans. ASM*, **50**, 856.
80. PARKER, E. R. and WASHBURN, J. (1953). Modern Research Techniques in Physical Metallurgy. American Society for Metals. p. 186.
81. HONEYCOMBE, R. W. K. (1951–2) *J. Inst. Metals*, **80**, 45.
82. HARPER, S. and COTTRELL, A. H. (1950). *Proc. phys. Soc.*, **63B**, 331.
83. GILMAN, J. J. and READ, T. A. (1952). *Trans. AIME*, **194**, 875.
84. BARRETT, C. S. (1953). *Acta metall.*, **1**, 2.
85. LÜCKE, K., MASING, G. and SCHRODER, K. (1955). *Z. Metallk.*, **46**, 792.
86. DAVIES, K. G. (1963). *Can. J. Phys.*, **41**, 1456.
87. SEEGER, A. and TRAUBLE, H. (1960). *Z. Metallk.*, **51**, 435.
88. CONRAD, H. and ROBERTSON, W. D. (1957). *Trans. AIME*, **209**, 503.
89. SCHMID, E. and SIEBEL, G. (1931). *Z. Elektrochem.*, **37**, 447.
90. YOSHINAGA, H. and HORUICHI, R. (1962). *Trans. Japan Inst. Metals*, **3**, 220,
91. HIRSCH, P. B. and LALLY, J. S. (1965). *Phil. Mag.*, **12**, 595.
92. CHURCHMAN, A. T. (1960). *Trans. AIME*, **218**, 262.
93. BERGHEZAN, A., FOURDEUX, A. and AMELINCKX, T. (1961). *Acta metall.*, **9**, 464.
94. STEEDS, J. W. (1966). *Proc. R. Soc.*, **A292**, 343.
95. DINGLEY, D. J. and HALE, K. F. (1966). *Proc. R. Soc.*, **A295**, 55.
96. ALTSHULER, T. L. and CHRISTIAN, J. W. (1967). *Phil. Trans. R. Soc.*, **A261**, 253.
97. STOLOFF, N. S. and DAVIES, R. G. (1964). *J. Inst. Metals*, **93**, 127.
98. MOTE, J. D., TANAKA, K. and DORN, J. E. (1961). *Trans. AIME*, **221**, 858.
99. DORN, J. E. and MITCHELL, J. B. (1965) in *High Strength Materials*. Ed. by V. F. Zackay. John Wiley, New York and London.
100. PARTRIDGE, P. G. (1967). *Metall. Rev.* **12**, No. 118, 169.
101. GEACH, G. A., JEFFERY R. A. and SMITH, E. (1962) in *Rhenium*. Ed. by B. W. Gonser, Elsevier, Amsterdam and New York.

Additional general references

1. HIRTH, J. P. and WEERTMAN, J. (Eds.) (1968). Work Hardening, Metallurgical Society (AIME) Conference Volume, Gordon & Breach, New York.
2. REID, C. N. (1973). *Deformation Geometry for Materials Scientists*. Pergamon, Oxford.
3. THOMPSON, A. W. (ed.) (1977). Work Hardening in Tension and Fatigue, Metallurgical Society of AIME, New York.
4. NABARRO, F. R. N. (1974). *The Theory of Crystal Dislocations* Vol. 4, Chapter by BASINSKI, S. J. and BASINSKI, Z. S. Clarendon Press, Oxford.

Chapter 5

Theories of Work Hardening of Metals

5.1 Introduction

While the low observed initial flow stresses of metal crystals can be shown to be due to slip by movement of dislocations, it is equally true that the subsequent work hardening arises when dislocations are hindered in their movement through the crystals, so that a higher stress must be imposed to continue the deformation. Many obstacles to dislocation movement exist, the most important being:

1. Other dislocations.
2. Grain and sub-grain boundaries.
3. Solute atoms.
4. Particles of a second phase.
5. Surface films.

However, when it is remembered that single crystals of pure metals show very marked work hardening, it is clear that other dislocations must be the most important obstacles to consider in the first instance. In Chapter 3 the stress fields around dislocations have been briefly examined and it has been shown that elastic interactions, either attractive or repulsive, arise when two dislocations approach each other. These interactions form the bases of the earlier general dislocation theories of work hardening.

5.2 Earlier theories

The earliest general theory of work hardening which involved dislocations was put forward by Taylor in 1934.[8] At that time the stress–strain curves of metal crystals such as aluminium were assumed to be, to a first approximation, parabolic, so Taylor set up a dislocation model from which such curves could be calculated. He appreciated that many dislocations do not reach the surface of a crystal but interact elastically with other dislocations, and become anchored within the crystal forming a network (Fig. 5.1). This is a progressive state of affairs so the dislocation concentration gradually increases as deformation proceeds, and the stress necessary to force a passage for the later dislocations is raised.

Taylor assumed that the average distance a dislocation moved was L before it was stopped. If the density of dislocations after a given deformation is D then the strain ϵ is given by

$$\epsilon = D \,.\, L \,.\, b \qquad \textbf{5.1}$$

where b is the Burgers vector.

The spacing of the dislocations (l) will be $1/\sqrt{D}$ and they will interact elastically with their neighbours. The effective internal stress τ as a result

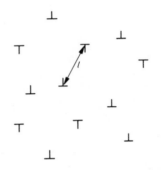

Fig. 5.1   Dislocation interaction—Taylor model

of these interactions is the stress just necessary to force the two dislocations past each other and is given by

$$\tau = \frac{Gb}{8\pi(1 - \nu)l} \qquad \textbf{5.2}$$

where l is the distance between the two slip planes. G and ν are the usual elastic constants.

$$\tau = k \cdot \frac{Gb}{l} = k \,.\, G \,.\, b \,.\, D^{1/2} \qquad \textbf{5.3}$$

where k is a constant. Substituting for D from eqn. **5.1**

$$\tau = kG \left(\frac{\epsilon b}{L}\right)^{1/2} \qquad \textbf{5.4}$$

This is a parabolic relationship between the stress τ and the strain ϵ which approximately describes the behaviour of many metals.

Taylor envisaged an array of dislocations forming a type of three-dimensional lattice. The rate of work hardening is dependent on L, which Taylor assumed to be determined by a mosaic structure which would not allow dislocations through. The magnitude of L is about 10^{-4} cm and if this is used in eqn. **5.1**, for relatively high strain of around 1, a dislocation density of $10^{12}/cm^2$ is obtained, which is in general agreement with experimental determinations of dislocation densities in heavily cold worked metals.

However, the Taylor theory is incompatible with experiments in so far as the deformation does not occur by movement of isolated dislocations, but as a result of numerous dislocations on slip bands arising from sources such as those of the Frank–Read type. As has been shown in Chapter 4, the microscopic changes as a result of plastic deformation are very well defined. Another objection is that pairs of dislocations while locked with respect to each other, can be forced to move by other dislocations.

Mott[9] overcame some of the objections to Taylor's theory by replacing individual dislocation interactions by interactions between piled-up groups of dislocations. These groups were visualized as dislocations on one slip plane emanating from a Frank–Read source, which became piled up against such obstacles as sessile dislocations (Fig. 5.2). This will result in a magni-

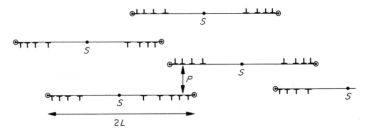

Fig. 5.2   Interaction of piled-up groups of dislocations on the primary system. Edge-on view. ⊙ = obstacle, S = source.

fied internal stress at the head of the pile-up which, amongst other effects, will cause sources on certain secondary systems to operate, and the dislocations formed to interact with the primary dislocations in the pile-up. These reactions would prevent the un-piling of these dislocations when the stress was removed.

The dislocation pile-ups are in a sense superdislocations with a strength of nb where n is the number of dislocations, and a stress at the head of the pile-up of n times the applied stress τ. Consequently, the distance between pile-ups can be much greater than between the dislocations in the Taylor model, and can be of the order of the slip band spacing. If it is assumed that dislocations originating from one source are piled up at each side, so that the separation distance is $2L$ (each section of dislocation line moves a distance L) and the distance between the slip planes is P, the density of piled up groups is $1/(L \cdot P)$ and the average distance between them is $(L \cdot P)^{1/2}$.

The mean stress τ acting on each pile-up is given by

$$\tau = \frac{Gnb}{2\pi(L \cdot P)^{1/2}} \quad \text{(cf. eqn. 5.2)} \qquad 5.5$$

The plastic strain ϵ is simply the summation of the strains from each pile-up which is the product of $nb \cdot L$ and the pile-up density $1/LP$ so

$$\epsilon = \frac{nb}{P} \qquad 5.6$$

Combining eqns. 5.5 and 5.6

$$\tau = \frac{G}{2\pi}\left(\frac{\epsilon nb}{L}\right)^{1/2} \qquad 5.7$$

And again this theory gives a parabolic relationship between stress and strain for single crystals; however, experimental work since 1950 has shown that most single crystals do not have parabolic stress–strain curves. Hexagonal metals in many cases give linear stress–strain curves to high strains, while those from face-centred cubic metals can frequently be divided into three stages with very different characteristics as shown in Chapter 4. It is thus necessary to look more closely at these stages which require more detailed theories.

5.3   More recent theories of work hardening

Before examining in detail the several stages of hardening which occur in metal crystals, it is logical to look at the stress necessary to make

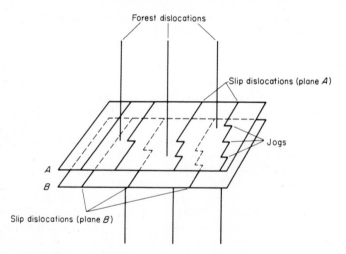

Fig. 5.3   Interaction of glide dislocations on plane A with, **a**, other glide dislocations on plane B; **b**, forest dislocations.

dislocations move in the first instance, i.e. the initial flow stress. In Chapter 2 it has been shown that the critical resolved shear stress for slip depends on certain important variables one of which is the temperature. Seeger[3] has pointed out that the initial flow stress will be determined not only by the interactions between the first dislocations to be generated on the chosen slip system, but also by the interaction of these dislocations with those which are present in the annealed state. These 'grown-in' dislocations can be assumed to be randomly distributed, many of them cutting across the primary slip planes; these are often referred to as the dislocation forest. An idealized model is shown in Fig. 5.3, where glide dislocations in two neighbouring slip bands are present as well as several forest dislocations.

5.4   Theory of the flow stress

Seeger[3, 15] has derived an expression for the flow stress in the following way. The strain ϵ which is observed is expressed in terms of the number of dislocations N per unit volume sweeping out an area A of the slip plane, and the Burgers vector b, thus

$$\epsilon = b \cdot A \cdot N \qquad\qquad 5.8$$

There will of course be obstacles to the movement of dislocations which will be surmounted either by increasing the applied stress or by thermal energy. The activation energy required $U(\tau)$ will thus depend on the applied shear stress. The strain rate is then expressed as

$$\dot{\epsilon} = b \cdot A \cdot N \cdot \nu_0 \exp\left[-\frac{U(\tau)}{kT}\right] \qquad\qquad 5.9$$

where ν_0 is a frequency factor determined by the nature of the obstacles and k is Boltzmann's constant.

If this equation is solved for τ we have an expression relating τ to T and $\dot{\epsilon}$, but before this can be done more information is needed about $U(\tau)$. Seeger assumed that τ is composed of two components τ_G and τ_S. The τ_G component arises from the interaction of parallel dislocations on the primary slip plane, all of which possess the same Burgers vector. The wavelength of the internal stress fields resulting from these interactions will be approximately equal to the dislocation spacing, viz. 10^{-4} cm in an annealed metal. This distance is atomically so large that thermal fluctuations would not make any significant contribution to assist the applied stress, consequently the τ_G component is independent of temperature.

The other component τ_S arises from the interaction of the dislocations causing the slip with pre-existing forest dislocations which cut through the slip plane. When the slip dislocations cut across the forest dislocations, jogs are formed which hinder subsequent movement. The τ_S component is strongly temperature sensitive because thermal activation will greatly assist the formation of jogs. Seeger assumes that there is a single rate-determining activation energy U_0 which depends linearly on the applied stress, thus

$$U = U_0 - v(\tau - \tau_G) \qquad \textbf{5.10}$$

where $U =$ the apparent activation energy, and $v =$ activation volume.

The activation volume $v = b \, . \, d \, . \, l_0'$ where d is the effective diameter of the intersected forest dislocations and is a measure of the distance through which work is done during the cutting process. l_0' is the distance between the forest dislocations which obstruct the glide dislocation and b is its Burgers vector.

This relation implies that the stress which is effective in lowering the activation energy is the difference between the applied stress τ and the internal stress τ_G. Finally an expression for the flow stress is obtained as a function of T and $\dot{\epsilon}$.

$$\tau(T) = \tau_G + \frac{U_0 - kT \ln (NAbv_0/\dot{\epsilon})}{v} \qquad \textbf{5.11}$$

$$= \tau_G + \tau_S \qquad \textbf{5.12}$$

At high temperatures the τ_S term disappears and thermal activation needs no assistance from the stress to move the primary slip dislocations through the forest. However, the τ_G term remains unaltered as the temperature is changed. The flow stress of a pure metal crystal would thus be expected to decrease rapidly with rising temperature at low temperatures, but to fall to a constant value when the temperature was raised sufficiently. This is in accordance with experimental results on a number of metals.[5] (See Chapter 4.)

5.5 Stage 1 hardening

In Chapter 4, the stress–strain curves of face-centred cubic metal crystals have been divided into three stages, of which Stage 1 represents the region of lowest hardening. The experimental observations of long uninterrupted slip lines in both cubic and hexagonal metals closely spaced on one slip system suggest that many dislocation sources are operating without much hindrance. The 'mean free path' of the dislocations is probably long, and a very high proportion reach the surface of the crystal. This view is strengthened by the fact that surface coatings of metal or oxide have a big effect on this stage of the stress–strain curve, probably causing dislocation pile-ups just beneath the coating.

Stage 1, at least for crystal orientations towards the (011) corner of the stereographic triangle, represents deformation on one slip system closely resembling the process in the soft hexagonal metal crystals at 293 K, where similar considerations apply. Probably the ideal Stage 1 deformation is produced in the simple shear experiments which have been carried out primarily on zinc crystals. However, for crystals of many orientations, particularly those towards the boundaries of the stereographic triangle, traces of slip on other systems can be detected during Stage 1. Moreover, the rate of hardening for these 'non-ideal' orientations is greater. This point is more forcibly brought home when a crystal of symmetrical orientation is tested; the Stage 1 hardening is entirely eliminated, the rate of hardening becomes very high and slip occurs on two or more systems depending on the specific orientation.

Seeger and co-workers[10, 15] have attributed Stage 1 hardening to long-range interactions between fairly widely spaced dislocation loops all on the primary slip system. It is assumed that there are N dislocation sources per unit volume, each of which has emitted n dislocation loops when the stress τ is reached. The experimental evidence indicates that each loop moves a large distance L through the crystal, while loops on adjacent slip planes are a distance d apart which is much smaller than L.

An increment in stress $\delta\tau$ will result in an increase in the number of loops by δn, which will give rise to an increment in strain $\delta\epsilon$ expressed thus:

$$\delta\epsilon = b \, . \, N \, . \, L^2 \, . \, \delta n \qquad \qquad \textbf{5.13}$$

N can also be defined in terms of d and L

$$N = \frac{1}{dL^2}$$

so that
$$\delta\epsilon = \frac{b \, . \, \delta n}{d} \qquad \qquad \textbf{5.14}$$

The generation of δn new dislocation loops will also increase the back stress τ_B on the dislocation sources by an amount

$$\delta\tau_B = \frac{Gb \, \delta n}{2\pi L} \qquad \qquad \textbf{5.15}$$

When the back stress becomes equal to the stress increment $\delta\tau$, no further loops are generated.

Combining eqns. **5.14** and **5.15**

$$\theta_1 = \frac{\delta\tau}{\delta\epsilon} = \frac{G}{2\pi}\left(\frac{d}{L}\right) \qquad \qquad \textbf{5.16}$$

A more detailed treatment gives a similar expression

$$\theta_1 = \frac{8G}{9\pi}\left(\frac{d}{L}\right)^{3/4} \qquad \qquad \textbf{5.17}$$

from which, when we substitute typical values, e.g. $d = 300$ Å, $L = 0\cdot5$ mm, a work-hardening rate of approximately $3 \times 10^{-4}G$ is obtained, which is close to the measured values.

5.6 Stage 2 hardening

The most important contribution to work hardening in face-centred cubic metals comes from Stage 2 which, like Stage 1, is linear, but with a much higher work-hardening coefficient approximately $2 \times 10^{-3}G$ (see, for example, Fig. 4.13). Moreover, the slope of Stage 2 is relatively independent of temperature, the main difference between curves at increasing temperatures being that Stage 3 or parabolic hardening sets in at an earlier stage, thus restricting the range of Stage 2. This suggests that the strong increase in flow stress in Stage 2 is predominantly due to the temperature independent contribution from the stress fields of dislocations on the primary slip plane, rather than the component arising from interactions with the forest which is temperature dependent.

Observations on the slip lines during Stage 2 indicate that they become progressively shorter, and Seeger[3] has interpreted this as a result of an increasing concentration of Lomer–Cottrell locks. Friedel[2] also has suggested that Lomer–Cottrell locks play a vital role, but he has developed a theory on the assumption that a large number is formed early in Stage 2, exhausting all the suitable sources of secondary slip. In Seeger's theory small groups of dislocations pile up against the barriers as Stage 2 proceeds and the slip distance becomes smaller. Friedel, on the other hand, assumes a constant slip distance but the number of dislocations in each piled-up group is proportional to the stress.

Seeger's original theory of Stage 2 hardening takes the following form. Dislocations are considered to be generated from Frank–Read sources, and to comprise simply edge and screw components which pile-up at Lomer–Cottrell sessile dislocations in three close-packed directions in the slip plane. The hardening then arises from the long range stresses of these piled-up groups of dislocations, which interact with each other (Fig. 5.4).

If there are n dislocations in each pile-up and the total number of such

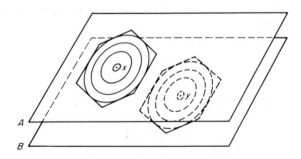

Fig. 5.4 Glide dislocations from sources x and y on planes A and B held up at obstacles in three directions. Consequent interaction between the pile-ups.

groups is N/unit volume, then if the slip lines are assumed to be of constant length R, an expression for the strain ϵ is

$$\epsilon = N \cdot \pi R^2 \cdot n \cdot b \qquad \textbf{5.18}$$

The stress is obtained by assuming the piled-up groups to be super dislocations. These groups are a distance l apart

$$l \simeq \frac{1}{(2NR)^{1/2}} \qquad \textbf{5.19}$$

and the stress τ required to move a dislocation through the stress fields is given by an expression of the form

$$\tau = \frac{1}{2\pi} \cdot \frac{Gbn}{l} \qquad \textbf{5.20}$$

The number of dislocations in the pile-ups depends on the stress according to the basic relationship of Eshelby, Frank and Nabarro,[13] thus

$$n = k \cdot \frac{\tau \cdot R}{b \cdot G} \qquad \textbf{5.21}$$

It can be shown from these equations that

$$\frac{\tau}{\epsilon G} = \text{constant} \qquad \textbf{5.22}$$

and that it is unaffected by the slip line length R.
Assuming a relationship for the slip distance of the type

$$R = \frac{\Lambda}{(\epsilon - \epsilon')} \qquad \textbf{5.23}$$

(see page 95, eqn. **4.3**) where ϵ' is the strain at the end of Stage 1, Seeger was able to evaluate the rate of hardening in stage 2 thus

$$\frac{d\tau}{d\epsilon} \simeq \frac{\beta G}{6\pi^2} \qquad \textbf{5.24}$$

where β is a constant $\simeq 0.5$,
which is relatively close to the experimentally determined values. The hardening is thus independent of the slip line length. When the experimental values of θ and G are put in the equations, the number of dislocations n in the pile-ups is found to be about 25–40.

A further attempt to give a theory of Stage 2 hardening in face-centred cubic metals has been made by Mott,[11] who has emphasized that it is a process in which temperature plays no part. In addition he has made three other points:

1. The process is similar in single and polycrystals.
2. The ratio $[(d\tau/d\epsilon)/G]$ differs little from one metal crystal to another. The value is approximately 0·03 but it varies by a factor of 2 with orientation.

3. In Stage 2 the slip lines are much shorter than in Stage 1, and as Seeger and co-workers have shown, their average length decreases with increasing strain.

Mott did not accept the Seeger theory for several reasons. First, it would be expected that the large stresses at the heads of piled-up groups should be dissipated by slip on other systems, and in any case, it is difficult to explain why the piled-up dislocations do not run back when the external stress is removed. Furthermore, a mechanism is needed to account for the sudden formation of slip bands observed on the surface of a deforming crystal, as well as the complicated tangled arrays of dislocations which are revealed by electron microscopy of thin metal foils.

Mott[11] and Hirsch[12] thus developed a theory which assumes that sessile jogs on dislocations cause the resistance to flow. In Chapter 3 it has been shown that jogs are created when dislocations intersect. In many cases these are able to move along with the rest of the dislocation, but in the case of screw dislocations, the jogs would have to move non-conservatively to keep up with the moving dislocation. This would be energetically unfavourable and vacancies or interstitial atoms would be created. The jogs are thus likely to remain sessile, although some authors have thought that the jogs could glide conservatively along the dislocation lines in the direction of the Burgers vector. Hirsch has, however, pointed out that jogs will tend to dissociate into partial dislocations separating stacking faults just as in the case of the piled-up dislocations in the Seeger model. Fig. 5.5

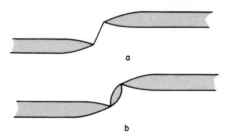

Fig. 5.5   Jogs on extended dislocations: **a,** constricted; **b,** extended

shows one type of extended jog which would be expected to form if the metal possesses a low enough stacking fault energy. If such jogs drag behind the dislocations, dipoles will form (see Chapter 3), and indeed these have been observed even in Stage 1 of the stress–strain curve. On the other hand, if the stress and/or temperature is high enough, the jogs can be associated or constricted, and then will be able to glide along the axes of screw dislocations in a conservative manner.

A detailed analysis by Hirsch has shown that jogs which would form interstitials if forced to move non-conservatively, can glide conservatively along the dislocations if a constriction is formed. On the other hand, jogs leading to vacancy formation are not constricted by stress and thus are more formidable obstacles, except at high temperatures when vacancies can diffuse away readily from the jogs.

Mott has examined the effect of this type of obstacle quantitatively. He considered that dislocation sources on primary slip planes were jogged by dislocations acting on secondary systems, and that the flow stress τ is needed to unlock a jogged dislocation source.

$$\tau = \alpha \,.\, G \,.\, b \,.\, m_v \qquad\qquad \textbf{5.25}$$

where α is a constant such that $\alpha G b^3$ is the energy to form a vacancy: $\alpha \approx 0{\cdot}2$ for most metals. m_v = no. of vacancy jogs/unit length.

The model is then applied to Stage 2 of the stress–strain curve.

If in an increment of strain $d\epsilon$, there are dN dislocation loops emitted from sources, each of mean radius R, then

$$d\epsilon = b \,.\, \pi R^2 \, dN \qquad\qquad \textbf{5.26}$$

Now turning to the slip on secondary planes, let this be g times that on the primary plane, then if m is the number of jogs per unit length

$$dm = g\pi R^2 \,.\, dN \qquad\qquad \textbf{5.27}$$

but only a small proportion dm_v of the jogs will be vacancy jogs, so

$$dm_v = f \,.\, dm$$

so from eqns. **5.25** and **5.27**

$$d\tau = \alpha G b \,.\, (f \,.\, g)\pi \,.\, R^2 \,.\, dN \qquad\qquad \textbf{5.28}$$

using eqn. **5.26**

$$\frac{d\tau}{d\epsilon} = \alpha G(f \,.\, g) \qquad\qquad \textbf{5.29}$$

Now from experimental observations on single crystals $[(d\tau/d\epsilon)/G] \approx 1/300$ for a number of metals and taking $\alpha \simeq 0{\cdot}2$ and the fraction of vacancy jogs as $0{\cdot}05$, then $g \simeq 0{\cdot}3$ which is a reasonable value. A further analysis shows that a slip line length of about 10 microns would be expected in Stage 2—this is rather too small but of the right order.

This theory succeeds in explaining a number of important features of Stage 2 hardening in face-centred cubic metals. Hardening arises from an increase in the density of jogs on the primary slip sources, and after the unlocking of the sources slip lines are produced. The jogs can be thermally activated, and so explain the lower flow stresses at higher temperatures. The theory is also in accord with the observation that edge dislocations move greater distances than screw dislocations which are susceptible to the formation of vacancy jogs.

On the other hand, the theory requires a rather substantial amount of secondary slip ($g \simeq \frac{1}{3}$); moreover, the theoretical slip distance is less than that observed. Further, in practice the slope of Stage 2, θ_{II} changes relatively slightly with orientation, although the tendency to secondary slip varies substantially, in which circumstances the theory would predict a marked change in θ_{II}.

A more recent attempt by Hirsch[4] to develop a general theory of Stage 2 hardening is a phenomenological theory in so far as it has concentrated on the observed slip distances, and the fact that dislocation tangles are observed in thin foils in the electron microscope. Like Seeger, he considers that piled-up groups of dislocations are formed, but these then relax by secondary slip to form the observed tangles which are the main obstacles to further slip. The new sources then operate and when the dislocations pile up near the obstacles, they too relax and contribute to the observed tangles. So the number of obstacles is directly related to the number of sources.

If the density of sources is N and the spacing of the dislocation tangles l, then $N \sim 1/l^3$ and as τ is inversely related to l

$$\tau \propto N^3 \qquad \textbf{5.30}$$

For a given increment of strain dϵ

$$d\epsilon = L_1 L_2 nb \, . \, dN \qquad \textbf{5.31}$$

where L_1 and L_2 are the slip distances of edge and screw components of dislocation loops. n = number of loops.

Now the condition for blocking a dislocation completely along its length is determined from the geometry of the model.

$$L_1 L_2 RN \sim 1 \qquad \textbf{5.32}$$

where R = radius of obstacle and is a fraction f of the stress field of a pile-up.
So

$$R = f \cdot \frac{nb \, . \, G}{2\pi\tau} \qquad \textbf{5.33}$$

From the above equations

$$\theta_{\text{II}} = \frac{f}{6\pi} \cdot G \qquad \textbf{5.34}$$

and for

$$\theta_{\text{II}} \simeq \frac{G}{300}, \quad f \approx \frac{1}{15}$$

This model has been shown to be compatible with the empirical relations $L_2 = \Lambda_2/(\epsilon - \epsilon_{\text{II}})$ and $\tau = \frac{1}{2}Gb\rho^{1/2}$ referred to in Chapter 4 (eqns. **4.3** and **4.1**).

The observation in electron micrographs of dislocation tangles (rather than pile-ups) at the end of Stage 1 has led Kuhlmann-Wilsdorf[14] to develop another theory of Stage 2 hardening. She assumes that Stage 2 work hardening occurs because the lengths of dislocation line, which act as Frank–Read sources, become progressively shorter due to the tangles. As the stress required to operate a source is inversely proportional to its length, this means that the stress for further deformation increases.

Despite the considerable theoretical and experimental work which has been done on Stage 2 hardening, a generally acceptable theory is unlikely in the near future. For a recent summary of the position, the reader is referred to reference 6, in particular the general discussion on pages 157–190.

5.7   Stage 3 hardening

The main characteristics of this stage are a decreasing rate of work hardening (parabolic) and a tendency to occur more readily, i..e at lower stresses, as the temperature is raised. The general implication is that barriers set up in Stage 2 are being broken during Stage 3. This view is confirmed by the work-softening experiment of Cottrell and Stokes in which aluminium crystals deformed into Stage 2 at 77 K, on restraining at room temperature showed a yield drop. The room temperature deformation allowed Stage 3 to be initiated with the sudden by-passing of obstacles created in the deformation at 77 K. If the general picture of the formation of Lomer–Cottrell locks as barriers in Stage 2 is accepted then Stage 3 occurs when the effects of the barriers are overcome by one of two methods.

(a) Collapse of the barriers by recombination of the partial dislocations of the sessile Lomer–Cottrell dislocation. Several possible mechanisms exist.

(b) By-passing of the barriers by slip on the cross-slip plane which has a common slip direction with the primary slip plane.

Seeger[3] has favoured the second alternative, and has pointed out that the beginning of Stage 3 is characterized by heavy slip bands with much cross-slip. The onset of this phenomenon is dependent on the stacking fault energy of the metal. Metals with relatively low stacking fault energy need a high activation energy for cross-slip, because before cross-slip can occur locally, the broad stacking fault between the partial dislocations must be constricted to form a length of unextended dislocation. Thus, in order to obtain a reasonable rate of cross-slip in copper at room temperature, the applied stress must be very high. On the other hand, a metal such as aluminium with a high stacking fault energy will soon exhibit Stage 3 hardening at room temperature, because the dislocations are undissociated and thus cross-slip can occur at much lower applied stresses.

Seeger[4] supported this view primarily because of the structural observations referred to in Chapter 4. The length of the slip lines does not increase in Stage 3 as could be expected if the Lomer–Cottrell barriers were collapsed. Seeger has derived an equation for the shear stress τ necessary to cause the partial dislocations of the leading dislocation in such a pile-up to coalesce:

$$\tau = \frac{1}{n}\left(\frac{\sqrt{2}}{4\pi}G - \frac{2\gamma}{b}\right)$$ 5.35

where γ = stacking fault energy. n = number of dislocations in the pile-up.

The stress τ is identified with the stress τ_{III} at which Stage 3 begins at a temperature sufficiently low to exclude thermal activation. Using values of τ_{III} for lead and aluminium at 4 K, and substitution in equation 5.35 together with appropriate values for G and γ gives $n \simeq 25$. The size of the pile-up corresponds closely with that which the Seeger theory yields for the Stage 2 hardening model.

5.8 Hexagonal metals

Most theories of work hardening have been concerned with face-centred cubic metals; however, some attention has been paid to the close-packed hexagonal metals. Seeger[3] has pointed out that the equation relating the flow stress to temperature and strain rate holds not only for face-centred cubic metals but also for close-packed hexagonal metals.

So

$$\tau(T) = \tau_G + \frac{U_0 - kT \ln\left(\frac{NAbv_0}{\dot{\epsilon}}\right)}{v} \quad \text{for} \quad T \leqslant T_0 \qquad 5.36$$

$$= \tau_G \quad \text{for} \quad T \geqslant T_0 \qquad\qquad\qquad 5.37$$

Above the critical temperature T_0 the flow stress consists entirely of the τ_G component. The work-hardening coefficient θ is obtained by differentiation of eqn. 5.36 with respect to the strain.

$$\theta = \frac{d\tau}{d\epsilon} = \frac{d\tau_G}{d\epsilon} + \frac{dv}{d\epsilon}\cdot\frac{kT}{v^2}\ln\frac{NAbv_0}{\dot{\epsilon}} \quad \text{for} \quad T \leqslant T_0 \qquad 5.38$$

$$= \frac{d\tau_G}{d\epsilon} \quad \text{for} \quad T \geqslant T_0 \qquad\qquad\qquad 5.39$$

Now at room temperature in zinc and cadmium crystals, basal slip is predominant to large strains, so the original dislocation forest will remain largely unchanged, and v will not vary with strain. So the second term of eqn. **5.38** is zero, and eqn. **5.39** applies. τ_G has been previously defined in terms of the dislocation density for unit area, N'.

$$\tau_G = \alpha b \, . \, G\sqrt{N'} \qquad\qquad \textbf{5.40}$$

so, as long as the variation of N' with strain is temperature independent, the implication is that the rate of work hardening is independent of temperature and strain rate.

The experimental results in Chapter 4 suggest strongly that this is so below the recovery range, i.e. below about 223 K in cadmium and zinc. In the recovery range the dislocation density will tend to be reduced as the temperature is raised, and during deformation there will be a balance between work hardening on one hand and recovery on the other. Consequently, in this temperature range θ will be both temperature and strain rate dependent. At still higher temperatures when recovery is the dominating process even during deformation, the rate of work hardening is very low. The rate of work hardening in zinc and cadmium crystals is low compared with the overall hardening of copper and aluminium crystals. The evidence suggests that the hardening is comparable to that of Stage 1 in face-centred cubic metals, and it is implied that dislocations move large distances. Mott has suggested that most dislocations in fact reach the surface, a hypothesis which finds support in the important effects surface films can play in the deformation process.

The picture is, however, not as simple as this. Basal slip has been shown not to predominate in other hexagonal metals such as beryllium and titanium, while even in zinc and cadmium there is much evidence of slip on pyramidal or prismatic planes. Moreover, the rate of hardening rises rapidly at low deformation temperatures, and it frequently increases substantially during deformation at a constant temperature (see, for example, Fig. 4.34). Further information is needed on dislocation reactions in hexagonal metals before a fully satisfactory theory can be put forward.

General references

1. COTTRELL, A. H. (1953). *Dislocations and Plastic Flow in Crystals.* Oxford University Press, London.
2. FRIEDEL, J. (1956). *Les Dislocations.* Gauthier-Villars, Paris. (English edition (enlarged), 1964, Pergamon Press, Oxford.)
3. FISHER, J. C., JOHNSTON, W. G., THOMSON, R. and VREELAND, T. (Eds.) (1957). *Dislocations and Mechanical Properties of Solids* (Lake Placid Conference). John Wiley, New York and London.
4. *The Relation between the Structure and Mechanical Properties of Metals.* (1963). National Physical Laboratory Conference. H.M.S.O.
5. MITCHELL, T. E. (1964). 'Dislocations and Plasticity in Single Crystals of Face-Centred Cubic Metals and Alloys.' In *Progress in Applied Materials Research*, Vol. 6. Ed. by E. G. STANFORD, J. H. FEARON and W. J. MCGONNAGLE.

6. 'Dislocations in Solids.' Faraday Society Discussion No. 38, 1964.
7. NABARRO, F. R. N., BASINSKI, Z. S. and HOLT, D. B. (1964). 'The Plasticity of Pure Single Crystals', *Adv. Phys.* **13**, (50).

References

8. TAYLOR, G. I. (1934). *Proc. R. Soc.*, **A145**, 362, 388.
9. MOTT, N. F. (1952), *Phil. Mag.*, **43**, 1151.
10. SEEGER, A., DIEHL, J., MADER, S. and REBSTOCK, K. (1957). *Phil. Mag.*, **2**, 323.
11. MOTT, N. F. (1960). *Trans. AIME*, **218**, 962.
12. HIRSCH, P. B. (1962). *Phil. Mag.*, **7**, 67.
13. ESHELBY, J. D., FRANK, F. C. and NABARRO, F. R. N. (1951). *Phil. Mag.*, **42**, 351.
14. KUHLMANN-WILSDORF, D. (1962). *Trans. AIME*, **224**, 1047.
15. SEEGER, A., MADER, S. and KRONMÜLLER, H. (1962), in *Electron Microscopy and Strength of Crystals*. Ed. by G. Thomas and J. Washburn. Interscience, New York and London.

Additional general references

1. HIRTH, J. P. and LOTHE, J. (1968). *Theory of Dislocations*. McGraw-Hill, New York.
2. HIRSCH, P. B. (ed.) (1975). *The Physics of Metals* Vol. 2, Defects. Cambridge University Press.
3. NABARRO, F. R. N. (Ed.) (1979). *Dislocations in Solids* Vol. 4, Chapter by BASINSKI, S. J. and BASINSKI, Z. S. on Plastic Deformation and Work Hardening, North Holland Publishing Co. Amsterdam.

Chapter 6

Solid Solutions

6.1 Interactions of solute atoms with dislocations

The nature of the elastic strain fields around dislocations has been discussed in Chapter 3. Isolated solute atoms, particularly if they are much smaller or larger than the solvent atoms, are centres of elastic distortion. Consequently, the solute atom and the dislocation will interact elastically and exert forces on each other. Other types of interaction are also possible, in particular electrical and chemical interactions, which will be briefly considered.

6.1.1. *Elastic interactions*

Cottrell and Bilby[8] first deduced that carbon atoms in interstitial solid solution would not be randomly distributed if a state of equilibrium existed. It has been shown in Chapter 3 that a dislocation has associated with it a strain field, which in the case of a positive edge dislocation causes a compressive stress above the core of the imperfection and a dilatation below. Likewise a solute atom randomly placed in a crystal has a strain field around it, which is minimized if the solute atom can move to the dislocation. With an interstitial atom such as carbon in iron the position of lowest energy in an edge dislocation is in the dilated region near the core (Fig. 6.1a) and provided adequate opportunity is given for diffusion, this is the most likely place to find such solute atoms in a dilute alloy at low temperatures. While the theory was worked out in detail for carbon in α-iron, it is important to appreciate that this is a general model for both interstitial and substitutional solutes. A substitutional solute atom smaller than the solvent atom would be expected to replace solvent atoms in the compressive part of the strain field while a larger solute atom would move to the dilated region (Fig. 6.1b).

There is one very significant difference between an interstitial atom such as carbon in iron, and a substitutional atom such as zinc in copper. The carbon causes not only a volume expansion, but a tetragonal distortion in α-iron. Both carbon and nitrogen atoms occupy interstitial positions at the face centres and the mid-points of the edges of the body-centred cube. If the atoms preferentially take up positions on the {001} faces and at the mid-points of ⟨001⟩ edges, the cell will become tetragonal along the [001] axis. So if the interstitial atoms carry out this ordering process, the tetragonal distortion produced will allow interaction with shear and hydrostatic stress fields. Consequently, the impurity atoms will interact and form atmospheres with both screw and edge dislocations. On the other hand, zinc in copper results in a fully symmetrical lattice distortion. This leads to fundamental differences in the interaction of the solutes and the dislocations, for the zinc atom with its symmetrical strain field will interact solely with hydrostatic stresses, and so can only interact strongly with edge dislocations which have a hydrostatic stress component (Chapter 3). Screw dislocations have stress fields of entirely shear character, and so no interaction with a substitutional solute such as zinc in copper would be antici-

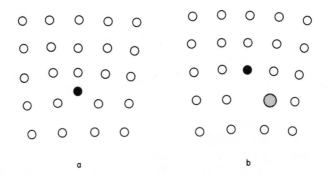

a   b

Fig. 6.1   **a,** Interstitial atom in a dislocation. **b,** Substitutional atoms of different sizes in a dislocation

pated. In consequence dislocations in α-iron are more strongly bound by carbon atoms than they are in a dilute copper–zinc alloy, with the result that a stronger yield point is obtained in iron than in substitutional solid solutions of comparable concentrations.

Let the hydrostatic pressure of a dislocation stress field at some point be p and the change in volume induced by the solute atom be Δv. Then the elastic interaction energy U which would bind a solute atom to that point is given by

$$U = p \cdot \Delta v = K\phi \cdot \Delta v \qquad\qquad 6.1$$

where K is the bulk modulus.

Elasticity theory gives the dilatation strain ϕ at a point with polar coordinates (r, θ) from the core of an edge dislocation (Fig. 6.2), viz.

$$\phi = \frac{b \sin \theta \, (1 - 2\nu)}{2\pi r(1 - \nu)} \quad \text{and} \quad K = \frac{2G(1 + \nu)}{3(1 - 2\nu)}$$

where G = shear modulus, and ν = Poisson's ratio; and after substitution in eqn. **6.1**

$$U_{(r,\,\theta)} = A \frac{\sin \theta}{r} \qquad\qquad \textbf{6.2}$$

where A is a parameter depending on the elastic constants, Δv, and b the strength of the dislocation. As eqn. **6.2** is derived using elasticity theory it breaks down for the core of the dislocation where linear elastic theory is no longer applicable. This is unfortunate, because this is the position of maximum binding energy ($U_{\max}$), and consequently the energy determined in this way is only an estimate.

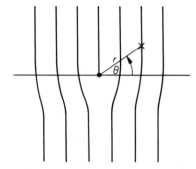

Fig. 6.2   Co-ordinates of a solute atom in a dislocation strain field

U is positive in the upper half of the dislocation ($0 < \theta < \pi$) for a large atom ($\Delta v > 0$), and negative on the lower side. This means that a solute atom larger than the solvent will tend to be repelled from the compression side of the dislocation and attracted to the side in tension where the interaction energy U will be negative. Intuitively the maximum binding energy $U_{\max}$ will occur at a point

$$\theta = \frac{3\pi}{2}, \quad r = r_0$$

where r_0 is approximately 2×10^{-8} cm.

Cottrell[1] has shown that for carbon in iron, $U_{\max} \approx 1$ eV. In the case of zinc in copper, the lattice distortion is smaller, with the result that $U_{\max}$ is much lower, about $\frac{1}{8}$ eV. However, these values are only to be regarded as rough approximations and subject to substantial modification. Typical experimental and calculated values are given in Table 6.1.

Table 6.1 ⋅Some interaction energies of solute atoms with dislocations

Alloy	U_{obs} (eV)	U_{calc} (eV)
Cu–Zn	0·12	0·11
Al–Cu	0·3	0·35
Al–Mg	0·2–0·27	0·28
Al–Zn	0·08–0·11	—
Al–Ge	0·17	0·21
Fe–C, N	0·55	1·0
Ni–H	0·08	—

6.1.2. *Modulus interactions*

There is a further contribution to the elastic interaction which arises from the different elastic properties of the matrix and the solute.[2] If an atom of the matrix is replaced by an impurity atom of the same size but with different elastic constants, there is then no pre-existing lattice distortion due to the solute atom, but an interaction between a dislocation and the solute atom occurs because the dislocation must do more work if it moves near an elastically harder impurity atom. Thus there is an attraction if the impurity is elastically softer than the matrix, and a repulsion if the solute atom is elastically harder. A solution for the change in energy ΔU_G resulting from this type of interaction has been given by Fleischer[25] as follows:

$$\Delta U_G = G \cdot \epsilon_G' b^2 \cdot R^3/6\pi r^2 \qquad\qquad 6.3$$

where ϵ_G' is a measure of the modulus difference between solvent and solute and is defined thus

$$\epsilon_G' = \frac{\epsilon_G}{1 - \epsilon_G/2} \quad \text{and} \quad \epsilon_G = \frac{1}{G} \cdot \frac{dG}{dc}$$

where G is the shear modulus, c the atomic concentration of solute, R the atomic radius of solute, and r the distance between solute atom and dislocation.

This relationship demonstrates that the elastic interaction energy between a solute atom and a dislocation is proportional to the difference in elastic constants between solute and solvent, and that it varies inversely as the square of the distance between the solute atom and the dislocation.

6.1.3. *Electrical interactions*

In ionic solids marked electrostatic interactions take place between solute atoms and dislocations. An edge dislocation will introduce locally an excess of positive or negative charge and this will occur along the whole dislocation line as an excess of alternately positive and negatively charged ions. As a result of this, the dislocation attracts electrostatically charged impurity atoms of both signs.[1]

In a metal the variation of hydrostatic pressure about an edge dislocation results in a rearrangement of the conduction electrons, which will
tend to move from the region of compression to the tension side, resulting
in the formation of an electrical dipole. The electronic charge of the impurity atom then interacts with the field of the dipole. The charge on the
dipole is proportional to the elastic distortion of the lattice, and the interaction with the solute will increase with the valency.

6.1.4. *Chemical interactions*

In Chapter 3 the structure of dislocations in close-packed crystals was
discussed, and it was shown that dissociation into partial dislocations
separated by a layer of stacking fault is energetically favoured. The stacking fault in a face-centred cubic crystal has the close-packed hexagonal
structure, and Suzuki[9] pointed out that in a solid solution there should be
a different equilibrium concentration of solute in the stacking fault from
that in the matrix, because the cohesive energy between atoms in the
stacking fault will differ from that in the face-centred cubic matrix. This
heterogeneous distribution of solute atoms should exert a locking force on
the dislocations.

6.2 Experimental evidence for solute–dislocation interactions: yield phenomena

It is as yet impossible to obtain direct evidence for the segregation of
solute atoms to dislocations, although this is a predictable result which
should emerge from the use of field ion microscopy by which means individual solute atoms can be seen. Indirect structural evidence is, however,
abundant. For example, the fact that dislocations and sub-boundaries are
preferred regions for precipitation implies that solute atoms are more
heavily segregated in these regions as a prerequisite to nucleation of the
second phase. In many metals dislocations cannot be revealed by etching
unless impurities are present,[10] and it is often necessary to conduct suitable
heat treatments to enable the solute atoms to migrate to the dislocations
before etching is effective.

Single crystals of most solid solutions, whether interstitial or substitutional if tested in a suitably hard* tensile machine, will exhibit a *yield point*,
i.e. at the end of the elastic part of the stress–strain curve the specimen
suddenly deforms plastically with a drop in stress which results in the
characteristic sharp *upper yield point* (*A*) and a *lower yield point* (Fig. 6.3).

Strong yield points are found in single crystals of iron containing carbon and nitrogen in interstitial solid solution (0·002–0·01 weight per
cent), and are also an important feature of the deformation of polycrystal-

---

* A 'hard' tensile machine is one which undergoes only a small elastic
distortion at high stresses. Such a stiff machine has a sensitive response to a
drop in load, and will reveal yield points which softer machines cannot detect.

line iron (see Chapter 9). Other interstitial solid solutions have shown similar though less pronounced yield points, e.g. cadmium[11] and zinc[12] crystals containing nitrogen, β-brass with nitrogen. Yield points have also been reported in numerous substitutional solid solution crystals, but yield points are not normally observed at the low concentrations typical of many interstitial solid solutions.

The upper and lower yield point is sometimes followed by a horizontal section of the stress–strain curve—the *yield elongation zone* (CD). In polycrystalline iron, this zone is well developed and has been shown to be the region in which the plastic deformation moves through the specimen on one or more sharp fronts called Lüders bands (see Chapter 9), the propagation occurring at constant load.[13] Dilute alloy crystals, particularly of substitutional alloys, do not normally exhibit this behaviour, but it has,

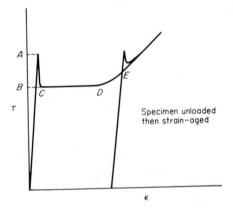

Fig. 6.3   Yield phenomena in a stress–strain curve (schematic)

however, been shown to be characteristic of the more concentrated solid solutions, e.g. 70/30 brass and copper–10 per cent indium single crystals.

If after stressing through the yield point, a crystal is unloaded (Fig. 6.3) and then restrained immediately, the yield point does not return. On the other hand if the crystal is rested for a long time at room temperature in the case of iron, or a shorter time at an elevated temperature (323–423 K), the yield point returns on subsequent restraining. This phenomenon is known as strain-ageing and involves the diffusion of solute atoms to new dislocations generated during the prior deformation which are thus rendered immobile.

As the temperature of deformation is raised, the sharp yield point is gradually eliminated, and instead the stress–strain curves exhibit small serrations (Fig. 6.4) which are not normally observed at low temperatures. These were first observed in iron containing small concentrations of carbon and known as the *Portevin–Le Chatelier effect*,[14] but the phenomenon is also common in non-ferrous solid solutions, e.g. brass and other copper base alloys, aluminium alloys. The phenomenon occurs in both single crystals and polycrystalline specimens. At the temperatures concerned 373–573 K for iron) the solute atoms are sufficiently mobile to be able to migrate to moving dislocations which in turn repeatedly break away from their atmospheres. Alternatively, the dislocations formed during the deformation readily collect solute atmospheres and remain locked in position so that new dislocations are being continuously created with the formation of repeated minute yield points.

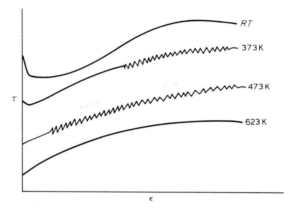

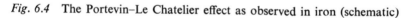

Fig. 6.4   The Portevin–Le Chatelier effect as observed in iron (schematic)

6.3   Theories of the yield point

Cottrell and Bilby[8] explained the sharp yield point in iron in terms of the locking of all mobile dislocations by carbon atoms.

The complete segregation of solute atoms to dislocations will only take place if opportunity is given for diffusion of the solute atoms. With carbon and nitrogen in iron, this happens readily at room temperature, but with many substitutional solutes elevated temperature treatments are necessary. In this way a solute atom atmosphere is built up, the concentration of which, c, at a point where the binding energy is U, can be expressed thus:

$$c = c_0 \, e^{U/kt} \qquad\qquad \textbf{6.4}$$

where c_0 = average concentration, U = elastic binding energy.

Such an equation predicts a Maxwellian-type distribution about the dislocation, but in the case of carbon and nitrogen in iron at room temperature, the elastic interaction is so strong that $U \gg kT$, and the atmosphere becomes completely condensed into a row of atoms along the core of the dislocation. The above equation can be used to predict the temperature below which this will happen.

In a fully condensed atmosphere $c = 1$ and $U = U_{max}$. From eqn. **6.4** above

$$T_{crit} = \frac{U_{max}}{k \ln 1/c_0}$$ **6.5**

With values of $c^0 = 10$ and $U_{max} = 0.5$ eV, a temperature of 700 K is obtained for T_{crit}. This accounts for the very strong yield points found in iron at room temperature and lower temperatures.

An important feature of Cottrell-locking is that a strong temperature dependence of the yield point is to be expected, the yield stress rising very steeply at low temperatures. At 0 K the atmosphere of solute atoms is fully condensed and there is no assistance from thermal vibrations, consequently the stress required to break a dislocation away from a row of solute atoms is high. As the temperature is raised, the local thermal fluctuations of atoms gradually increase and augment the effect of the applied stress to achieve the breakaway of dislocations. Only a short length of dislocation line need escape from the atmosphere of solute atoms for a catastrophic break-away to set in. The Cottrell–Bilby analysis gives a relationship between activation energy W and τ of the form in Fig. 6.5,

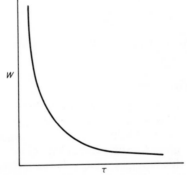

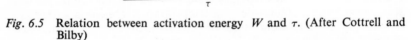

Fig. 6.5 Relation between activation energy W and τ. (After Cottrell and Bilby)

and when τ the applied stress approaches τ_l, the local stress necessary to cause a break-away of a dislocation loop, the required activation energy can be defined thus:

$$W(\tau) = W_0 \left(1 - \frac{\tau}{\tau_l}\right)^{3/2}$$ **6.6**

where W_0 is the activation energy required in the absence of an applied stress.

The probability of a dislocation escaping locally from its atmosphere can then be given as

$$\exp\left[\frac{W(\tau)}{kT}\right]$$

The dislocation is assumed to vibrate under the influence of thermal movement at a frequency $V\,(\sim 10^{11}\ \text{s}^{-1})$, then the stress τ necessary to liberate the dislocation at a temperature T in unit time is

$$\tau = \tau_l - \left[\frac{kT}{W_0}\ln Vt\right]^{2/3} \qquad \textbf{6.7}$$

This relationship gives a strong $T^{2/3}$ dependence of the break-away stress, i.e. the yield stress. Experiments confirm that a marked temperature dependence exists, but it does not always obey a $T^{2/3}$ relationship.

While it was originally thought that this yield drop could be generally characterized by the unpinning of the locked dislocations, in recent years it has been found that many *new* dislocations are often generated during yielding. The original dislocation–solute atom theory has thus been modified, although the locking of *pre-existing* dislocations by solute atoms is still an essential feature of the yield point.

The strain rate $\dot{\epsilon}$ of a crystal can be expressed in terms of the movement of dislocations thus,

$$\dot{\epsilon} = Nvb \qquad \textbf{6.8}$$

where N is the number of dislocations moving per unit area, v is their average velocity and b the Burgers vector.

To maintain a given strain rate, N dislocations have to move at velocity v, but if the number increases say to $2N$, the velocity needed is only $v/2$. If we now assume that the lower velocity can be achieved with a lower stress, it follows that in the stress–strain curve a drop in stress will take place. Johnson and Gilman [15, 58] carried out stress–strain measurements on lithium fluoride crystals, and by using an etch pit technique for revealing dislocations were able to show that two criteria were necessary for a sharp yield point.

1. An increase in the number of moving dislocations;
2. a direct relationship between stress and velocity of dislocations.

They found that the yield point of a lithium fluoride crystal became less marked as the number of free dislocations increased; this was demonstrated by repeated straining through the yield point followed by reloading and restraining (Fig. 6.6). Calculated curves assuming different dislocation densities were found to be in close agreement with the experimental results.

When the dislocation density is studied in the early stages of deformation by etch pits or by thin-foil electron microscopy[15, 17] it is found for $\epsilon < 0\cdot1$ that

$$\rho = \rho_0 + c\epsilon^\alpha \qquad\qquad 6.9$$

where ρ = dislocation density, ρ_0 = initial dislocation density (between $10^3/cm^2$ and $10^7/cm^2$), c = constant $\approx 10^8\ cm^{-2}$, α = constant $\approx 1 \pm 0\cdot5$.

We can now define the strain rates at the upper yield point ($\dot{\epsilon}_U$) and the lower yield point ($\dot{\epsilon}_L$) according to eqn. 6.8.

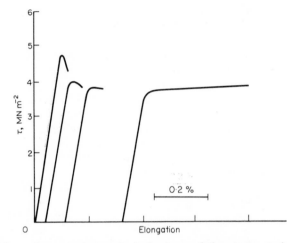

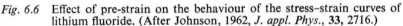

Fig. 6.6   Effect of pre-strain on the behaviour of the stress–strain curves of lithium fluoride. (After Johnson, 1962, *J. appl. Phys.*, **33**, 2716.)

$$(\dot{\epsilon})_U = \rho_U v_U b$$

$$(\dot{\epsilon})_L = \rho_L v_L b$$

where ρ_U and ρ_L are the densities of *mobile* dislocations at the upper and lower yield points, and v_U and v_L their velocities, so

$$\frac{v_U}{v_L} = \frac{\rho_L}{\rho_U} \qquad\qquad 6.10$$

The velocity of the dislocations is stress dependent and has been shown to obey the following relationship:

$$v = k\tau^m \qquad\qquad 6.11$$

where m can vary between 1 and 100. Consequently,

$$\frac{\tau_U}{\tau_L} = \left(\frac{\rho_L}{\rho_U}\right)^{1/m} \qquad\qquad 6.12$$

where τ_U = upper yield stress, and τ_L = lower yield stress.

It follows that both m and ρ_U are criteria for the extent of the yield drop. With small m values, the ratio τ_U/τ_L will be large, representing a pronounced yield drop, whereas when m is between 100 and 300 the yield point is barely perceptible. Lithium fluoride crystals ($m \simeq 16\cdot5$) give pronounced yield points, while germanium crysals deformed at 773 K also have low m values and large yield drops. In Fig. 6.7 calculated stress–strain curves are plotted for a range of m values. The initial density of *mobile* dislocations must also be small, for example in the case of α-iron where

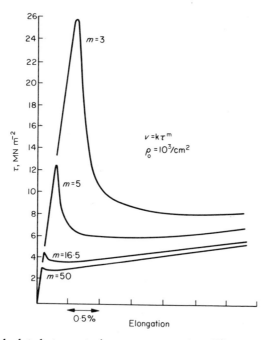

Fig. 6.7   Calculated stress–strain curves, assuming different relationships for the dependence of dislocation velocity on stress. (After Johnson, 1962, *J. appl. Phys.*, **33**, 2716.)

$m \simeq 35$, the yield point is substantial only when the initial density of free dislocations is less than about $10^3/cm^3$. As the observed density of dislocations in annealed iron and steel is greatly in excess of this value, it follows that the majority of the original dislocations are firmly pinned.

The above mechanism can occur in a pure metal if the necessary criteria are met; however, it is an experimental fact that sharp yield points are found when solute atoms have maximum opportunity of interacting with dislocations. It now appears that an important role of solute atom locking in the development of the yield point is to control the number of mobile dislocations. Furthermore, the actual locking mechanism is no longer vital to the theory, and locking of dislocations by precipitate particles can also lead to yield points (Chapter 7).

Cottrell[4] has emphasized that the yield point is a type of plastic instability which is usually only detected when the rate of work hardening over the critical strain range is low. This explains why it is easier to detect yield points in face-centred cubic single crystals than in polycrystals, because the single crystals are in Stage 1 of the stress–strain curve where the work hardening is low. On the other hand, polycrystals exhibit Stage 2 hardening from the earliest stages of deformation, so the yield point is hidden except at relatively high solute concentrations: Cottrell distinguishes three conditions for the production of yield points.

1. Initial dislocation density = zero. In this case we start with a perfect crystal, and yielding will occur at the very high stress (approaching the theoretical strength) necessary to generate dislocations in the crystal. The stress falls drastically to a low value needed to move the new dislocations through the crystal. This type of behaviour is sometimes observed in metal whisker crystals, an example of which is shown in Fig. 6.8 for a copper whisker.[18]

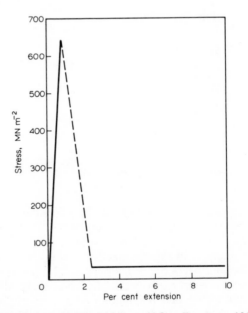

Fig. 6.8 Yielding of a copper whisker. (After Brenner, 1958, *Growth and Perfection of Crystals*. Ed. by R. R. Doremus *et al.* Wiley, New York and London. p. 157)

2. Initial dislocation density greater than zero, and N the number of mobile dislocations is zero. The dislocations are no longer absent but are locked by impurities. In this case the yield point can occur either as a result of unlocking the dislocations (weak pinning) or where new dislocations are formed (strong pinning).

3. $N > 0$, and the stress increases with the velocity v of the dislocations. In this case the mobile dislocations already exist and the yield point develops as they multiply. This type of yielding is often met in non-metallic crystals

6.4 Deformation of solid solution crystals

6.4.1. *Critical shear stress of solid solutions*

All elements in solid solution raise the critical shear stress for slip above that of the pure metal. Reliable data are, however, scarce, partly because

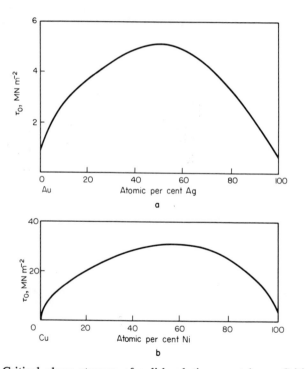

Fig. 6.9 Critical shear stresses of solid solution crystals. **a**, Critical shear stress of Au–Ag solid solution crystals. (After Sachs and Weerts, 1930, *Z. Phys.*, **62**, 473). **b**, Critical shear stress of Cu–Ni solid solution crystals. (After Osswald, 1933, *Z. Phys.*, **83**, 55)

of experimental difficulties, such as segregation during the growth of crystals, and the problem of the exact measurement of the critical shear stress in the presence of yield points.

The classical work of Von Goler and Sachs[19] on the copper–zinc α solid solutions showed that the critical resolved shear stress τ_0 increased almost linearly for small zinc contents, but deviations from linearity occurred at higher concentrations. Osswald,[20] who examined the whole range of copper nickel solid solutions, showed that the τ_0-composition curve had a maximum at 60 atomic per cent nickel (Fig. 6.9). Similarly, Sachs and Weerts[21] examined silver–gold solid solutions over the whole range of composition, and found a well-defined maximum at 50 atomic per cent corresponding to a strength of 5 MN m^{-2} an order of

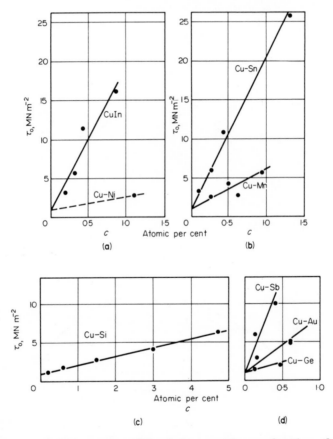

Fig. 6.10 τ_0 of dilute copper solid solution crystals as a function of atomic concentration of solute. (After Linde *et al.*, 1954, *Ark. Fys.* **8**, 511)

magnitude higher than the critical shear stresses of the component pure metals.

Experiments with more dilute alloys have shown that solid solution additions have a particularly marked effect at low concentrations, for example Greenland[22] showed that a concentration of 10^{-4} of silver in mercury raised the critical shear stress of the metal by a factor of 5 when crystals were tested at 223 K.¡ The first systematic work was done by Linde and co-workers[23, 24] on dilute solid solutions of copper with antimony, germanium, gold, indium, manganese, nickel, silicon and tin. Alloy single crystals were prepared with concentrations of solute mainly up to 1 atomic per cent and the critical shear stresses determined at room temperature (Fig. 6.10a–d). The τ_0-composition curves were all linear in this range, but the slope varied markedly from one element to another. The

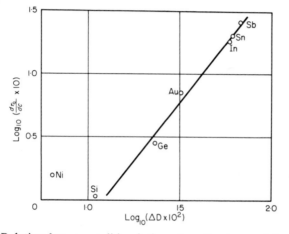

Fig. 6.11 Relation between solid solution strengthening and lattice distortion in copper-base solid solutions. ΔD is the difference in atomic diameters of solvent and solute

smallest effect on the flow-stress resulted from the addition of nickel and silicon which have atomic sizes close to that of copper, while antimony, indium and tin which have substantially larger atoms had a more pronounced effect. The role of relative atomic size of solvent and solute is more clearly shown when the log. of the slope of the τ_0-composition curve $d\tau_0/dc$ is plotted against the log. of the difference in Goldschmidt's atomic diameter* between solvent and solute (ΔD) (Fig. 6.11). Alternatively $d\tau_0/dc$ can be plotted against

$$\frac{1}{a}\frac{da}{dc} = \theta$$

* The atomic diameter determined from the interatomic distances in the crystals of the elements (close-packed structures).

where a is the lattice parameter. In each case a linear plot is obtained for dilute alloys and a relation of the following type applies:

$$\frac{d\tau_0}{dc} = K\theta^n$$

where $n \simeq 2$.

Fleischer[25] has investigated the strengthening of copper-base alloys, taking into account both the size factor effect and the effect of difference in modulus between solvent and solute atoms. The results are plotted in Fig. 6.12 as a function of a combined mismatch parameter ϵ_s which is defined as

$$\epsilon_s = (\epsilon_G' - 3\epsilon_b) \qquad\qquad 6.14$$

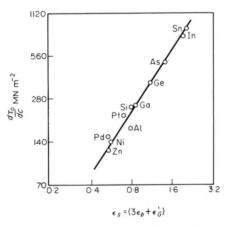

$$\epsilon_s = (3\epsilon_b + \epsilon_G')$$

Fig. 6.12 Relation of solid solution strengthening to a combined mismatch parameter. (After Fleischer, 1963, *Acta Metall.*, **11**, 203)

where ϵ_b is the size factor parameter and ϵ_G' the modulus parameter, the expression being that appropriate for screw dislocations. These results emphasize the importance of both atomic size and modulus. Fleischer concluded that both effects are substantial, but that they will contribute a varying proportion to the increased strength dependent on the solute metal.

Similar systematic experiments[26] have been carried out with magnesium crystals to which had been added small concentrations of aluminium, cadmium, indium, thallium and zinc. The results which are summarized in Fig. 6.13a again illustrate the wide variations arising from different solutes. The effects of the solutes on the lattice parameter of magnesium are plotted in Fig. 6.13b. These results show that while relative atomic size or effect on the lattice parameter is significant, it is not the only consideration. For

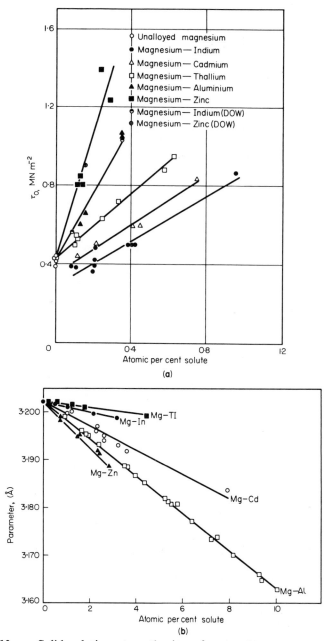

Fig. 6.13 **a**, Solid solution strengthening of magnesium crystals; **b**, effect of solutes on the lattice parameter of magnesium. (**a**, After Levine *et al.*, 1959, *Trans. AIME*, **215**, 521; **b**, After Busk, 1950, *Trans AIME*, **188**, 1400).

example, although thallium has a very small effect on the lattice parameter of magnesium, it has a greater strengthening effect than cadmium.

It is relevant to refer to experiments by Ainslie and co-workers[27] on polycrystalline binary copper-base solid solutions of the same grain size and free from texture. In this work, the electron/atom ratio $e/a*$ was taken as a criterion, and it was found that three alloys of copper with aluminium, zinc and silicon of similar e/a had equal yield strengths. However, this criterion failed for copper–tin alloys which had much higher initial yield strengths, yet much lower e/a ratios than the other alloys.

It seems clear that both the relative atomic size of solvent and solute atoms, their elastic properties and the electron/atom ratio of the alloy are significant variables in determining the flow stress of solid solutions. This problem will be considered further in the section dealing with theories of solution strengthening (Section 6.6).

6.4.2. Temperature dependence of τ_0 of solid solutions

There is substantial experimental evidence which indicates that the change of the flow stress with temperature of alloy crystals is greater than that for pure metals.[28] The results of several investigations on copper and copper–zinc alloys are summarized in Fig. 6.14 where the temperature

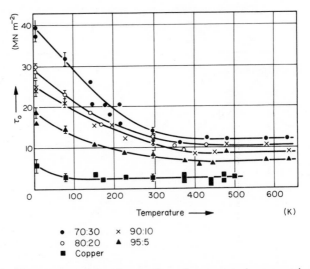

Fig. 6.14   Temperature dependence of τ_0 of copper and copper–zinc crystals. (After Mitchell, 1964, *Progress in Applied Materials Research*, **6**)

* Assuming the usually assigned valencies, this is the ratio of valency electron concentration to concentration of atoms in an alloy.

dependence of τ_0 increases with increasing zinc concentration.[3] The main effect is observed below 300 K.

Detailed analysis of the results reveals that the increased temperature dependence of τ_0 is only marked at the higher alloy concentrations. This is confirmed by work on copper and its dilute alloys with silver and germanium, where the ratio of τ_{293K}/τ_{80K} is practically the same for copper and the three dilute alloys.[28] Similar results have been obtained with aluminium and copper–aluminium alloys up to 14 atomic per cent Al,[29] for copper–germanium and copper–gallium alloys[30], nickel–cobalt solid solutions,[31] and silver–indium alloys.[55]

It has been found that the temperature dependence of the flow stress in the early stages of deformation (Stage 1), determined by the temperature change method on a single specimen, is identical to the temperature dependence of the critical resolved shear stress τ_0[30] found from a group of crystals. This indicates that the main contribution to τ_0 is a friction stress on moving dislocations, rather than a stress needed to unlock dislocations from their solute atmospheres.

6.4.3. Stress–strain curves of solid solution crystals

1. YIELD POINTS

In most alloy crystals the transition from elastic to plastic behaviour in the tensile test is a sharp one if proper care is taken. At the yield point there is often a sharp drop in the load to the lower yield stress, which in ideal circumstances is followed by a flat region or yield elongation zone. With polycrystalline steels, the yield elongation zone represents the zone in which yielding propagates at uniform macroscopic stress along the specimen, the moving front or Lüders band giving rise to surface relief effects. Similar behaviour is obtained in single crystals of solid solutions when the solute concentration is fairly high.

The first yield points in metal crystals (other than iron) were observed with zinc and cadmium containing nitrogen in interstitial solid solution.[33] Ardley and Cottrell[32] showed that as little as 1 per cent zinc in substitutional solid solution in copper produced a sharp yield point (Fig. 6.15), but it was often necessary to strain age the specimen before the yield point was obtained. Similarly, yield points have been observed in single crystals of copper–germanium, –gallium,[30] –aluminium,[34] –indium[35] and –arsenic,[36] also silver–aluminium.[37] There is some evidence that atoms which cause greater distortions in the copper lattice give larger yield drops, for example a copper–indium crystal gives a much larger yield point than a copper–zinc crystal of similar atomic concentration.[35]

As in the case of iron, straining beyond the yield point causes it to disappear on subsequent unloading and restraining; however, a low-temperature anneal causes the phenomenon to return (Fig. 6.15). Recent work confirms the viewpoint that yield points of varying intensity are a normal

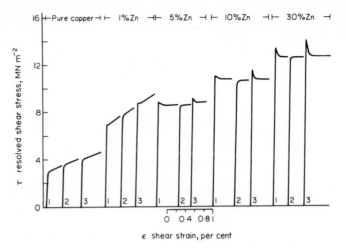

Fig. 6.15 Stress–strain curves of copper and copper-zinc crystals at room temperature. Curve 1: first loading; curve 2: immediate reloading; curve 3: after 2 hr at 473 K. (After Ardley and Cottrell, 1953, *Proc. R. Soc.*, **A219**, 328)

characteristic of the deformation of solid solutions, whether of the substitutional or interstitial type. The absence of a yield point during a tensile test can usually be attributed to poor experimental technique, for example non-axiality of loading, damage to the specimen, unsuitable specimen dimensions, or to the characteristics of the testing machine. A very 'hard' machine in which the load is rapidly relaxed when a sudden strain occurs is essential for this type of experiment.

After the yield point, in dilute solid solutions Stage 1 hardening dominates the stress–strain curve, but crystals of more concentrated alloys exhibit a wide variety of Lüders band propagations. Recent work has shown that the type of Lüders band obtained is dependent on the orientation of the crystal. For example, copper–10 atomic per cent indium crystals of soft orientations showed a marked 'geometrical' type of propagation in which such a large amount of slip occurred at the Lüders front that the cross-section of the crystal changed drastically from circular to elliptical (Fig. 6.16).[35] This phenomenon is associated with a large yield drop, and

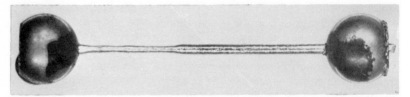

Fig. 6.16 Lüders band propagation in a copper–indium single crystal

is in part due to the fact that geometrical softening takes place, i.e. as the crystal axis rotates during tension, the resolved shear stress on the primary slip system increases, causing slip to occur under a diminishing load (i.e. the Schmid factor, $\sin \chi \cos \lambda$ increases). However, this geometrical effect can only be observed if it is not offset by a high rate of work hardening.

This is undoubtedly not the complete explanation for many crystals lying outside the range in which geometrical softening occurs still exhibit the same type of Lüders band propagation.

On the other hand, crystals of harder orientations, i.e. crystals which will work harden more rapidly in the early stages of deformation, show a simpler form of Lüders band where the amount of slip taking place at the front is very much less, and does not lead to marked changes in cross-section. The front is more diffuse as indicated by the gradual fading out of slip bands as it is crossed. It is likely that these two types of Lüders band are only extremes of the same behaviour, the differences being dictated primarily by the rate of work hardening during the early stages of deformation. In concentrated alloys the bands so intrude on the stress–strain behaviour that the intrinsic hardening is not readily determined.

2. THE STRESS–STRAIN CURVE

As with pure metals it is convenient to divide the stress–strain curves of solid solution crystals into three stages, 1, 2 and 3, but it should be emphasized that in many cases yield elongation zones may obscure part or even all of Stage 1 of the curve. Fig. 6.17 shows some typical curves for cop-

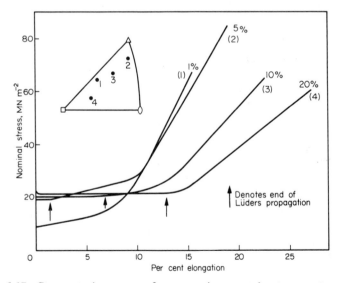

Fig. 6.17   Stress–strain curves of copper–zinc crystals at room temperature (Brindley, *et al.*, 1962, *Acta Metall.*, **10**, 1043)

per–zinc crystals between 1 and 20 atomic per cent zinc; Stage 1 and Stage 2 hardening is revealed in all curves except that for 20 per cent zinc, where the Lüders extension has completely masked the Stage 1 hardening.[35]

Stage 1, as in pure metals, corresponds to deformation on the primary slip plane with only very limited amounts of slip on other systems, and as a consequence, the rate of work hardening is low. Stage 1 in alloy crystals is, however, more extensive and increases with increasing solute concentration (Fig. 6.17). As much as 60 per cent shear strain has been measured at room temperature for copper–0·50 atomic per cent silver crystals with $\tau \simeq 6\ \mathrm{MN\,m^{-2}}$.[28] A pure copper crystal of similar orientation had an easy glide range of about 6 per cent and $\tau_0 \simeq 0·6\ \mathrm{MN\,m^{-2}}$, so the easy glide is increased in the same ratio as the critical shear stress for glide. A similar trend over a much wider compositional range is shown by the nickel–cobalt single crystals[31, 38]; at 90 K, nickel crystals have only a small Stage 1 strain, whereas Ni–20 atomic per cent Co crystals exhibit 40 per cent shear strain in this stage, and crystals with 40 atomic per cent Co deform to about 80 per cent shear strain before entering Stage 2 of the stress–strain curve. The extent of Stage 1 at two temperatures as a function of cobalt content is shown in Fig. 6.18; the length reaches a maximum at 60 atomic per cent cobalt.

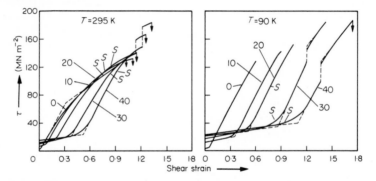

Fig. 6.18 Shear stress/shear strain curves of nickel–cobalt single crystals at 295 K and 90 K. S indicates the stress at which the crystal crossed the symmetry line. Numbers represent the cobalt concentration. (Breaks in the curves arise from the use of a different criterion to calculate τ from the point when duplex slip commences.) (After Pfaff, 1962, *Z. Metallk.*, **53**, 411, 466)

Alloying raises the critical shear stress for glide on all possible systems *proportionately*, so to produce the patches of secondary slip which characterize the end of Stage 1, higher localized stresses must be created compared with the pure metal. If these stresses are considered to arise from the pile-up of dislocations on the primary system then larger pile-ups will be needed, and these will only occur if easy glide is more prolonged in the alloy, than in the pure metal crystal.

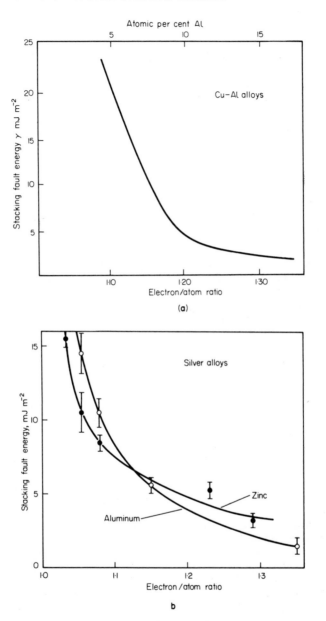

Fig. 6.19 **a,** Effect of aluminium on the stacking fault energy of copper. (After Swann, in ref. 50). **b,** Stacking fault energies of silver solid solutions. (After Thornton *et al.*, 1962. *Phil. Mag.*, **7**, 1349)

The hardening rate for pure metals in Stage 1 is approximately $3 \times 10^{-4}G$ for soft orientations. Similar measurements in alloy crystals indicate a slightly lower hardening rate of about $1-2 \times 10^{-4}G$. This would be expected because by raising the critical shear stress alloying elements in solid solution will make it more difficult for dislocations on secondary systems to move.

Stage 2 is characterized by a marked increase in the rate of hardening, which is of the same order as that of pure metals but extends to much higher stresses; shear strains of 75–80 per cent are commonly achieved before the rate of hardening diminishes. It is reasonable to consider that the beginning of Stage 2 is associated with the formation of Lomer–Cottrell locks or other effective barriers, and that the end of this stage is characterized by the massive avoidance of the obstacles by cross-slip of screw dislocations leading to the formation of slip bands. Coarse slip bands tend to develop less readily in dilute alloys than in pure metals.

The delay in the transition to Stage 3 seems at least in part to be connected with the effect of the solute additions on the stacking fault energy.[47] Recent work has shown that such additions usually lower the stacking fault energy, in some cases to very low values (Fig. 6.19). For example, 15 atomic per cent aluminium lowers the stacking fault energy of copper and silver from 40 and 20 mJ m^{-2} respectively to about 2 mJ m^{-2} (Fig. 6.19)†. This means that dislocations in a face-centred cubic solid solution will show more tendency to dissociate into partials separated by stacking faults than the pure metals.

In Chapter 4 the lower stacking fault energy of copper with respect to aluminium was shown to postpone cross-slip and Stage 3 of the stress–strain curve. Likewise, copper base solid solutions would be expected to exhibit Stage 3 at an even later stage than pure copper, because of the difficulty of forcing widely dissociated dislocations to associate locally. Such localized constrictions will require a high stress to form, and they are a necessary prerequisite for cross-slip to occur. However, alloys of aluminium with copper or magnesium which have relatively high stacking fault energies nevertheless show much greater work hardening in Stage 2 than does pure aluminium. In contrast, aluminium–silver solid solutions are closer in their behaviour to pure aluminium. The copper and magnesium atoms, which differ substantially in size from those of aluminium and silver, introduce greater lattice strains in aluminium than do silver atoms, with the possible result that the cross-slip mechanism is hindered and so the formation of coarse slip bands is less pronounced.[39]

In view of the late development of Stage 3 hardening in solid solution crystals at low temperatures, the transition is best demonstrated by raising the temperature of deformation. Figure 6.20 shows stress–strain curves at different temperatures of a copper–5 per cent zinc crystal in which Stage

† See also Gallagher, P. C. J. (1970) *Met Trans.* **1**, 2429 for further data.

3 is only evident at 673 K. The specimen deformed at 77 K is still under-going Stage 1 hardening after 20 per cent elongation.

Another factor which tends to eliminate any Stage 3 hardening is the onset of duplex slip. Unlike pure metals, solid solutions show extensive 'overshooting' of the symmetry boundary; that is, during deformation the specimen axis rotates well past the symmetry boundary [001]–[Ī11] before slip on the secondary system occurs. This phenomenon is usually most pronounced in solid solutions of low stacking fault energy, because the widely extended dislocations on the primary slip system make it difficult

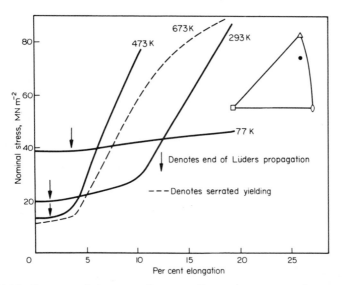

Fig. 6.20   Stress–strain curves of copper–5 atomic per cent zinc crystals at several temperatures. (Brindley *et al.*, 1962, *Acta metall.*, **10**, 1043)

for dislocations to move on the secondary systems. Alternative explana-tions involve the hardening of the secondary slip system (latent hardening), i.e. the raising of the critical stress for secondary slip, by various mechan-isms, for example dislocation pile-ups on the primary system. The occur-rence of duplex slip generally causes a sharp flattening of the stress–strain curves, which is frequently associated with the propagation of the duplex slip through the crystal behind a type of Lüders front.

A further feature of the stress–strain curves of alloy crystals tested at elevated temperatures is the occurrence of fine serrations throughout the test. This phenomenon is similar to the Portevin–Le Chatelier effect ob-served in mild steel and in some polycrystalline aluminium alloys at ele-vated temperatures. It is one of the typical phenomena in alloys which exhibit yield points at lower temperature. As in steels the serrations dis-

appear if the temperature of deformation is sufficiently high. The phenomenon is undoubtedly associated with the diffusion of solute atoms to
glide dislocations, and it seems likely that this process is assisted by vacancies produced during the deformation which associate with the solute atoms.

6.4.4. *Dislocations in solid solutions*

Measurements on dislocation nodes in solid solution alloys indicate
that γ, the stacking fault energy of a metal, is lowered by many elements
in solid solution. If the stacking fault energy of a solid solution system is
plotted against electron to atom ratio, there is initially a fairly sharp drop
in the curve[3] (Fig. 6.19a and b). Consequently, solutes of high valency
have a greater effect on γ for the same atomic concentration than solutes
with low valencies. In concentrated solid solutions of copper and silver
with metals of high valency, such as aluminium, gallium or germanium,
stacking fault energies of the order of $1-5\,\mathrm{mJ\,m^{-2}}$ are found.[40] Nevertheless, metals of lower valency, e.g. zinc, have a pronounced effect
on the stacking fault energy (Fig. 6.19b), and as they often form more
extensive solid solutions, low values of γ are achieved in these cases if the
solute concentration is high. Consequently, as the solute concentration in a
solid solution is increased, the dislocations become more widely dissociated.
Fig. 6.21a and b show the effect of increasing concentrations of gallium in silver[54]. At 10 atomic per cent (6·7 weight per cent) the dissociation
of some dislocations is just detectable, but when the concentration is increased to 17 atomic per cent (11·7 weight per cent) the dissociation distance
is so great, that long ribbons of stacking fault are seen in thin foil specimens. These photographs both indicate that at small deformations the
dislocations are distributed on well defined slip bands (they are *co-planar*),
and that there is little tendency to form sub-boundaries or dense networks
of dislocations as for example in aluminium.

The gradual change in the dislocation patterns in a typical series of
copper–aluminium solid solutions has been studied by Howie.[5] During
easy glide, he found that with increasing aluminium content there was an
increasing tendency for dislocations to line up on slip planes. In addition,
there was a marked tendency to form dipoles, that is, paired positive and
negative dislocations. These were also observed during Stage 1 hardening
in nickel–cobalt single crystals by Mader.[5] Typical examples in a deformed
Cu–In solid solution crystal can be seen in Fig. 3.19. Mader *et al.*[4] adopted
the technique of sectioning parallel to the operative slip plane to examine
the dislocation arrays in nickel–cobalt single crystals. By the use of dark
field electron microscopy they were able to show that a high proportion of
these dislocations belonged to the primary glide system. In Stage 2 of the
stress–strain curves, long straight dislocation lines were observed which
were along the intersection of the primary plane with the conjugate plane.
At least some of these dislocations were considered to be Lomer–Cottrell
dislocations, which in the Seeger theory of work hardening are assumed to

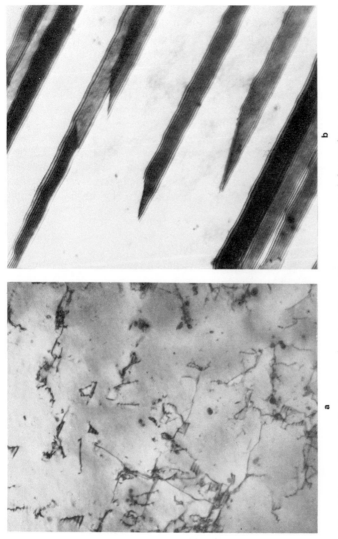

b

a

Fig. 6.21 **a**, Silver–10 atomic per cent gallium alloy, 5 per cent deformation at room temperature; **b**, silver–17 atomic per cent gallium alloy, 5 per cent deformation at room temperature. Thin-foil electron micrographs ×51,000 (Hutchison)

develop during Stage 2. At this stage of the deformation, the dislocation distribution is more irregular than in Stage 1 and becomes increasingly so at higher strains. The dipoles are less predominant but the density of elongated dislocation loops increases. The influence of stacking fault energy is seen by comparing pure nickel crystals (high γ) with crystals containing 67 per cent Co (fairly low γ). In the alloy crystal, the dislocations are predominantly straight, while in the pure nickel they are more curly as would be expected because of the greater ease of cross slip.

Stage 3 in alloy crystals has similar characteristics to that in pure metals. The slip bands are thick and short and connected to each other by cross-slip. The lower stacking fault energy of the alloys means that the stresses at which this phenomenon is observed are substantially higher than in pure metals. Cross-slip can, however, be generated in Stage 2 by another mechanism where the pile-ups of dislocations in the primary slip plane lead to the activation of nearby sources on cross-slip planes. As the dislocations are created on the cross-slip plane, no constriction of extended dislocations is needed, so the process can occur in the absence of thermal activation. This type of cross-slip is common during Stage 2 deformation of α-brass crystals.[41]

6.5 Deformation of ordered solid solutions

In substitutional solid solutions involving two atomic species, the atoms are usually arranged in a random fashion on the lattice sites, thus forming a disordered solid solution.[5] However, a number of solid solutions, particularly those with near-stoichiometric compositions such as AB, A_2B, etc., undergo a structural rearrangement or ordering below a certain critical temperature T_{crit} to produce a *superlattice*. In the ordered condition the two atomic species are each arranged in regular configurations, hence the term superlattice. For example, in the ordered CuZn solution, which is body-centred cubic, the zinc atoms are in the centres of the unit cells, while in Cu_3Au the copper atoms occupy the face-centred positions in the face-centred cubic lattice, while gold atoms occupy the corners.

While disordered solid solutions deform plastically in the manner which has been described, in ordered solid solutions the dislocations have special properties. These arise from the fact that normally an ordinary dislocation moving in this type of crystal would leave behind it a trail of disorder across the slip plane which is referred to as an anti-phase boundary. It was first realized by Koehler and Seitz[48] that it would be energetically more favourable for dislocations to move in pairs or groups bounded by antiphase boundaries, such that the total Burgers vector of the group was equivalent to the identity distance of the superlattice in the slip direction. These superlattice dislocations (super-dislocations) were first observed by Markinkowski *et al.*[49] in Cu_3Au, using thin-foil electron microscopy, and subsequently in a number of other ordered alloys.[5, 50] They are often seen characteristically as closely spaced pairs, the separation being of the order 100–200 Å. The structure of such a super-dislocation in an ordered solid solution of the *AB* type is shown in Fig. 6.22, in which the pair of dislocations are shown with an anti-phase boundary in between. The width of the anti-phase boundary is the result of a balance between the elastic repulsion of the two dislocations (of the same sign) and the energy of the anti-phase boundary (or fault-in-order). There is a close analogy with the dissociation of dislocations in close-packed lattices such as the face-centred cubic.

It has been known for a long time[51] that there is a marked difference between the stress–strain curves for ordered and disordered crystals of the same alloy. For example, a disordered crystal of Cu_3Au has a high critical shear stress, a long region of Stage 1 hardening and exhibits the phenomenon of overshooting of the $[001]$–$[\bar{1}11]$ side of the basic stereographic triangle by the crystal axis. In contrast a highly ordered solid solution

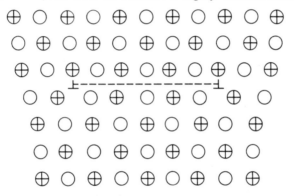

Fig. 6.22 A super-dislocation in a cubic ordered lattice of the *AB* type. The dashed line is an anti-phase boundary. (After Brown, 1959, *Phil. Mag.*, **4**, 693)

crystal of the same alloy becomes more like a pure metal with a low stacking fault energy; Stage 2 hardening is well developed and the transition to Stage 3 is at high stresses. An ordered alloy containing super-dislocations would find cross-slip as difficult as a pure metal of low stacking fault energy, because in each case, association of the partial dislocations must occur before the slip dislocations can move to the cross-slip plane.

The presence of order plays an important part in the deformation behaviour of many intermetallic compounds, the plastic behaviour of which is often very temperature dependent. Westbrook has published detailed surveys.[52, 53]

6.6 Theories of solid solution strengthening

There are a number of theories of solid solution hardening which can be roughly divided into two categories.[6] Firstly, there is a group of theories depending on various models of dislocation locking where the dislocation is at rest, while the other theories are concerned with the frictional resistance of solute atoms to the movement of dislocations. A number of dislocation locking mechanisms have already been referred to earlier in this chapter, so we shall only deal briefly with two theories in this category, those due to Cottrell and to Suzuki.[9, 46]

6.6.1. *Cottrell locking theory*

The formation of Cottrell atmospheres both in interstitial and substitutional solid solutions is well established, but while this type of dislocation locking plays a part in the development of sharp yield points, it is difficult to see that it can account for more than a small part of solid solution strengthening. The modern theory of the sharp yield point assumes that dislocations with atmospheres remain locked in position, while it is the newly created dislocations which move. Even if the locked dislocations are torn from their atmospheres, the drop in stress observed at the yield point is usually only a small fraction of the initial flow stress. One is thus forced to the conclusion that theories dependent primarily on dislocation locking are not sufficient, and that the major contribution to the flow stress comes from resistance to *moving* dislocations by various configurations of solute atoms. This view is supported by many experimental results which indicate that addition of solute raises the level of the stress–strain curve as a whole[56, 57]. Stress–strain curves for increasing solute concentration are parallel but progressively displaced to higher stresses. This is most readily explained in terms of a friction stress or impedance which is independent of strain but increases with solute concentration. Recent work indicates that for a range of solutes in silver, the friction stress increases linearly with the electron/atom ratio.

6.6.2. *Chemical interaction theory*

The detailed theory of strengthening due to segregation to stacking faults is complex but Suzuki[9, 46] gives the following relationship for the dependence of critical shear stress on solute concentration and temperature:

$$\tau_0 = \frac{2h}{vb} H c_0 (1 - c_0) \frac{1 - e^{-H/RT}}{1 - c_0(1 - e^{-H/RT})} \qquad \textbf{6.15}$$

where τ_0 = maximum locking stress, c_0 = average concentration of solute, v = molecular volume (volume of alloy/mole), and

$$H = (\gamma_b - \gamma_a)v/2h,$$

where γ_a and γ_b are the stacking fault energies in metals a and b; h = half width of a fault. H is not dependent on temperature, and as it is the main variable in the equation, the flow stress due to this mechanism is likely to be relatively insensitive to temperature, but very dependent on solute concentration. Suzuki has determined experimentally the variation of the critical shear stress with temperature for a range of copper–nickel solid solutions. These results show that above 500 K, τ_0 is insensitive to temperature, but sensitive to solute atom concentration; Suzuki has suggested that stacking fault segregation plays a dominant part in determining the strength in this range of temperature. Below 500 K, τ_0 rises rapidly with decreasing temperature, and Suzuki attributes this to Cottrell locking, which is temperature sensitive.

The Suzuki theory can, like the atmosphere theory, be criticized on the grounds that dissociated dislocations, to which segregation takes place in the stacking fault, will not move when the stress is applied. They will be so firmly locked that new dislocations will be generated, and the stress to move these dislocations is the solution hardening stress. It is, however, possible that the interaction of solute atoms with extended dislocations will contribute to the frictional stress during dislocation movement.

6.6.3. Mott and Nabarro theory

Mott and Nabarro [42] developed a theory which assumed that solid solution strengthening arose from the interaction of randomly distributed solute atoms with *moving* dislocations. The isolated solute atoms give rise to small local internal stress fields, the magnitude of which was assumed to depend only on the difference in size between the atoms of solvent and solute. The dislocation line was assumed to be flexible, so that the whole line did not have to move simultaneously, and was restrained periodically by the interaction of its stress field with those of solute atoms.

There are thus two important parameters, firstly the distance apart of the solute atoms or the spacing of the internal stress fields. This is a 'wavelength' Λ which is simply

$$\Lambda = \frac{a}{c^{1/3}} \qquad\qquad 6.16$$

where a = interatomic spacing, and c = atomic concentration of solute.

The second parameter θ is a measure of the magnitude of the internal stress field which depends on the difference in size between solvent and solute atoms. It is defined as

$$\theta = \frac{1}{a} \cdot \frac{da}{dc} \qquad\qquad 6.17$$

where da/dc is the change in lattice parameter of the solid solution with atomic concentration of solute.

Now the spacing Λ of the local solute stress fields in a solid solution will be much smaller than the radius of curvature to which the dislocation line can be bent. Equation 3.23 relates the yield stress inversely to the spacing Λ and it can be seen that the stress necessary at small Λ is very high, approaching G. It is thus clear that in these circumstances the dislocation line will move in lengths much greater than Λ. The difficulty is to estimate the length of the moving loop; Mott and Nabarro assumed that this would have to be long enough for the centre of the line to move a distance Λ.

Their first calculation for the critical shear stress gave

$$\tau_0 = G \cdot \theta^2 \cdot c \qquad\qquad \textbf{6.18}$$

but a re-examination using a longer coherent length of dislocation led to the result

$$\tau_0 = 2{\cdot}5G \cdot \theta^{4/3} \cdot c \qquad\qquad \textbf{6.19}$$

Cottrell[1] applied this relationship to the results of Linde and co-workers,[23] and found that the hardening predicted was approximately ten times too large. He suggested that the model would be more accurate if a longer coherent dislocation line was used in the model. Moreover, the theory took into account only the size factor, while the experimental results on solid solutions indicate that the elastic constants and valencies of solute and solvent are also of significance.

6.6.4. *Theory combining size and modulus effects* (Fleischer)

It has already been shown that both differences in atomic size and moduli of solute and solvent atoms result in substantial solute atom dislocation interactions. Fleischer[25] considers the flow stress to be proportional to a quantity $(\epsilon_G{}' - \alpha\epsilon_b)$, where $\epsilon_G{}'$ is a measure of the modulus difference (see sections 6.1.2 and 6.4.1)

and
$$\epsilon_b = \frac{1}{b}\frac{db}{dc}$$

where b is the slip vector, and c is the atomic concentration of solute. (This quantity is equivalent to θ in the Mott and Nabarro theory.) $\alpha = $ constant (different for edge and screw dislocations).

In Fig. 6.12 the solution hardening $d\tau/dc$ of a large number of copperbase solid solutions is plotted against ϵ_s, the combination of the mismatch parameters $(\epsilon_G{}' - 3\epsilon_b)$. All the results lie fairly close to a straight line and the small scatter indicates that the two interactions considered are likely to be the principal ones. Nevertheless, the proportion of the flow stress resulting from each interaction will differ with the particular solvent. With big size differences, e.g. Pd, each interaction accounts for about half, but with atoms like Si and Zn the size difference may account for as little as 10 per cent. Fleischer thus considers that the modulus interaction is normally the greater effect. He finally derives an expression for the flow stress τ_0

$$\tau_0 = \left(\frac{G\epsilon_s{}^{3/2} \cdots c^{1/2}}{700}\right) \qquad\qquad \textbf{6.20}$$

The slope of the line in Fig. 6.12 is in good agreement with the $\frac{3}{2}$ exponent, but the variation of τ_0 with $c^{1/2}$ is not always observed in practice.

Summarizing, it appears that the biggest contribution to the flow stress of solid solutions arises not from the locking of dislocations but from resistance to their movement, a frictional force, the magnitude of which is sensitive both to atomic size differences, and differences in elastic properties between solute and solvent. Valency differences are also likely to be significant, but it is not a simple matter to estimate the relative contributions of these phenomena to the flow stress.

General references

1. COTTRELL, A. H. (1953). *Dislocations and Plastic Flow in Crystals*. Oxford University Press, London.
2. FRIEDEL, J. (1956). *Les dislocations*. Gauthier-Villars, Paris. English edition (enlarged) 1964, Pergamon Press, Oxford.
3. MITCHELL, T. E. (1964). 'Dislocations and Plasticity in Single Crystals of Face-Centred Cubic Metals and Alloys.' In *Progress in Applied Materials Research*, Vol. 6. Ed. E. G. STANFORD, J. H. FEARON and W. J. MCGONNAGLE.
4. *The Relation between Structure and Mechanical Properties of Metals* (1963). National Physical Laboratory Conference. H.M.S.O.
5. MASSALSKI, T. B. (Ed.) (1965). 'Alloying Behaviour and Effects in Concentrated Solid Solutions.' *AIME*. Met. Soc. Conference, Vol. 29.
6. CAHN, R. W. and HAASEN, P. (Eds.) (3rd ed. 1983). *Physical Metallurgy*. North Holland Pub. Co., Amsterdam. Chapter by P. HAASEN, p. 134.
7. Strengthening Mechanisms in Solids, (1960). The American Society for Metals, article by P. A. FLINN.

References

8. COTTRELL, A. H. and BILBY, B. A. (1951). *Proc. phys. Soc.*, **A62**, 490.
9. SUZUKI, H. (1957). *Dislocation and Mechanical Properties of Crystals*. Edited by J. C. Fisher *et al*. John Wiley, New York and London. p. 361.
10. WYON, G. and LACOMBE, P. (1954). Bristol Conference on Crystal Defects. Physical Society, London.
11. COTTRELL, A. H. and GIBBONS, D. F. (1948). *Nature, (Lond.)*, **162**, 488.
12. WAIN, H. L. and COTTRELL, A. H. (1950). *Proc. phys. Soc.*, **B63**, 339.
13. SYLWESTROWICZ, W. and HALL, E. O. (1951). *Proc. phys. Soc.*, **B64**, 495.
14. PORTEVIN, A. and LE CHATELIER, F. (1923). *C. R. Acad. Sc.*, **176**, 507.
15. JOHNSON, W. G. and GILMAN, J. J. (1959). *J. appl. Phys.*, **30**, 129; (1962). *ibid.*, **33**, 2050; (1962). *ibid.*, **33**, 2716.
16. CHAUDHURI, A. R., PATEL, J. R. and RUBIN, L. G. (1962). *J. applied Phys.*, **33**, 2736.
17. HAHN, G. T. (1962). *Acta metall.*, **10**, 727.
18. BRENNER, S. S. (1958). *Growth and Perfection of Crystals*. Ed. by R. R. DOREMUS *et al.*, John Wiley, New York and London. p. 157.
19. VON GOLER, F. and SACHS, G. (1929). *Z. Phys.*, **55**, 581.
20. OSSWALD, E. (1933). *Z. Phys.*, **83**, 55.
21. SACHS, G. and WEERTS, J. (1930). *Z. Phys.*, **62**, 473.
22. GREENLAND, K. M. (1937). *Proc. R. Soc.*, **A163**, 28.
23. LINDE, J. O., LINDELL, B. O. and STADE, C. H. (1950). *Ark. Fys.*, **2** (11), 89.

24. LINDE, J. O. and EDWARDSON, T. (1954). *Ark. Fys.*, **8** (51), 511.
25. FLEISCHER, R. L. (1963). *Acta metall.*, **11**, 203.
26. LEVINE, E. D., SHEELY, W. F. and NASH, R. R. (1959). *Trans. AIME*, **215**, 521.
27. AINSLIE, N. G., GERARD, R. W. and HIBBARD, W. R. (1959). *Trans. AIME*, **215**, 42.
28. GARSTONE, J. and HONEYCOMBE, R. W. K. (1957). *Dislocations and Mechanical Properties of Crystals*. Ed. by J. C. FISHER *et al.* John Wiley, New York and London.
29. KOPPENAAL, T. J. and FINE, M. E. (1962). *Trans. AIME*, **224**, 347.
30. HAASEN, P. and KING, A. (1960). *Z. Metallk.*, **51**, 722.
31. PFAFF, F. (1962). *Z. Metallk.*, **53**, 411, 466.
32. ARDLEY, G. and COTTRELL, A. H. (1953). *Proc. R. Soc.*, **A219**, 328.
33. a. OROWAN, E. (1934). *Z. Phys.*, **89**, 634.
    b. WAIN, H. L. and COTTRELL, A. H. (1950). *Proc. Phys. Soc.*, **B63**, 339.
34. KOPPENAAL, T. J. and FINE, M. E. (1951). *Trans. AIME*, **221**, 1178.
35. BRINDLEY, B. J., CORDEROY, D. J. H. and HONEYCOMBE, R. W. K. (1962). *Acta metall.*, **10**, 1043.
36. SCHRODER, K. (1959). *Proc. phys. Soc.*, **73**, 674.
37. HENRICKSON, A. A. and FINE, M. E. (1961). *Trans. AIME*, **221**, 103.
38. MEISSNER, J. (1959). *Z. Metallk.*, **50**, 207.
39. THOMAS, G. and NUTTING, J. (1956). *J. Inst. Metals*, **85**, 1.
40. HOWIE, A. and SWANN, P. R. (1961). *Phil. Mag.*, **6**, 1215.
41. FOURIE, J. T. (1960). *Acta metall.*, **8**, 88.
42. MOTT, N. F. and NABARRO, F. R. N. (1948). Report on Conference on the Strength of Solids. Physical Society, London. p. 1.
43. MOTT, N. F. (1952). *Imperfections in Nearly Perfect Crystals*. Edited by W. SHOCKLEY. John Wiley, New York and London.
44. FISHER, F. C. (1955). *Acta metall.*, **3**, 413.
45. FLEISCHER, R. L. (1964). Chapter in *The Strengthening of Metals*. Ed. by D. PECKNER. Reinhold, New York.
46. SUZUKI, H. (1952). *Sci. Rep., Res. Inst.*, Tohoku University, **A4**, 455.
47. CHRISTIAN, J. W. and SWANN, P. R. (1965). General Reference 5, p. 105.
48. KOEHLER, J. S. and SEITZ, F. (1947). *J. appl. Mech.*, **14**, A217.
49. MARKINKOWSKI, M. J., BROWN, N. and FISHER, R. M. (1961). *Acta metall.*, **9**, 129.
50. THOMAS, G. and WASHBURN, J. (Eds.) (1963). *Electron Microscopy and the Strength of Crystals*. Interscience, New York and London. (Article by M. J. Markinkowski.)
51. SACHS, G. and WEERTS, J. (1931). *Z. Phys.*, **67**, 507.
52. WESTBROOK, J. H. (Ed.) (1960). *Mechanical Properties of Intermetallic Compounds*. John Wiley, New York and London.
53. WESTBROOK, J. H. (Ed.) (1967). *Intermetallic Compounds*. John Wiley, New York and London.
54. HUTCHISON, M. M. and HONEYCOMBE, R. W. K. (1967). *J. Mat. Sci.*, **1**, 186.
55. HAASEN, P. (1964). *Z. Metallk.*, **55**, 55.
56. BULLEN, F. P. and HUTCHISON, M. M. (1963). *J. Aust. Inst. Metals*, **8**, 33.
57. HUTCHISON, M. M. and BULLEN, F. P. (1962). *Phil. Mag.*, **7**, 1535.
58. GILMAN, J. J. and JOHNSON, W. G. (1962). *Solid State Phys.*, **13**, Academic Press, New York and London.

Additional general references

1. KEAR, B. H., SIMS, C. T., STOLOFF, N. S., WESTHOOK, J. H. (Eds.) (1970). *Ordered Alloys* Proc. Third Bolton Landing Conference 1969, Claitor's Publishing Division, Baton Rouge.
2. HIRSCH, P. B., (Ed.) (1975). *The Physics of Metals* Vol. 2, Defects, Cambridge University Press.
3. NABARRO, F. R. N. (Ed.) (1979). *Dislocations in Solids* Vol. 4, Chapters by Haasen, P. and Suzuki, H. on solid solution hardening, North Holland.

Chapter 7

The Deformation of Crystals
Containing a Second Phase

7.1 Introduction

The first step in the strengthening of a pure metal by alloying is to form a solid solution. The second is to supersaturate the solid solution, and by suitable heat treatment or ageing cause the excess solute to be precipitated as a second phase. This process, which is known as age hardening or precipitation hardening, can be carried out in a large number of alloy systems, but the detailed behaviour varies from alloy to alloy. This is particularly true of the structure and morphology of the precipitate and its relationship to the matrix. On the other hand, it is possible to generalize concerning the behaviour on deformation of precipitation-hardened alloys, for the interaction of a dislocation with a precipitate is much more dependent on particle size, spacing and density than on the composition. We shall firstly consider the various ways in which dislocations and precipitate interact, and the theories which have been developed to explain strengthening by finely dispersed phases. The deformation behaviour of single crystals containing a finely divided second phase will be then discussed.

7.2 The interaction of dislocations with precipitates

When a fine precipitate is formed in a solid solution, an additional barrier to the movement of dislocations is created. The precipitate particles will lie across the slip planes along which dislocations move, so that the dislocations must behave in one of two ways:

1. cut through the particles of precipitate,
2. take a path around the obstacles.

While it is difficult to observe these interactions even in the electron microscope, the first mechanism is thought to predominate in lightly aged alloys with very fine coherent precipitates or zones, while the second mechanism is more characteristic of over-aged alloys with coarser dispersions of pre-cipitate. The first dislocation theory of aged alloys was that of Mott and Nabarro,[5] which, in its broadest aspect, included the case of solid solution strengthening which has already been considered. We shall now consider the cases where particles of precipitate replace the individual solute atoms as the obstacles to dislocation movement.

7.3   Mott and Nabarro theory for aged alloys

Mott and Nabarro considered an alloy with spherical solute atoms or groups of solute atoms, which by virtue of their different atomic size from that of the solvent atoms, caused internal stresses in the matrix. If the atomic radius of the solvent atom is R_s then the atomic radius of the solute atom is $R_s(1 + \theta)$ where θ is the misfit parameter defined as $1/a \cdot da/dc$, a being the lattice parameter and c the atomic concentration of solute. The same concept can be applied to *groups* of solute atoms such as occur in Guinier–Preston zones* in precipitation hardened aluminium alloys. The internal stress fields caused by the elastic strain between the group of solute atoms and the matrix are an average distance Λ apart which is the 'wavelength' of the internal stress field (Fig. 7.1). From elasticity theory,

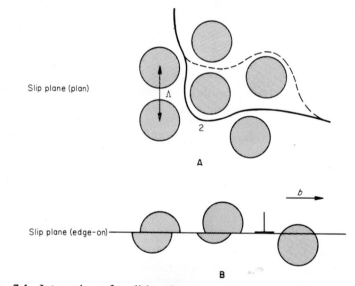

Fig. 7.1   Interaction of a dislocation line with zones in an alloy. **a,** Plan view; **b,** edge-on to slip plane.

* These are plate-like or spherical regions rich in solute atoms which form on crystallographic planes of the matrix. In Al–Cu alloys the Guinier–Preston zones are copper-rich plates on {100} planes, and in the early stages are between 50 Å and 200 Å in diameter and a few Ångström thick.

the shear strain ϵ in the internal stress field at distance l from the centre of a spherical particle of radius r_0 when $l \geqslant r_0$ is given as:

$$\epsilon = \frac{\theta r_0{}^3}{l^3} \qquad\qquad 7.1$$

They then calculated a mean shear strain ϵ_M from eqn. 7.1 by assuming that l could be defined as half the distance between particles: i.e.

$$l = \tfrac{1}{2}N^{-1/3}$$

where N is the number of particles per unit volume, so

$$\epsilon_M = (\theta r_0{}^3)(2N^{1/3})^3 = 8\theta r_0{}^3 N \qquad\qquad 7.2$$

The concentration of solute $c_0 = \tfrac{4}{3}\pi r_0{}^3 N$. The critical shear stress of the alloy is defined in terms of the mean elastic strain thus:

$$\tau_0 = \epsilon_M \cdot G$$
$$= 8G\theta r_0{}^3 N \simeq 2G\theta c_0 \qquad\qquad 7.3$$

This expression is independent of the particle spacing, the yield stress depending only on the mismatch function and the solute concentration, a result which is incompatible with the experimental results, which show that for incoherent particles the yield stress is related inversely to the particle spacing.

In the above analyses a rigid dislocation line was first assumed; however, Mott and Nabarro later introduced the idea that the dislocation was flexible, and could move locally independent of the line as a whole. The movement would then be dependent on the distance between internal stress centres, i.e. the wavelength Λ. We must thus know what are the factors limiting the radius to which a dislocation line can be bent. A dislocation line always tends to reduce its energy by shortening, i.e. it tries to straighten itself. So we introduce the concept of a *tension T* along the line, analogous to surface tension (section 3.11).

Mott and Nabarro have shown that $T \simeq Gb^2$. So if we have a curved dislocation, it can only be in equilibrium if acted on by a stress. Let us assume that τ_0 is the stress needed to maintain the dislocation in a curvature of radius r.

We consider a small arc δs of a dislocation of strength b (Fig. 7.2). The angle subtended by the arc at the centre of curvature O is $\delta\phi = \delta s/r$. There is an outward force along OA due to the applied stress equal to $\tau_0 b\,\delta s$, and an opposing inward force (radially inward) due to the line tension of

$$2T \sin\frac{\delta\phi}{2} \simeq T\delta\phi$$

In equilibrium

$$T \delta\phi = \tau_0 b \cdot \delta s$$

so

$$\tau_0 = \frac{T \delta\phi}{b \delta s} = \frac{T}{b \cdot r} = \frac{Gb}{r} \qquad \textbf{7.4}$$

Thus the radius of curvature of a dislocation is inversely proportional to the applied stress. This additional concept helps to explain the effect of the dispersion of the precipitate, because the extent of the localized movement of a loop in the dislocation line is determined by the spacing of the precipitate, i.e. Λ the wavelength of the internal stress field. With a solid solution $r \gg \Lambda$, but there are two other cases relevant for alloys containing precipitates.

1. $r \approx \Lambda$. This applies to the optimum aged condition where a high stress would be needed to force the dislocation loops between the closely spaced precipitate, e.g. Guinier–Preston zones in aluminium–copper alloys.

Fig. 7.2   Limiting radius of a dislocation line (schematic).

In this case Mott and Nabarro showed that the critical stress τ_0 needed to force a loop to move was defined by eqn. **7.3**.

$$\tau_0 = 2G\theta c_0 \qquad \textbf{7.5}$$

where θ is the misfit parameter.
Now

$$\Lambda = \frac{\alpha Gb}{\tau_0}$$

so the optimum dispersion size Λ_c is obtained by substituting

$$\Lambda_c = \frac{\alpha b}{2\theta c_0} \qquad \textbf{7.6}$$

With the appropriate data substituted in these relationships high yield stresses are obtained ($\sim 10^{-2}G$), and the critical dispersion of the particles is about 50–100 Å: Thus it is likely that yielding occurs, not by the dislocations bending around the precipitate particles, but by the shearing through of the particles by the dislocations (Fig. 7.1).

2. $r \ll \Lambda$. In this case ageing has proceeded further and the particles are much wider apart. It is thus easier for the dislocations to bend around the particles, and so the yield stress is lower. The yield stress of the alloy can thus be defined as that necessary to bend the dislocation line into loops of radius $\frac{1}{2}\Lambda$ and thus

$$\tau_0 = \frac{2\alpha Gb}{\Lambda} \qquad 7.7$$

The yield stress has also been calculated by Orowan[6] using the model shown in Fig. 7.3, when the particles are by-passed but leave residual dis-

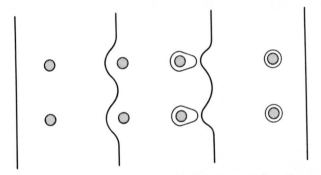

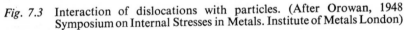

Fig. 7.3 Interaction of dislocations with particles. (After Orowan, 1948 Symposium on Internal Stresses in Metals. Institute of Metals London)

location loops around each particle. A simplified form of the relationship for the initial flow stress on this model is

$$\tau_0 = \tau_s + \frac{T}{b\Lambda/2} \qquad 7.8$$

where τ_s is the critical resolved shear stress of the matrix and T is the line tension of the dislocation.

The yield stress thus varies inversely as the spacing between the precipitate particles, and the alloy softens as it over-ages. The relationship is not concerned with the magnitude of the internal stress fields, and is more properly applied to systems where the precipitate particles are incoherent with the matrix, i.e. there are no coherency stresses caused by the lattices of precipitate and matrix distorting at or near the interface in an attempt to preserve the lattice continuity which is a characteristic of coherent precipitates.

Guinier–Preston zones in aluminium–copper alloys are coherent with the aluminium-rich matrix in so far as they are segregations of copper atoms on {100} planes of the matrix, and still conform to the crystal structure of the solid solution.[2] Eventually a distinct new phase θ' forms with a structure which is in the early stages coherent with the matrix. This leads to a close crystallographic relationship of the structures of precipitate and matrix, with certain planes and directions in the matrix being in continuity with other planes and directions in the precipitate.

7.4   The cutting of zones and precipitates by dislocations

It is generally accepted that coherent zones and precipitates can be sheared by the dislocations, and consequently the work done in forcing the earlier dislocations through the particles will be important in determining the flow stress. Kelly and Nicholson[2] have pointed out that the resistance to shear is governed by several factors.

1. The interaction of the dislocation with the stress field of the precipitate.

2. If the precipitate has an ordered lattice, work will be done in creating a disordered interface across the slip plane.

3. If the lattice parameters of matrix and precipitate differ, then during shearing of the particles, misfit dislocations must be created at the precipitate–matrix interface. The magnitude of the Burgers vector of the interface dislocation will be the difference between the Burgers vector of the slip dislocation in the matrix, and in the precipitate, i.e. $(b_m - b_p)$.

4. Differences in the elastic moduli of matrix and precipitate. In Mott and Nabarro's expression for the flow stress (eqn. **7.3**), G is not a constant. If the shear modulus of the particle is greater than that of the matrix, a larger stress will be needed to force dislocations through the particles than through the matrix.

5. If the matrix and precipitate possess different atomic volumes, a hydrostatic interaction would be expected between a moving dislocation and a precipitate similar to that between dislocations and individual solute atoms.

Kelly and Fine[7] have made an estimate of the yield stress in the case of internally ordered particles which become disordered when the dislocations cut through. A new interface is formed which has an energy of γ_p/unit area.

$$\text{Approximate area of newly formed interface} = \pi r_i^2$$

where r_i = average radius of particle interface.

Now an additional area of particle–matrix interface is formed as a result of the shear (Fig. 7.1B); let the energy of this be γ_s/unit area.

$$\text{Additional area} \simeq 2r_i b$$

A dislocation is moved over unit area of the glide plane and the work done leads to an increase in energy ΔE when the particles are cut, which is

$$\Delta E = n(\pi r_i^2 \gamma_p + r_i b \gamma_s) \qquad\qquad 7.9$$

where n is the number of particles crossing unit area of the slip plane.

The applied stress τ_0 necessary to move a dislocation is then given as

$$\tau_0 = \frac{n}{b}(\pi r_i^2 \gamma_p + 2r_i b \gamma_s) \qquad\qquad 7.10$$

$$n = \frac{3f}{2\pi r^2}$$

where f = volume fraction of precipitate.

Substituting in eqn. **7.10** for n

$$\tau_0 = \frac{f\gamma_p}{b} + \frac{3}{\pi}\sqrt{\frac{2}{3}\frac{f}{r}\cdot\gamma_s} \qquad\qquad 7.11$$

where $r = \sqrt{\frac{2}{3}}r_i$ = mean radius of precipitate particles.

If $\gamma_p \gg \gamma_s$, then $\tau_0 \geqslant f\gamma_p/b$, which is independent of r, the particle size. If, however, the particles are disordered, $\gamma_p = 0$, as no energy is needed to create disorder across the slip plane, then

$$\tau_0 \geqslant \frac{\sqrt{6}}{\pi}\cdot\frac{f\gamma_s}{r}$$

These estimates of the flow stress only apply to particles which are reasonably large compared with the Burgers vector of the dislocation.

7.5   The interaction of dislocations with incoherent precipitates

When the precipitate ceases to be coherent, and its crystal structure is very different from that of the matrix, the particles are no longer sheared by the dislocations, which have to find ways of moving around the particles. We have already discussed the Orowan mechanism, which predicts that the yield stress varies inversely with the interparticle spacing.

Fisher, Hart and Pry[8] take the Orowan model somewhat further, and emphasize the importance of the dislocation loops which this model indicates will form successively around the particles as each dislocation line bows around them. These dislocation rings exert stresses on the particles which are resisted because the particles are often small perfect crystals of high strength. Likewise, the stresses from the loops oppose further slip on the slip plane by acting on the dislocation sources. The overall result is that aged crystals in which such a mechanism operates work harden very rapidly in the early stages of deformation. This is seen in over-aged copper-aluminium alloys.

The shear stress due to the loops $\simeq \dfrac{NGb}{r}$

where N = number of loops about each particle of radius r.

The increment in flow stress τ_p due to the work hardening resulting from the loops was found to be

$$\tau_p = af^{3/2} \qquad\qquad\qquad 7.12$$

where $f = (r/\frac{A}{2})^2 =$ volume fraction of precipitate, where A = particle spacing, a = constant.

Thus the work hardening increases both with increasing fineness of dispersion, and as the volume fraction of precipitate is made larger.

Ansell and Lenel[9] have also looked more closely at the Orowan model, and came to the conclusion that appreciable plastic flow will take place only when the particles are fractured as a result of the stress concentration caused by dislocations piling up against them. These pile-ups must be in the form of multiple loops or rings of dislocations around the particles. The theory gives a relationship for the flow stress τ_0 in terms of the volume fraction of precipitate f.

$$\tau_0 = \tau_s + \frac{G'}{4a}\left(\frac{f^{1/3}}{0\cdot 82 - f^{1/3}}\right) \qquad\qquad 7.13$$

where τ_s is the yield stress of the particle-free matrix, a is a constant and G' = shear modulus of the particles.

This theory raises a number of difficulties, the prime one being that no plastic flow is assumed to occur prior to the fracture of the particles. This is demonstrably not the case in a number of matrix–particle systems.

7.6   Stress–strain curves of aged alloys

By using identically oriented crystals of aluminium–3·5 weight per cent (1·5 atomic per cent) copper in three different conditions, namely the solution treated, aged to peak hardness and the over-aged conditions, Carlsen and Honeycombe[10] showed that marked differences occurred in the stress–strain curves. While the solution treated crystals exhibited easy glide, the other two conditions led to parabolic stress–strain curves. A more detailed investigation has been made on aluminium–4·5 weight per cent (2 atomic per cent) copper crystals[11] heat-treated in four ways,

1. air quenched superaturated solid solution;
2. aged 2 days at 403 K—G.P.1 zones about 100 Å diameter on {100} planes of the matrix;
3. aged 27½ hours at 463 K—G.P.2 zones and θ' precipitate; optimum hardness;

4. over-aged, slowly cooled from 623 K—coarse θ, $CuAl_2$ precipitate in depleted matrix.

Fig. 7.4 shows a typical set of shear stress/shear strain curves at 77 K for similarly oriented crystals in the four conditions. The most striking feature of these curves is the combined effect of solute and ageing on the yield stress. Pure aluminium crystals have a τ_0 of about $1\,MN\,m^{-2}$, whereas Fig. 7.4 shows that a supersaturated solution of 4·5 weight per cent copper in aluminium raises this figure to around $30\,MN\,m^{-2}$. On allowing ageing to take place, τ_0 is further raised to over $80\,MN\,m^{-2}$. This illustrates very effectively the progressive role firstly of solute atoms, then

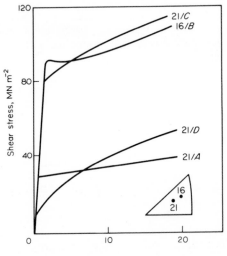

Fig. 7.4 Shear stress/shear strain curves of Al–4·5 weight per cent Cu crystals. 21/*A* As air cooled; 16/*B* aged 2 days at 403 K; 21/*C* aged 27½ hr at 463 K; 21/*D* over-aged at 623 K and slowly cooled (Greetham and Honeycombe)

aggregates of solute atoms, and finally precipitates in the strengthening of a relatively weak metal.

The supersaturated solid solution behaves at low temperatures of deformation in a similar way to other solid solutions, exhibiting a large Stage 1 hardening preceded by a sharp yield. The curves are sensitive to the orientation of the crystal; symmetrically oriented crystals work harden much more rapidly than other crystals. The slip lines are fine and characteristic of the easy glide region. Crystals aged at 403 K have a much higher yield stress and in most cases exhibit a yield phenomenon followed by a yield elongation zone. The subsequent hardening is low and comparable to that in the supersaturated solid solution, while the slip lines are indistinguishable from those in the solid solution. This suggests that the zones are cut through once the stress reaches a high enough level, and subsequently the alloy behaves like a solid solution.

The crystals aged to peak hardness, although they often have slightly lower yield stresses than the crystals aged at 403 K, work harden much more rapidly, giving a parabolic form of stress–strain curve. The appearance of the slip markings confirms that a different type of deformation process is taking place, in so far as long straight slip lines are no longer visible, but are replaced by indistinct markings. In some crystals in this condition slip markings are almost impossible to detect at low strains. Such evidence suggests that the dislocations are no longer cutting the second phase particles to form well-defined slip bands, but are moving around them in an irregular manner. This mechanism is undoubtedly operative in the over-aged condition, where the yield stress is very low, but the subsequent rate of work hardening is higher than in any of the other three conditions of heat treatment. Slip traces can be detected at this stage, but slip on secondary systems is frequently observed in the neighbourhood of the coarse hard particles of $CuAl_2$ (the equilibrium θ phase), because the distortions associated with the large particles activate dislocations on other slip planes (particle-induced polyslip).

Further work on single crystals of aluminium–copper[12] and copper–2 atomic per cent cobalt crystals[2] has confirmed that the behaviour of single crystals containing coherent G.P. zones is similar to a pure metal deforming in Stage 1, in so far as the rate of hardening and the slip line pattern are comparable, but the yield stress is very much higher. It should, however, be emphasized that the comparative tests must be carried out at low temperatures to avoid the complicating effect of further ageing taking place during deformation.

Experiments involving progressive ageing of aluminium–copper crystals at 463 K have further confirmed that the low rate of hardening typical of alloys containing zones is replaced by the more rapid parabolic hardening characteristic of the presence of θ' precipitate. This precipitate is coarser and only partially coherent with the matrix, consequently the particles are now obstacles which the dislocations cannot cut through, so they must avoid them. Dew-Hughes and Robertson[13] have demonstrated this transition in an aluminium–copper alloy, and it has also been observed in aluminium–silver and copper–beryllium alloys.[2]

7.7 Stress–strain curves of crystals with incoherent precipitates

The over-aged aluminium–copper alloy referred to above is an example of a matrix with an incoherent precipitate, where the rate of work hardening is relatively high. This behaviour has also been found in aged copper–10 atomic per cent indium crystals[14] which give parabolic stress–strain curves (Fig. 7.5) over a wide range of testing temperatures, as the Cu_9In_4

precipitate becomes incoherent at a very early stage. The precipitate is in these cases causing *dispersion* hardening, just as dispersions of silica and alumina produced by internal oxidation of dilute alloys can be made to strengthen copper. Ashby[15] has grown single crystals of such dispersion hardened copper alloys, and found that high rates of work hardening are characteristic of these alloys.

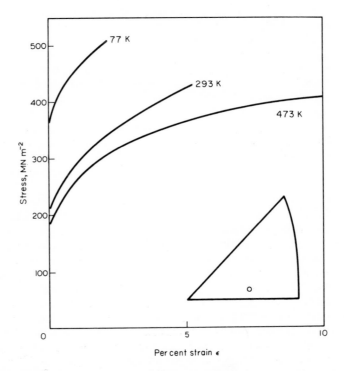

Fig. 7.5 Stress–strain curves of identically oriented copper–10 at. per cent indium crystals aged 500 hr at 573 K

Recent work by Ebeling and Ashby[18] on copper crystals dispersion hardened with spherical particles of silica has clearly demonstrated the replacement of Stage 1 hardening by a parabolic stage which becomes more extensive with increasing volume fraction (f) of silica, until at a volume fraction of 1 per cent the whole curve is parabolic (Fig. 7.6). Comparison of the stress–strain curves with those of pure copper oriented for multiple slip, suggest that particle-induced polyslip is an important strengthening mechanism. A detailed theory of this type of work hardening has been put forward by Ashby.[19]

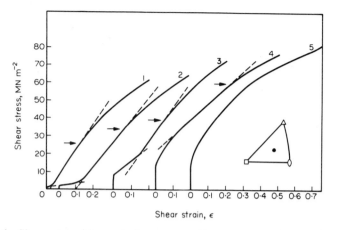

Fig. 7.6 Shear stress/shear strain curves of copper crystals with different volume fractions of SiO_2 particles, particle diam. about 900 Å. 1, Pure Cu; 2, Cu, oxidized then deoxidized, $f = 0$; 3, $f = 0.33 \times 10^{-2}$; 4, $f = 0.66 \times 10^{-2}$; 5, $f = 1 \times 10^{-2}$. (After Ebeling and Ashby, 1966, *Phil. Mag.*, **13**, 805)

7.8 Effect of dispersion on the yield stress

We have discussed theories by Orowan and by Ansell and Lenel which express the yield stress of a dispersion hardened crystal in terms of the spacing of the particles and the volume fraction of the dispersed phase. It is interesting now to look at some systematic experiments which can be used to test the theories.

Ashby[15] has investigated the yield stress of internally oxidized single crystals of copper with silicon, aluminium and beryllium. He determined the mean particle radius r (the SiO_2 particles are spherical but those of BeO and Al_2O_3 are non-spherical) by electron microscopy, then knowing the volume fraction of oxide f, found the mean particle spacing Λ from the relationship

$$f = \frac{r^2}{(\Lambda/2)^2}$$

A precise form of Orowan's equation was then used which takes into account the effect of the particle radius r on the interparticle spacing Λ. τ_s is the yield stress of the matrix without dispersion, ϕ is a constant.

$$\tau_0 = \tau_s + \frac{Gb}{4\pi} \phi \ln \left(\frac{\Lambda - 2r}{2b}\right) \frac{1}{(\Lambda - 2r)/2} \qquad \textbf{7.14}$$

Ashby calculated the yield stress of his dispersion strengthened alloys at 77 K and 293 K to obtain the results in Table 7.1.

Table 7.1 Measured and calculated yield stresses of internally oxidized copper alloy crystals[15]

Alloy, weight per cent	Volume fraction oxide	Particle radius, Å r	Particle spacing, Å Λ	Yield stress τ_0 MN m^{-2}			
				77 K		293 K	
				obs.	calc.	obs.	calc.
0·3 Si	0·026	485	3000	34	33	25	38
0·25 Al	0·011	100	900	80	112	64	105
0·34 Be	0·028	76	450	157	207	112	194

The best agreement was obtained with the crystals containing the spherical particles of SiO_2; greater discrepancies were found at the higher temperature when the dislocations are more readily able to move out of their original slip planes by cross-slip or by climb.

Another check on the Orowan relationship has been made by Dew-Hughes and Robertson,[13] who determined the critical resolved shear stresses of aluminium–copper crystals containing between 1·3 and 2·1 atomic per cent copper, which had been over-aged so that the equilibrium incoherent precipitate θ CuAl$_2$ was present in a series of different degrees of dispersion. They found that the shear stress plotted against the reciprocal of the mean interparticle spacing Λ^{-1} gave a straight line, which the simple

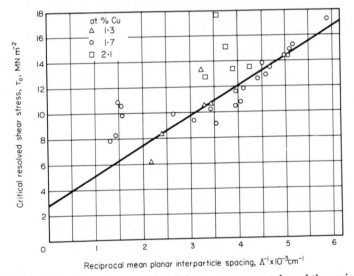

Fig. 7.7 Relation between τ_0 of aluminium–copper crystals and the reciprocal of the interparticle spacing. (After Dew-Hughes and Robertson, 1960, *Acta metall.*, **8**, 147)

Orowan equation predicts (Fig. 7.7). Moreover, a plot of shear stress against the logarithmic term of the Orowan equation (eqn. **7.14**) gives a straight line with the slope correct to within a factor of 2.

An ingenious two-phase system was made by Meiklejohn and Skoda[16] who dispersed very fine spherical particles of iron in a mercury matrix which was tested in the solid state at 213 K. For particle sizes in the range 50–800 Å they found that the Ansell and Lenel equation (eqn. **7.13**) fitted the results; however, Ashby has recently criticized its validity.[15]

7.9   The effect of temperature on the deformation of aged alloy crystals

It has been found that lightly aged crystals of aluminium–2 atomic per cent copper containing G.P. 1 zones showed a marked temperature dependence of the yield stress, while the yield stress of crystals aged to peak strength was very insensitive to temperature.[11] Byrne and co-workers[12] have examined the temperature dependence of the critical resolved shear stress of Al–1·7 atomic per cent Cu crystals over a wide range of temperature from 4 K to 400 K. Their work shows that the temperature dependence of τ_0 for crystals containing G.P.1 and G.P.2 zones is much greater than for θ', and still greater than for θ (Fig. 7.8). Likewise, Al–6 per cent

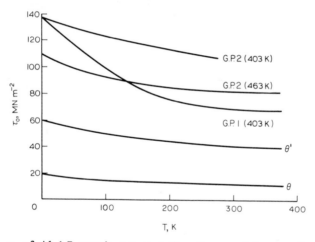

Fig. 7.8 τ_0 of Al–1·7 atomic per cent Cu alloy crystals as a function of temperature for different aged structures. (After Byrne *et al.*, 1961, *Phil. Mag.*, 6, 1119)

Ag and Al–5 per cent Zn crystals aged to contain G.P. zones show a similar temperature dependence of the critical resolved shear stress.[17] It is interesting to note that aged crystals subjected to *reversion*, i.e. a brief treatment at a higher temperature after low temperature ageing, which tends to reduce the zone size, give the greatest dependence of τ_0 on temperature.

These results can be interpreted in terms of the help which thermal activation can give dislocations in cutting through zones. When the zones become semi-coherent precipitates, the dislocations avoid the particles and the deformation mechanism would not be expected to be particularly dependent on temperature. A small effect would arise from the temperature dependence of the shear modulus, and at high temperatures thermally activated cross-slip would help the movement of dislocations. On the other hand Corderoy and Honeycombe[14] found with copper–10 atomic per cent indium alloys, that the fully aged crystals (having an incoherent dispersion of Cu_9In_4) showed a similar temperature dependence for the critical resolved shear stress as crystals in the supersaturated solid solution condition. In this case the temperature dependence arises from the solid solution which even in the aged alloy must contain several per cent of indium. It seems clear that the temperature dependence of the critical resolved shear stress depends not only on the stage of ageing, but on the detailed relationship between the precipitate and the matrix,

A study of the temperature dependence of the stress–strain curves of solution treated and aged crystals is likely to lead to difficulties unless the investigation is limited to low temperatures, because further precipitation can occur during deformation at temperatures as low as 293 K. Aluminium–4·5 per cent copper crystals, both in the supersaturated and aged at 403 K conditions, give steeper stress–strain curves at 293 K than at 77 K because of this strain-induced precipitation.[11] On the other hand, crystals aged at 463 K and higher temperatures seemed to be stable when deformed at 293 K. The role of deformation in causing further precipitation is also revealed when a stress–strain curve at 77 K is interrupted and the crystal rested at 293 K. Subsequent testing at 77 K leads to a marked increment in the curve due to strain-induced precipitation (Fig. 7.9). This phenomenon is very pronounced in the supersaturated solution, but also occurs

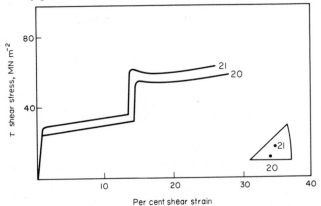

Fig. 7.9   Strain-induced ageing of Al–4·5 wt per cent (2 at. per cent) Cu crystals, solution treated and air cooled.

to a decreasing extent in aluminium–copper crystals aged at 403 K and 463 K[11]. The temperature dependence of the shear stress/shear strain curves of aluminium–copper crystals aged at 463 K is not very pronounced in the range 77–473 K. However, when the precipitate becomes fully incoherent (θ), the temperature dependence of work hardening becomes greater. This has been found generally in alloys containing a non-coherent precipitate, for example internally oxidized copper-alloy crystals, and in copper–10 atomic per cent indium crystals in the fully aged condition. Fig. 7.5 shows stress–strain curves of three identically oriented copper–indium crystals deformed at 77 K, 293 K and 473 K; there is a substantial difference in the extent of work hardening achieved.

7.10   Microstructure of Precipitate–dislocation interactions

It has already been pointed out that, as the ageing temperature is raised, the pattern of slip markings on a polished surface undergoes marked changes. Crystals containing zones deform on the slip plane on which the resolved shear stress is highest, and the zones are sheared by the passage of the dislocations. The crystals change shape in the way expected from slip on one system of slip planes.[11] On the other hand, aluminium–copper crystals aged to peak hardness and containing θ' precipitate do not always change shape like single crystals deforming on one system and behave more like polycrystalline material. This suggests that a number of slip systems are equally favoured, and that they operate on such a fine scale that no well-defined slip can be observed. In the over-aged condition, slip is somewhat easier to observe, and in these circumstances slip on several systems can be seen. Slip lines are also not observed on internally oxidized copper crystals and in Al–8 per cent Mg crystals when the second phase is finely divided.

Since the introduction of the technique of thin-foil electron microscopy it has become easier to make significant observations on the interaction of dislocations with precipitate, although the interpretation of the micrographs is frequently difficult. Electron microscopic examination has shown conclusively that zones are cut by dislocations. The following zones and phases have been reported as being sheared by the deformation[2]: G.P.1, G.P.2 and θ' in Al–Cu, G.P. zones in Al–Zn, γ' in Al–Ag. However, whether or not a phase is cut by dislocations will depend on the extent of deformation; it is likely that semi-coherent precipitates such as θ' and γ' are not cut during small deformations, but only after heavier deformation. It seems generally agreed that incoherent precipitates such as θ in Al–Cu are not cut by dislocations, but may fracture after heavy deformations. While it is not simply a question of size, in aluminium alloys at least, particles of 200 Å diameter or less which are coherent, will deform with the matrix from the start, but particles up to 1000 Å which are at least partly coherent may be sheared by dislocations at heavier deformations.

Turning to incoherent particles where the Orowan mechanism might be expected to operate, there is substantial evidence for dislocation lines being held up by particles, for example at A (Fig. 7.10a), then bowing between them (Fig. 7.10b). It is not easy to detect the residual dislocation loops the

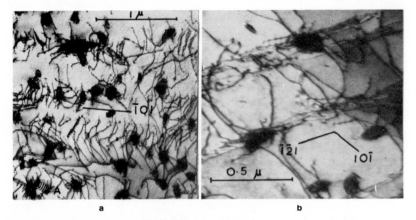

a   b

Fig. 7.10 Interaction of dislocations with precipitates in copper–10 atomic per cent indium alloy aged 10 hr at 573 K then deformed 5 per cent in tension. Thin-foil electron micrographs (After Corderoy).

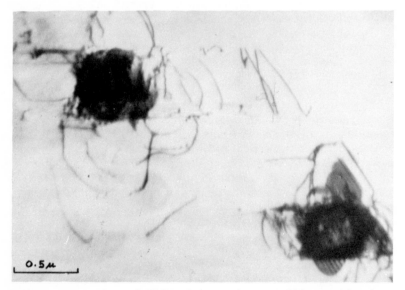

Fig. 7.11 Generation of dislocations at large carbide particles in an austenitic steel. Thin-foil electron micrograph. (After Harding)

Orowan mechanism predicts around such particles, but these have occasionally been observed.[14] Internally oxidized copper alloys after deformation have prismatic loops around the oxide particles, which have been explained in terms of local cross-slip to avoid the obstacles.[15] Such a mechanism would not be expected to operate at low temperatures in alloys with low stacking fault energies. In any case, whichever mechanism operates, the dislocation density increases rapidly with strain, and the work hardening rate is thus high. One complication is that large particles in a matrix can generate fresh dislocations on cooling from the ageing temperature as a result of different coefficients of thermal expansion of matrix and precipitate, or even during growth of the particles at constant temperature as a result of the different specific volumes of matrix and precipitate. This is a common source of dislocations in many alloys; Fig. 7.11 shows the generation of dislocations around large particles of niobium carbide in an austenitic steel probably initiated by quenching stresses.

General references

1. FRIEDEL, J. (1956). *Les Dislocations*, Gauthier-Villars, Paris. (English Edition (enlarged), 1964, Pergamon Press, Oxford.)
2. KELLY, A. and NICHOLSON, R. B. (1963). 'Precipitation Hardening.' *Progress in Materials Science*, **10**, 151.
3. THOMAS, G. and WASHBURN, J. (Eds.) (1963). *Electron Microscopy and the Strength of Crystals*. Interscience, New York and London.
4. MCLEAN, D. (1977). *Mechanical Properties of Metals and Alloys*. John Wiley, New York and London.

References

5. MOTT, N. F. and NABARRO, F. R. N. (1948). Report on the Strength of Solids. Physical Society, London. p. 1.
6. OROWAN, E. (1948). Symposium on Internal Stresses in Metals. Institute of Metals, London. p. 451.
7. KELLY, A. and FINE, M. E. (1957). *Acta metall.*, **5**, 365.
8. FISHER, J. C., HART, E. W. and PRY, R. R. (1953). *Acta metall.*, **1**, 336.
9. ANSELL, G. A. and LENEL, F. V. (1960). *Acta metall.*, **8**, 612.
10. CARLSEN, K. and HONEYCOMBE, R. W. K. (1954–55). *J. Inst. Metals*, **83**, 449.
11. GREETHAM, G. and HONEYCOMBE, R. W. K. (1960–61). *J. Inst. Metals*, **89**, 13.
12. BYRNE, J. G., FINE, M. E. and KELLY, A. (1961). *Phil. Mag.*, **6**, 1119.
13. DEW-HUGHES, D. and ROBERTSON, W. D. (1960). *Acta metall.*, **8**, 147.
14. CORDEROY, D. H. J. and HONEYCOMBE, R. W. K. (1964–65). *J. Inst. Metals*, **93**, 432.
15. ASHBY, M. F. (1964). *Z. Metallk.*, **55**, 5.
16. MEIKLEJOHN, W. K. and SKODA, R. E. (1959). *Acta metall.*, **7**, 675.
17. DASH, J. and FINE, M. E. (1961). *Acta Metall.*, **9**, 149.
18. EBELING, R. and ASHBY, M. F. (1966). *Phil. Mag.*, **13**, 805.
19. ASHBY, M. F. (1966). *Phil. Mag.*, **14**, 1157.

Additional general references

1. CAHN, R. W. and HAASEN, P. (Eds.) (1983). *Physical Metallurgy*. Chapter by Ansell, G. S. on Mechanical Properties of Two-Phase Alloys, North-Holland Amsterdam (3rd ed.)
2. HIRSCH, P. B. (Ed.) (1975). *The Physics of Metals* Vol. 2, Defects, Cambridge University Press.
3. NABARRO, F. R. N. (Ed.) (1979). *Dislocations in Solids* Vol. 4, Chapter by Gerold, V. on Precipitation Hardening, North-Holland Amsterdam.
4. HAASEN, P., GEROLD, V. and KOSTORZ, G. (Eds.) (1980). *Strength of Metals and Alloys*, Proc. Fifth International Conference on the Strength of Metals and Alloys, Aachen. Review: Precipitation and Dispersion Hardening, Brown, L. M. pp. 1551–1571.

Chapter 8

Other Deformation Processes in Crystals

8.1 Introduction

We have seen how the geometry of slip can explain the rotation of crystal axes during deformation in tension or compression. In using such a model it is assumed that the deformation is homogeneous throughout the crystal and that the changes in orientation occur uniformly. When the dislocation distributions in deformed crystals are studied, it is found that there is no uniformity on a microscopic scale because of the development of dislocation networks and sub-boundaries which involve local disorientations up to several degrees. However, this does not complete the picture because, even with simple modes of deformation such as tension, much larger disorientations can occur within crystals, and when more complex deformation processes are used, e.g. rolling, torsion, severe inhomogeneities are unavoidable.

8.2 Evidence for inhomogeneous deformation in crystals

The X-ray Laue technique provides a simple means of studying in single crystals disorientations produced by cold work. The method is widely used to determine the orientations of single crystals which give well-defined diffraction spot patterns from the various crystal planes, which select the appropriate wavelength from the polychromatic X-ray beam. An undeformed perfect crystal gives a pattern of sharp spots, and if the crystal is then uniformly deformed, the spots will remain sharp, but the arrangement of the pattern will reflect the small change in orientation of the crystal as a result of the deformation. Single crystals of zinc and cadmium will still give sharp patterns after substantial (100 per cent) elongation in tension; on the other hand if the crystal is then bent, the sharp diffraction spots will be smeared out into arcs[4]—this phenomenon is referred to as *asterism*. It shows that in the crystal a unique orientation has been replaced by a range of orientations.

Crystals of face-centred cubic metals show asterisms generally at much lower strains, for example, asterism has been observed in aluminium crystals after 1 per cent elongation in tension. Correlation with the stress–strain curves shows that asterism can occur in the easy glide region, but it becomes much more pronounced in Stage 2. Such evidence suggests that the deformation of crystals in tension is not uniform. The asterisms frequently show intensity maxima which indicate that a type of deformation sub-structure is being formed, of a grosser nature than the dislocation sub-boundaries which have been observed in thin-foil electron micrographs.

The inhomogeneities or variations in orientation which occur with more complex types of deformation, e.g. rolling or compression of thin crystals, can be readily revealed by etching when the changes in orientation show up as differently etched bands. These were first studied in detail by Barrett[5] and given the name *deformation bands*. A technique of X-ray microscopy first described by Berg[6] has been used to study these inhomogeneities which occur both in the deformation of single crystals and polycrystalline aggregates.[4, 5] Two main types of inhomogeneity have been established, which are usually referred to as kink bands, and bands of secondary slip.

8.3 Kinking in hexagonal crystals

A cadmium or zinc crystal with the c axis almost parallel to the rod axis, if loaded in compression, will undergo local collapse to form a kink[7] (Fig. 8.1). Such configurations have been known in some minerals for many years, but they are now recognized to occur widely in metals.

Fig. 8.1 Kinked cadmium crystal. $\times 17$

It has been shown that the kinks form gradually during compression of the crystal by a progressive rotation of the lattice, which may be as little as a few degrees or as large as 80°.[8] In general, kinking does not occur in compression if the angle between the slip plane and the axis, χ, is less than 2·5°[9] but it occurs readily in the range 2·5–24°; however, at the higher angles the form of the kinks is not well-defined. The process can be reversed by applying a tensile stress to the kinked crystal.

The structure of kinked regions in compressed zinc crystals in the ideal case is shown in Fig. 8.2, where two regions of severe lattice curvature are separated from each other and the unkinked crystal by well-defined kink planes *AB* and *CD*, which can be simply described as walls of edge dislocations. The curved regions have excess dislocations of one sign, and are responsible for the marked X-ray asterisms observed in Laue photographs; on annealing they are also preferred sites where polygonization and recrystallization take place (Chapter 11).

Kink bands also occur during tensile deformation of hexagonal metals, particularly in regions near the grips or in the vicinity of heavy scratches. It is thought that the bands nucleate at inhomogeneities which block dis-

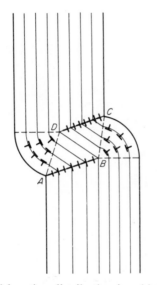

Fig. 8.2   Dislocation distribution in a kinked crystal

locations on several adjacent slip lines, leading to local curvatures, which in turn interact with dislocations on other slip lines and cause the lateral growth of the band.[10]

The kink planes which form part of the kinked structure in a zinc crystal are a basic feature of deformation in hexagonal metals. The dislocation structure discussed above is similar to that of a tilt polygonization boundary, but they can form and cause large disorientations at temperatures where rapid climb of dislocations is not possible. It is thus concluded that the kink planes are formed by the rapid feeding in and planar arrangement of glide dislocations on the basal slip plane. Formation of kink planes (accommodation kinking) also helps to relieve stresses set up by twin formation (see later).

8.4 Deformation bands in face-centred cubic crystals

X-ray microscopy shows that both aluminium and copper single crystals deformed in tension can develop deformation bands in the earliest stages of plastic deformation (1 per cent strain), and that these become more pronounced as the deformation proceeds. They are revealed by virtue of the fact that they cause local rotation of the lattice, so that these regions do not contribute to the X-ray image (Fig. 8.3). Two types have been found.

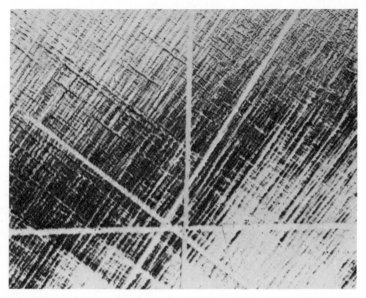

Fig. 8.3 X-ray micrograph of an aluminium crystal deformed 2 per cent in tension. White lines are reference scratches. × 25

8.4.1. *Kink bands*

These are *normal* to the operative slip plane and slip direction, in which the lattice is rotated with respect to the rest of the crystal about an axis lying in the slip plane and perpendicular to the slip direction. These have been studied in detail by Cahn,[11] Honeycombe[4] and Calnan,[12] and found to have features similar to kinks in hexagonal metals, except that sharp kink planes across which the lattice orientation changes abruptly are not formed. On the other hand, optical examination confirms that these bands contain localized regions of lattice curvature (Fig. 8.4) which comprise two zones of opposite curvature. At heavier strains localized secondary slip is frequently seen in the kink bands and slip bands tend to stop in the bent regions, i.e. the kink bands become barriers to the movement of dislocations.

The spacing and size of kink bands is dependent on the purity of the metal, the orientation of the crystal and the temperature of deformation. With aluminium of 99·99 per cent purity, the kink bands in a tensile specimen are uniformly spaced about 0·05 mm apart, while the bands are about 0·01 mm across; they can be seen in an optical microscope under slightly oblique illumination. It is, however, more difficult to detect them optically in copper under similar conditions; nevertheless they are visible in X-ray micrographs. The bands are particularly pronounced in crystals of 'soft' orientations and appear to be absent in symmetrically oriented crystals, when several slip systems operate throughout the crystal. As the temperature of deformation is raised, the individual kink bands become

Fig. 8.4 Kink bands in an aluminium crystal deformed 17·5 per cent in tension. Micrograph × 100

wider and involve larger lattice rotations, while the spacing between bands increases. Slip on a second system is more often observed in the bands under these conditions. These observations have been made on aluminium[4], gold[13] and copper[14] single crystals.

8.4.2. *Bands of secondary slip*

In the early stages of deformation of a face-centred cubic single crystal, gaps appear in between patches of primary slip lines in which primary slip is apparently absent. These bands are initially almost parallel to the active slip plane, and thus can be readily distinguished from kink bands. As the deformation proceeds, it becomes apparent that these regions are preferred sites for limited slip on other systems (Fig. 8.5). X-ray micrographs reveal that the lattice has rotated in a different sense to the rest of the crystal. Calnan[12] investigated the rotations and found that, unlike the kink bands, they were variable and could not be systematically analysed.

The intensity of the bands of secondary slip is rather dependent on orientation; and as they involve slip on systems other than the primary, it is perhaps not surprising that such bands are often very coarse and heavily developed in symmetrically oriented crystals, e.g. [001]. They are also readily observed in lightly deformed coarse-grained polycrystalline specimens.

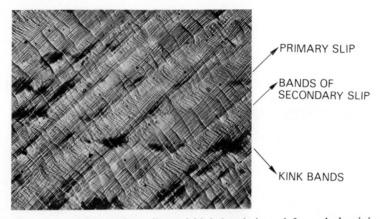

PRIMARY SLIP

BANDS OF
SECONDARY SLIP

KINK BANDS

Fig. 8.5   Bands of secondary slip and kink bands in a deformed aluminium crystal. Micrograph × 66

8.5   Deformation by twinning

A number of metals when deformed undergo sudden localized shear processes called twinning, which involve a small but well-defined volume within the crystal. This is in contrast to an individual slip process which, although it occurs by shear, is limited to one crystal plane, and is thus two-dimensional in character. Twinning is frequently a catastrophic process which is accompanied by energy release in the form of sound. The cry of tin is associated with the copious formation of twins which results from the bending of a rod of this metal, a phenomenon also found in zinc, cadmium and magnesium. Microscopic examination of a metal which has twinned reveals that the twinned region is often bounded by parallel or nearly parallel sides which correspond with planes of low indices; these are the *twin habit plane* or *twinning plane* (Fig. 8.6). It is also apparent that the twinned region differs appreciably in orientation from the grain in which it formed, a fact which is confirmed by the behaviour on etching, or more quantitatively by X-ray crystallographic techniques. If a crystal surface is polished flat prior to twinning, it is easy to show that the twinned region is tilted uniformly with respect to the original surface. The displacement of scratches can be used to demonstrate that shear of the twinned volume

Twin parallel to
active slip direction

Fig. 8.6 Twins in a zinc crystal which has fractured in a tensile test. Optical micrograph of fracture surface. (Deruyterre and Greenough, 1956, *J. Inst. Metals*, **84**, 337)

has taken place. The basic difference in the lattice translations by slip and by twinning is illustrated in Fig. 8.7.

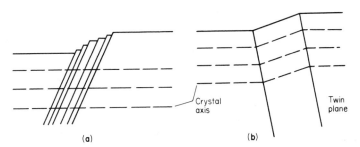

Fig. 8.7 Difference between, **a**, slip and, **b**, twinning (schematic)

8.6 Crystallography of twinning

Twinning is not a dominant mode of deformation in metals which possess many possible slip systems, e.g. face-centred cubic metals; however, it is very significant in metals where the possible slip systems are severely limited, e.g. the close-packed hexagonal metals, where slip is frequently limited to the unique basal system. Zinc single crystals deform readily by basal slip if the basal plane is suitably oriented, but if, for example, it is parallel to the tension axis, the crystals will deform preferentially by twinning. In polycrystalline aggregates the changes in shape accompanying deformation require the operation of several slip systems;

however, this is not possible in the hexagonal metals zinc, cadmium and magnesium, so twinning becomes an important deformation mechanism. In recent years it has been found that deformation twinning can also occur in face-centred cubic metals, particularly at low temperatures, and in metals and alloys which possess low stacking fault energies.

The process of twinning is a co-operative movement of atoms in which individual atoms move only a fraction of the interatomic spacing relative to each other, but the total result is a macroscopic shear which can often be observed with the naked eye. For example, it is easy to produce a twin shear of macroscopic dimensions in a crystal of calcite by pressure from a sharp knife blade. The crystallographic result of such a transition is that the lattice in the twinned region is related to the untwinned crystal, usually by reflection across the twinning plane which results in the twin lattice being a mirror image of the lattice of the matrix in which it forms. This can result from several different crystallographic transformations. A simple example is given in Fig. 8.8, where the lattice has been twinned along a

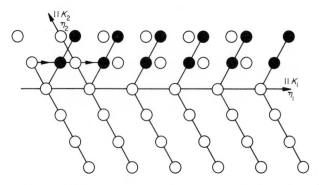

Fig. 8.8   A twinning transformation. (After Hall, 1953, *Twinning*. Butterworth, London)

plane K_1 the twinning plane in the twinning direction η_1. The open circles are the original atomic positions while the new positions following the twin displacements are indicated by the black circles. The twinning lattice is the mirror image of the parent lattice. In contrast to slip, which involves displacements which are equivalent to one or more interatomic spacings, twinning results from smaller displacements which successively occur on adjacent planes.

The best way to envisage the transformation involved in twinning is to consider a spherical single crystal (Fig. 8.9) in which the twinning plane K_1 is shown as an equatorial plane and η_1 is the direction of shear. We also need to know the magnitude of the shear S, which is defined as the distance moved by a point unit distance from the twinning plane. If now the upper hemisphere is allowed to twin, it is distorted to an ellipsoid (Fig. 8.9) in which only *two* planes remain undistorted, that is they are still semi-circular in shape, the twinning plane K_1 and another plane K_2 which is shaded in the diagram. The plane of shear is defined as that plane which contains both the direction η_1 and the normal to the twinning plane K_1, while the line of intersection of the plane of shear with the second undistorted plane K_2 is identified as η_2.

A further criterion of the twinning transformation is that the second undistorted plane K_2 has the same angle with K_1 before and after the transformation. If we assume that the sphere is of unit radius, then the

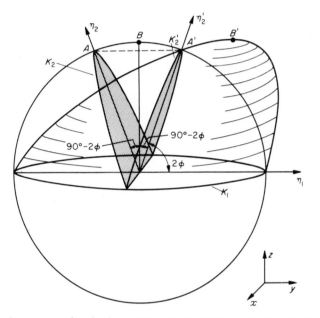

Fig. 8.9 Geometry of twinning. (After Hall, 1953, *Twinning*. Butterworth, London)

magnitude of the shear S is represented by the distance BB' which can be expressed in terms of the angle 2ϕ between the two undistorted planes

$$S = 2 \cot 2\phi \qquad \qquad \textbf{8.1}$$

Consequently, if the two undistorted planes are found, the shear strain in the twin is easily determined after measuring the interplanar angle. Typical values are given in Table 8.1, p. 206.

The twinning transformation, which leads to a change in orientation of the twinned region, does not alter the symmetry or structure of the crystal. This is readily shown by X-ray diffraction investigations. Consequently, the shape and size of the unit cell is unaltered by the transformation, and as a result there must be three rational lattice vectors (non-coplanar) which have the same magnitudes and the same angular relationships before and after twinning. As K_1 and K_2 are the only distorted planes, the three vectors must lie in these two planes.

If we now consider any vector ϵ in plane K_1 (Fig. 8.10), it follows that η_2 is the only vector in plane K_2 which makes the same angle θ with ϵ before and after twinning. This arises because η_2 is the only vector perpendicular to the intersection of the two planes K_1 and K_2. This condition applies to ϵ and any other vector lying in K_1. If it is now assumed that K_1 is a rational plane containing rational directions, and that η_2 is rational, we have the requirement that three non-coplanar vectors are unchanged in magnitude and angular relationship by the twinning shear. Similarly it can be shown that η_1 which is perpendicular to the intersection of K_1 and

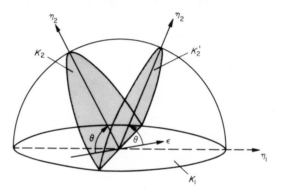

Fig. 8.10   Constant vectors in twinning

K_2, is the only vector in plane K_1 which makes the same angle with arbitrary vectors in K_2 before and after twinning. So again the requirements for twinning can be met if the direction η_1 and the plane K_2 are rational. There is also a third possibility that both planes K_1 and K_2 and the two vectors η_1 and η_2 are all rational. Summarizing, there are three modes of shear which preserve the symmetry and crystal structure of the material, and thus satisfy the requirements for twinning.

1. Twins of first kind:
   K_1 plane rational, η_2 direction rational
2. Twins of second kind:
   K_2 plane rational, η_1 direction rational.
3. Rational or compound twins:
   K_1 and K_2 rational, η_1 and η_2 rational.

It will be obvious from the above discussion that the exact nature of the lattice rotation during twinning will depend on the type of twinning which takes place.

Most metals, as they possess crystal structures of high symmetry, fall into the third class; however, α-uranium which is orthorhombic exhibits twins of the first and second kind.[15] Uranium also shows another type of twinning, called *reciprocal twinning*, where K_1 and η_2 of a twin of the first kind possess the same indices as K_2 and η_1 of a twin of the second kind.

To define completely a twinning process, the indices of the planes K_1 and K_2 and the directions η_1 and η_2 should all be specified together with the twinning shear S. Table 8.1 lists some typical data for a number of metals, from which it can be seen that metals within one crystallographic system behave similarly except that in the hexagonal metals, the twinning modes are influenced by the c/a ratio which varies from 1·886 for cadmium to 1·56 for beryllium.

Table 8.1. Some Crystallographic Data on Twinning

| Metal | Crystal structure | Twinning indices | | | | Shear |
		K_1	η_1	K_2	η_2	S
Copper and other f.c.c. metals	f.c.c.	(111)	[11$\bar{2}$]	(11$\bar{1}$)	[112]	0·707
α-Iron	b.c.c.	(112)	[11$\bar{1}$]	(11$\bar{2}$)	[111]	0·707
Magnesium	c.p.h. $c/a = 1\cdot624$	(10$\bar{1}$2)	[10$\bar{1}\bar{1}$]	(10$\bar{1}$1)	[10$\bar{1}$1]	0·129
Zinc	c.p.h. $c/a = 1\cdot856$	(10$\bar{1}$2)	[10$\bar{1}\bar{1}$]	(10$\bar{1}$1)	[10$\bar{1}$1]	0·139
Cadmium	c.p.h. $c/a = 1\cdot866$	(10$\bar{1}$2)	[10$\bar{1}\bar{1}$]	(10$\bar{1}$1)	[10$\bar{1}$1]	0·171
β-tin	tetragonal $c/a = 0\cdot541$	(301)	[$\bar{1}$03]	($\bar{1}$01)	[101]	0·119
Bismuth	rhombohedral	(110)	[00$\bar{1}$]	(001)	[110]	0·118

8.7 Twinning of hexagonal metals

Twinning in the hexagonal metals is a common form of deformation in view of the restrictions on deformation by slip imposed by the occurrence of basal slip. We shall primarily consider zinc, cadmium and magnesium, in which basal slip predominates, because the c/a ratios are close to or greater than the value for ideal close packing. Single crystal experiments on these metals have demonstrated that twinning occurs at substantially higher values of resolved shear stress than does deformation by basal slip, e.g. a shear stress of $1-7\,\mathrm{MN\,m^{-2}}$ is needed to cause twinning in pure cadmium, whereas a stress of $0\cdot2-0\cdot3\,\mathrm{MN\,m^{-2}}$ will cause basal slip. Consequently, twinning will only take place if the basal plane is unfavourably oriented, e.g. within 5–10° of the tension axis.

Work with zinc crystals has proved that no critical resolved shear stress criterion can be used for the onset of twinning.[16] Crystal wires with the basal plane parallel to the axis were found to twin at resolved shear stresses between 25 and $50\,\mathrm{MN\,m^{-2}}$, but twinning was preceded by slip on pyramidal planes $\{11\bar{2}2\}$.

There is evidence to suggest that slip is essential prior to twinning, because arrays of dislocations must be formed to create stress concentrations which can then cause the nucleation of twins. It is a familiar aspect of twinning studies that badly handled crystals, which have been bent or distorted, twin at much lower stresses than undamaged crystals. This accounts for the wide variations in resolved shear stresses for twinning reported in the literature. Once a twin forms during a test, it is followed by many other twins comprising a burst which is often accompanied by appreciable noise. The stress–strain curves frequently show marked serrations when twinning commences and the load relaxes. On reloading, deformation by twinning is resumed at a much lower stress level, and is accompanied by further deformation by slip within the twins, which is now possible because of the re-orientation of the crystal in the twinned regions.

The mode of deformation, for example, whether tension or compression, has a decisive role in the occurrence of twins in hexagonal metals. As we have seen, the direction and magnitude of the twinning shear is determined by the movement of the second undistorted plane K_2, moreover the shear occurs in only one direction so that twinning will cause important dimensional changes in the specimen. Referring to Fig. 8.9, any vector to the left of the initial position of the second undistorted plane K_2 will decrease in length as a result of the twinning transformation, while any vector to the right of K_2 will increase in length. In terms of actual crystals, this means that twinning will cause either extension or contraction of the specimen depending on the orientation of the specimen axis with respect to the crystal axes. If we consider a crystal of zinc with its basal plane parallel to the axis of

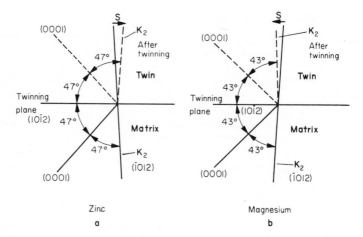

Fig. 8.11   Geometry of twinning in, **a**, zinc; **b**, magnesium. (After Reed-Hill, 1964, *Principles of Physical Metallurgy*, Van Nostrand)

the specimen, twinning on any $\{10\bar{1}2\}$ twin plane will cause extension of the specimen. Consequently, a *tensile* deformation will result in twinning, but a compressive deformation will not. In the latter case, the crystal deforms by kinking.

The situation changes, however, if we consider a metal of different c/a ratio, e.g. magnesium. As the axial ratio changes from 1·856 for zinc to 1·624 for magnesium, the angle between the basal plane and the twinning plane $(10\bar{1}2)$ changes from 47° to 43°. Fig. 8.11a and b represents the situation in the two metals before and after twinning. With zinc the crystal is increased in length parallel to the basal plane within the twinned material as the undistorted plane K_2 rotates in a clockwise direction. On the other hand, in magnesium the undistorted plane K_2 must rotate anti-clockwise to conform to the angular requirements, so the crystal is shortened in a direction parallel to the basal plane. Consequently, with magnesium twinning is to be expected when the deformation is in compression rather than in tension, whereas in zinc the reverse is the case.

When metals of smaller axial ratio are considered, the twinning behaviour becomes more complicated.[3] For example, while titanium and zirconium twin on the $\{10\bar{1}2\}$ plane, twinning has also been observed on the $\{11\bar{2}2\}$ plane and on other planes of this type.

8.8   Twinning of body-centred cubic metals

When iron is shock loaded, for example during explosive forming,† very thin crystallographic lamellae are formed which are known as Neumann bands after their discoverer. Despite the fact that they have been known since 1850, it was only comparatively recently that they were positively identified as twins by crystallographic methods.[17, 18] It is now known that twins are formed during deformation in many body-centred cubic metals including molybdenum, tungsten,[19] chromium, niobium[20] and tantalum,[21] however, most of the systematic information has come from studies of iron.

Twins in body-centred cubic metals are usually long and thin, rarely thicker than 5×10^{-4} cm because there is a large shear strain associated with the twinning (Table 8.1). The twinning plane is in most cases $\{112\}$ and the direction of shear $\langle 111 \rangle$. The geometry of the process is shown in Fig. 8.12a–d, where the figures represent the structure of a, b and c layers in the plane of the twinning shear i.e. a $\{110\}$ plane. The open circles represent atoms in the plane of the paper, while the black circles represent atoms in the adjacent planes above and below the plane of the paper. The twinning or composition plane is (112) and is normal to the plane of the figure, while the direction $[\bar{1}\bar{1}1]$ represents the twinning direction. The twinning planes are packed in the sequence $ABCDEFABCDE\ldots$ and if a

† Rapid plastic deformation initiated by an explosion.

displacement of $\frac{1}{6}a[\bar{1}\bar{1}1]$ is introduced (Fig. 8.12c) a stacking fault is formed $ABCDEFEFABC$.... To produce a twinned crystal the operation must be carried out on each successive (112) plane to produce the stacking sequence $ABCDEFEDCBA$ (Fig. 8.12d). Such a reaction produces a shear of 0·707, which is the value obtained from experimental observations.[18] The same orientation change would be achieved by a larger shear in the opposite direction, namely $-\frac{1}{3}a[\bar{1}\bar{1}1]$, but this is energetically unlikely and has not been observed. This underlines a point made earlier, that the twinning shear is unidirectional. So, as in the case of hexagonal structures,

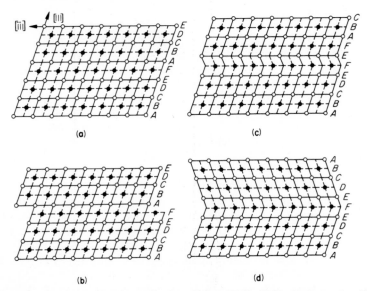

(a)

(c)

(b)

(d)

Fig. 8.12   Twinning in a b.c.c. lattice. (After Hull, 1964, *Deformation Twinning*. Edited by Reed-Hill, Hirth and Rogers, AIME)

the orientation of the crystal relative to the stress axis, and the mode of stressing whether tensile or compressive will determine which twinning systems operate.

Twins form more readily in body-centred cubic metals as the rate of deformation is increased, and are not normally encountered in pure iron in a tensile test at room temperature. However, the tendency to twin increases with decreasing temperature, and at 4 K pure iron will deform by twinning in a tensile test. On the other hand, deformation by impact will cause twinning at room temperature. The effect of impurities can also be substantial; for example, addition of silicon to iron causes twinning to be more prominent during deformation. The effect of interstitial atom impurities such as oxygen, nitrogen and carbon on the twinning behaviour of niobium has been studied in some detail.[20] Twinning as a result of slow deformation can be suppressed completely when the concentration of interstitials reaches a particular level (2000 p.p.m. oxygen, 500 p.p.m. nitrogen and 200 p.p.m. carbon).

However, impact deformation produced twins at these impurity levels. Solid solution alloys of molybdenum and tungsten with rhenium, and of niobium with vanadium twin very readily at room temperature even at slow rates of deformation. It has been suggested that the increased frequency of twinning arises from a lowering of the stacking fault energy, but no accurate measurements of stacking fault energies have yet been made on these alloys.

8.9   Twinning of face-centred cubic metals

Indirect evidence has long indicated that deformation twins form in α-brass.[22,23] X-ray diffraction data obtained from powder patterns from brass filings led to values for stacking fault and twin densities, but it is only recently that confirmation of twin formation as a result of deformation has been obtained by precision X-ray methods, electron microscopy and electron diffraction. Copper crystals deformed at 4 K were found to form deformation twins at high stress levels[24] and crystals of some orientations also twinned at 77 K. Silver,[25] gold[25] and nickel[26] also twin, but in all cases very high resolved shear stresses are needed, e.g. 150 MN m^{-2} for copper, 300 MN m^{-2} for nickel, which are only reached at low temperatures and/or heavy deformations. As in hexagonal metals, twinning at low temperatures in face-centred cubic metals often results in serrations in the stress–strain curve. The twin plane is invariably the close-packed {111}, which is also the slip plane in all face-centred cubic metals, while the twinning direction is ⟨112⟩.

Experiments with copper crystals at 4 K demonstrate that twinning propagates through the crystal often commencing at one grip, rather like a Lüders band propagation. The twinning plane is usually the primary (111) slip plane which becomes the twin plane after the deformation has raised the shear stress to a high enough level. The onset of twinning is then revealed by serrations and a Lüders-type plateau in the stress–strain curve (Fig. 8.13). Microscopically the crystal has the appearance of having twinned completely, but electron microscopic examination reveals that the twinned region contains large numbers of fine twin bands, up to 5000 Å wide and amounting to about 50 per cent of the volume of the region.

Solid solution alloys of face-centred cubic metals, for example silver–gold,[25] copper–zinc, –aluminium, –gallium, –germanium, –indium,[27] twin much more readily during deformation than do the pure metals. This is shown by the lower resolved shear stresses at which twins form, e.g. 40–120 MN m^{-2} in the case of copper-base solid solutions. The whole range of solid solution has been studied in the Ag–Au alloys in which twinning was encountered after deformation at room temperature.[25] The details of the twin propagation depend on the temperature of deformation in a rather complex way, and the load drops associated with twinning *increase* with increasing temperature, despite the increased incidence of twinning as the temperature is lowered. Twin formation has also been found in copper-base solid solutions above room temperature,[27,28,29] if the solute atom concentration is high, thus raising the value of the shear stress which the alloy can reach by cold work. For example, copper–10 at. per cent indium crystals twin at 473 K while 5 per cent indium crystals twin at 293 K and

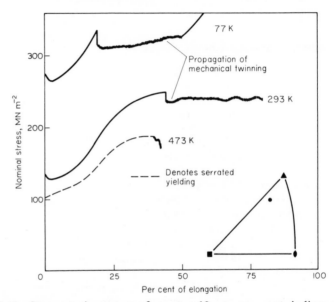

Fig. 8.13   Stress–strain curves of copper–10 at. per cent indium crystals showing the onset of twinning (Corderoy)

1 per cent indium crystals must be deformed at 77 K before they twin. In all cases the shear stress at which twinning commences is approximately 100 MN m^{-2}.

Recent work on solid solutions of face-centred cubic metals has demonstrated that the stacking fault energy of the solvent metal can be substantially lowered by the presence of solute atoms, and that the fall in stacking fault energy is related to the electron/atom ratio of the alloy. In view of the fact that in face-centred cubic metals twins form by changes in the stacking sequence of the close packed {111} planes, on which stacking faults can be easily created by dislocation dissociations, a correlation between stacking fault energy and the tendency to form deformation twins has been sought. Venables[3] has taken the resolved shear stress at which twins occur in a series of copper-base solid solutions and plotted the values against γ the stacking fault energy (Fig. 8.14) to obtain a reasonably smooth curve. This means that an increased solute concentration by depressing γ, lowers the stress at which twinning will start and, moreover, that polyvalent metals, e.g. germanium, will be more effective than monovalent metals such as nickel in causing deformation by twinning in solid solutions. A similar correlation exists between the occurrence of annealing twins and the stacking fault energy. Annealing twins occur as parallel-sided bands in recrystallized face-centred cubic metals and alloys, particularly in alloys of low stacking fault energy, e.g. 70/30 brass.

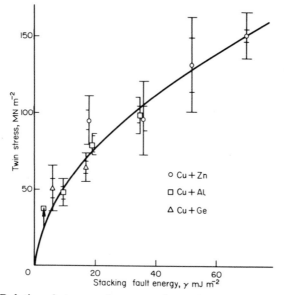

Fig. 8.14   Relation between the stress for twinning, and stacking fault energy for copper-base alloys. (After Venables, 1964, *Deformation Twinning*, AIME)

The geometry of the twinning process in face-centred cubic metals is well established. The shear is in $\langle 112 \rangle$ directions in $\{111\}$ planes which normally have the packing

ABCABCABC

The formation of a twin alters the sequence thus

ABCABACBA
↑

the arrow indicating the twin boundary beyond which the crystal has the opposite stacking to that in the untwinned region. The translation is best shown in a lattice section normal to the twinning plane, that is a (211) plane (Fig. 8.15) in which rows of atoms represent the edges of the close-packed (111) planes. The twinning shear in the direction [11$\bar{2}$] has created a sequence of planes in the opposite sense to the original sequence, so that the twin lattice is a mirror image of the matrix. The second undistorted plane is also octahedral, namely (11$\bar{1}$) and $\eta_2 = [112]$. Thus the twinning shear is calculated from eqn. **8.1** to be 0·707.

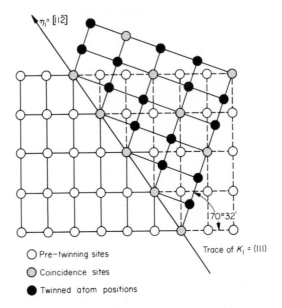

$\eta_1 = [11\bar{2}]$

170°32'

Trace of $K_1 = (111)$

○ Pre–twinning sites

◐ Coincidence sites

● Twinned atom positions

Fig. 8.15   Geometry of twinning in an f.c.c. lattice. (After Hall, 1953, *Twinning*. Butterworth, London)

8.10   Dislocation movements in twinning

It seems clear that dislocations must be directly responsible for twinning as they are for slip. While twinning usually requires a higher shear stress to initiate it than does slip, these stresses are still well below the theoretical strength, consequently a mechanism involving the simultaneous movement of all atoms in the twin is unrealistic. Therefore, we shall briefly examine several proposed dislocation mechanisms to emphasize their main features.

8.10.1. *Body-centred cubic structure*

The earliest model of a twinning dislocation postulated a partial dislocation rotating around a sessile dislocation in such a way as to cause shear on successive planes.[30] In Fig. 8.16 we have two intersecting {112} planes with a common [11$\bar{1}$] direction. There is a dislocation AO with $b = \frac{1}{2}a[111]$ lying in the (112) plane, which has dissociated at O to give two partial dislocations OB and $OEDB$, a reaction which is described as follows

$$\frac{a}{2}[111] \rightarrow \frac{a}{3}[112] + \frac{a}{6}[11\bar{1}]$$

The dislocation OB is, however, sessile because its Burgers vector is normal to the (112) slip plane. On the other hand, the dislocation $OEDB$ can move, but forms a stacking fault when it does so. The section OE is parallel to the

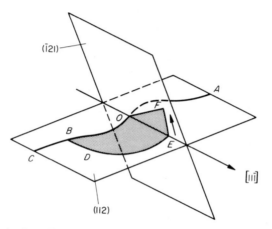

Fig. 8.16   Twinning dislocation in the b.c.c. lattice. (After Cottrell and Bilby, 1951, *Phil. Mag.*, **42**, 573)

Burgers vector $\frac{1}{6}a[111]$ and is thus screw; consequently, it can cross-slip on to the ($\bar{1}$21) plane, forming again a region of stacking fault OEF as it moves. The essential feature of the mechanism is that OF rotates about O forming a faulted or two-dimensional twin layer. However, there is a displacement in the direction normal to ($\bar{1}$21) so that as the dislocation OF completes a revolution it will have moved forward along OB, with the result that the second revolution of OF causes the two-dimensional twin to thicken. There is no theoretical reason why the process should not continue until a twin of substantial thickness is formed. Cottrell and Bilby pointed out that the outer end of the dislocation OF would tend to lag behind the centre which would have the highest angular velocity; this would lead to a dislocation of spiral form (cf. the dislocation sources discussed in Chapter 3 and the formation of growth spirals on crystal faces). The whole process could occur very rapidly, which would be in accord with observed times of formation of twins which can be as short as several micro seconds.[31] No direct experimental evidence has been obtained for this type of dislocation reaction; moreover, the evidence for dissociation of dislocations on (112) planes in body-centred cubic metals to form stacking faults is as yet slender. While such faults have been observed for example in niobium,[32] they are rare and may depend critically on the purity of the material.

8.10.2. *Dislocation mechanisms for the face-centred cubic structure*

The pole mechanism was applied by Cottrell and Bilby to the face-centred cubic structure by means of the dislocation dissociations shown in Fig. 8.17, where a dislocation AC dissociates into a Frank partial $A\alpha$ and

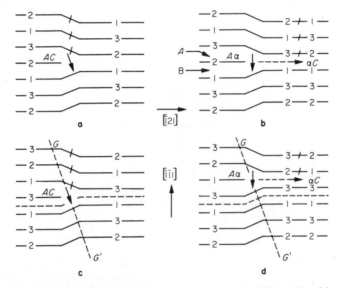

Fig. 8.17 Twinning mechanism in an f.c.c. crystal. (After Venables, 1964, *Deformation Twinning*, AIME)

a twinning dislocation αC which rotates about the point at the end of AC. After one rotation, αC again meets $A\alpha$ at either level A or C (Fig. 8.17b) thus producing one layer of fault. Venables[3] has suggested that after one rotation, recombination occurs, and the dislocation AC glides along GG' only to re-dissociate again at the next atomic plane, whereupon the partial αC can undergo another rotation and create another faulted layer. In this way a succession of faults amounting to a twin will be formed. The detailed dislocation geometry is complex, so the reader is referred to reference 3 for further details of this, and other pole mechanisms which have been proposed.

8.10.3. *Hexagonal structures*

A difficulty in hexagonal crystals is that there are several twin systems which can occur. Several mechanisms have been proposed, but the detailed geometry is complex and beyond the scope of the present treatment. Further information will be found in reference 3.

General references

1. JAOUL, B. (1965). *Etude de la Plasticité et Application aux Métaux.* Dunod, Paris.
2. HALL, E. O. (1953). *Twinning.* Butterworth, London.
3. REED-HILL, R. E., HIRTH, J. P. and ROGERS, H. C. (Eds.) (1964). *Deformation Twinning.* AIME, Conference Vol. 25.

References

4. HONEYCOMBE, R. W. K. (1951). *J. Inst. Metals,* **80**, 49.
5. BARRETT, C. S. (1940). *Trans. AIME,* **137**, 128.
6. BERG, W. (1934). *Z. Krist.,* **89**, 286.
7. OROWAN, E. (1942). *Nature, Lond.,* **149**, 643.
8. HESS, J. B. and BARRETT, C. S. (1949). *Trans. AIME,* **185**, 599.
9. GILMAN, J. J. (1954). *Trans. AIME,* **200**, 627.
10. FRANK, F. C. and STROH, A. N. (1952). *Proc. phys. Soc.,* **65B**, 811.
11. CAHN, R. W. (1951). *J. Inst. Metals,* **79**, 129.
12. CALNAN, E. A. (1952). *Acta crystallogr.,* **5**, 557.
13. SAWKILL, J. and HONEYCOMBE, R. W. K. (1954). *Acta metall.,* **2**, 854.
14. ANDRADE, E. N. DA C. and ABOAV, D. A. (1957). *Proc. R. Soc.,* **A240**, 304.
15. CAHN, R. W. (1953). *Acta metall.,* **1**, 49.
16. BELL, R. L. and CAHN, R. W. (1957). *Proc. R. Soc.,* **239**, 494.
17. KELLY, A. (1953). *Proc. phys. Soc.,* **A66**, 403.
18. PAXTON, H. W. (1953). *Acta metall.,* **1**, 141.
19. SCHADLER, K. W. (1960). *Trans. AIME,* **218**, 649.
20. MCHARGUE, C. J. (1962). *Trans. AIME,* **224**, 334.
21. BARRETT, C. S. and BAKISH, R. (1958). *Trans. AIME,* **212**, 122.
22. MATHEWSON, C. H. (1928). *Trans. AIME,* **78**, 7.
23. SAMANS, C. K. (1934). *J. Inst. Metals,* **55**, 209.
24. BLEWITT, J. K., COLTMAN, R. R. and REDMAN, J. K. (1957). *J. appl. Phys.,* **28**, 651.
25. SUZUKI, H. and BARRETT, C. S. (1958). *Acta metall.,* **6**, 156.
26. HAASEN, P. (1958). *Phil. Mag.,* **3**, 384.
27. HAASEN, P. and KAY, A. (1960). *Z. Metallk.,* **51**, 722.
28. CORDEROY, D. H. J., HONEYCOMBE, R. W. K. and BRINDLEY, B. W. (1962). *Acta metall.,* **10**, 1043.
29. VENABLES, J. A. (1962). *Proc. 5th International Conference on Electron Microscopy,* **1**, J8.
30. COTTRELL, A. H. and BILBY, B. A. (1951). *Phil. Mag.,* **42**, 573.

31. THOMPSON, N. and MILLARD, D. J. (1952). *Phil. Mag.*, **43**, 422.
32. FOURDEUX, A. and BERGHEZAN, A. (1960–1). *J. Inst. Metals*, **89**, 31.
33. WESTLAKE, D. G. (1961). *Acta metall.*, **9**, 327.
34. COSSLETT, V. E. and NIXON, W. C. (1960). *X-ray Microscopy*. University Press, Cambridge.

Additional general reference

1. KLASSEN-NEKLYNDOVA, M. V. (1960). *Mechanical Twinning of Crystals*. Translated from the Russian by Bradley, J. E. S. 1964, Consultants Bureau, New York.

Chapter 9

The Deformation of Aggregates

9.1 The role of grain boundaries in plastic deformation

A single crystal deformed in tension is usually free to deform on a single slip system for a large part of the deformation, and to change its orientation by lattice rotations as extension of the crystal takes place. If such a crystal is replaced by an aggregate of randomly oriented crystals, the situation changes drastically. Individual grains are no longer subjected to a uniaxial stress system even when the specimen is deformed simply in tension. The boundaries between the deforming crystals remain intact, and as each crystal is disoriented from its neighbours, it will tend to behave differently from them, with the result that neighbouring grains will impose restaints on the deformation of each other.

We shall first briefly discuss the nature of grain boundaries and the roles they can play during deformation processes. Basic experiments will be described which give information about the behaviour of individual boundaries in bicrystals and in coarsely grained polycrystals. The theories which have been developed to explain the behaviour of polycrystalline aggregates in terms of the properties of single crystals will then be summarized.

9.2 The nature of grain boundaries

It has been pointed out that dislocation models can be used to describe small angle boundaries (Chapter 3). The tilt boundary produces a disorientation about an axis normal to the plane of the paper and in the plane of the boundary (Fig. 3.16). It is composed entirely of edge dislocations necessary to create continuity at the interface. Likewise, a *twist* boundary can be formed by rotating the two grains about an axis in the plane of the paper and normal to the plane of the boundary, when the disorientation introduced by the relative rotation of the grains can be accommodated by an array of screw dislocations.

Clearly it is also possible to produce disorientations between grains by combined tilts and twists, in which case the dislocation structure of the boundary becomes a network of edge and screw dislocations. Each dislocation in such an array has an elastic energy and the energy of the boundary is the total of the individual energies of the component dislocations. Thus the energy will increase with the disorientation θ and it has been shown by Read and Shockley to be†

$$E = E_0\theta[A - \ln \theta] \qquad\qquad 9.1$$

where E_0 is a parameter dependent on the elastic distortion around the dislocations, and A is a constant dependent on the core energy of the dislocation. When E is plotted against θ a curve with a maximum is obtained (Fig. 9.1), θ_m being about 0·5 radians. However, the dislocation model of

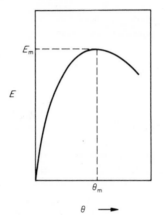

Fig. 9.1 Variation of grain boundary energy E with increasing disorientation θ

a grain boundary breaks down at much smaller disorientations, and, already at 10°, the individual dislocations cannot be distinguished in a bubble model. The higher angle boundaries can only be described as narrow regions where atoms are disordered. Recently direct observations have been made on grain boundaries in tungsten using the field ion microscope which reveals that high angle boundaries possess large regions of good atomic fit separated by regions of bad fit which are associated with ledges.[7] The width of the boundary region is no more than one or two atomic diameters.

It might be expected that, as the number of dislocations in a boundary increases, the effectiveness of the boundary as an obstacle to the movement of dislocations would increase. In certain cases, e.g. the tilt boundary, a dislocation would move readily from one grain to the next *between* the grain boundary dislocations at small disorientations, but the ease of movement decreases as the spacing of dislocations in the boundary decreases.

† See Reference 1 for a derivation of this equation.

The boundary becomes a complete barrier to dislocation movement at quite small disorientations, and the stress concentrations which then arise as the dislocations pile up at the boundaries cause dislocation sources to operate in neighbouring grains.

9.3   Effects of boundaries on the strength of crystals

9.3.1. *Bicrystals*

Chalmers[8] systematically investigated the effect of a unique grain boundary by growing bicrystals of tin with a longitudinal boundary which contained the specimen axis. The two crystals comprising the specimens had the same orientation with respect to the specimen axis, but the angle between the tetragonal (*c* axis) in the two grains was varied. It was found that the elastic limit of the bicrystals increased almost linearly with the angle between the tetragonal axes of the two crystals, and extrapolation to zero angle gave a value for the elastic limit close to that of a single crystal. Clark and Chalmers[9] made a similar study on bicrystals of aluminium with the boundary again parallel to the tensile axis, while each crystal had a common {111} plane at 45° to the tension axis. Orientation differences between the two crystals were of the type produced by rotation about the common {111} plane normal. These authors found that, as the angle of disorientation increased to 30°, the yield stress increased from 0·9 to 1·3 MN m^{-2} and then remained constant up to the largest disorientation (76°). The rate of strain hardening during the early stages of deformation was linear, but the slope increased until the angle of disorientation be-

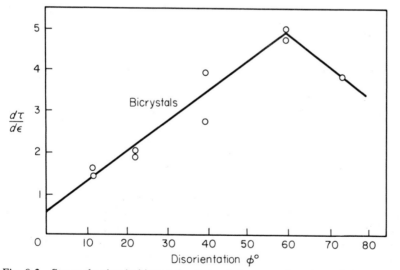

Fig. 9.2   Strengthening in bicrystals of aluminium as a function of disorientation. (After Clark and Chalmers, 1954, *Acta. metall.*, **2**, 80)

tween the crystals was 60°; at this stage the rate of strain hardening was five times that of a single crystal (Fig. 9.2).

Gilman[10] has examined the deformation of bicrystals of zinc with the common boundary parallel to the tensile axis. Little strengthening effect was observed when the basal slip plane was inclined to the boundary at the same angle in each crystal, but when the orientations were altered by a rotation about the tension axis, the stress–strain curves became very much steeper.

9.3.2. *Deformation of coarse-grained aggregates—variations in strain*

While experiments on bicrystals have yielded useful information, it is also necessary to study directly specimens in which the grain boundary distributions are typical of polycrystalline aggregates. A number of investigations have been made with coarse-grained aggregates where variations in strain and in work hardening could be explored within the individual grains.

It has been known for a long time that the plastic deformation of individual grains in aggregates is by no means uniform. Boas and Hargreaves,[11] using coarse-grained commercial aluminium, made a quantitative study of local variations in strain and hardening along linear traverses parallel to the tension axis, which covered a number of grain boundaries. They made diamond hardness tests at regular intervals along these lines to provide not only a measure of the hardness, but also small gauge lengths between indentations to enable the local strains to be determined (Fig. 9.3). The specimens were then deformed in tension and the distance between the

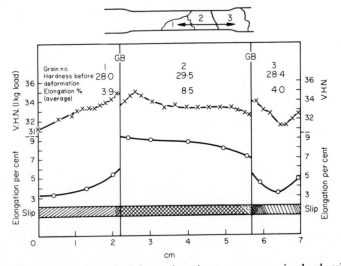

Fig. 9.3 Inhomogeneity of deformation in a coarse-grained aluminium specimen. Overall deformation 5% in tension. (After Boas and Hargreaves, 1948, *Proc. R. Soc.*, **A193**, 89)

original hardness impressions re-measured. In this way it was shown that for an overall elongation of 5 per cent, elongations varying between 2 and 14 per cent were found in various grains. Not only did different grains deform to greatly different degrees, but there were marked changes in strain within one grain. For example, the strain in the vicinity of a grain boundary often differed markedly from the strain at the centre of the grain; if the adjacent grain had deformed less than the first grain, the strain near the boundary of the first grain tended to be less than at the centre. If, on the other hand, the neighbouring grain had deformed more, the strain in the first grain near the common boundary was usually greater than in the centre of the grain. A further important point is that the strain is continuous across the boundary, although there may be a sharp strain gradient in these regions.

This is more clearly revealed by the work of Urie and Wain,[12] who printed photo-grids on the polished surfaces of coarse-grained high purity aluminium, in order to achieve a local gauge length of 0·5 mm instead of the 5–10 mm between hardness impressions in the work of Boas and Hargreaves. With the more sensitive techniques, they found that the elongation could vary by a factor of 10, and in addition they frequently observed minima in elongation near the boundary in the less deformed grain of a pair differing greatly in their overall elongation. There was also marked restriction of elongation in the region of the boundary of the more heavily deformed grain. These results indicated that the strain was continuous across grain boundaries, but that the boundary had a considerable influence on deformation even at macroscopic distances from it. Carreker and Hibbard[13] have examined the inhomogeneity in coarse-grained copper, using the change in angle of a crystallographic axis as a measure of the extent of plastic deformation, and have made similar observations to those referred to above. They also found minima in deformation near the boundaries at low deformations, but at heavier deformations these were replaced by maxima which were explained in terms of forced slip on unpredicted systems.

9.3.3. *Deformation of coarse-grained aggregates—microstructure*

Micro-examination of previously polished surfaces of deformed coarse-grained polycrystalline aggregates shows immediately that the circumstances are greatly different from those in single crystals deformed in tension. Even at small strains (1–5 per cent) each grain is clearly deforming on several slip systems (Fig. 9.4) so that the only appropriate comparison which can be made is with a single crystal of symmetrical orientation, e.g. [001] or [$\bar{1}$11], where four or three slip systems are operating simultaneously. In very large grains after small deformations, there may be a region near the centre which is deforming essentially on a one slip system, but this will soon change with increasing deformation. At smaller grain size such as those normally found in practice, the effects of the grain boun-

Fig 9.4. Multiple slip in lightly deformed polycrystalline aluminium.
Micrograph × 200

daries are felt throughout the crystals, and the deformation is essentially by multiple slip.

The slip systems operating near a grain boundary are frequently different from those elsewhere, because of the need for the grain to accommodate the deformation behaviour of the neighbouring grain. Boas and Ogilvie[14] examined this aspect in aluminium polycrystals, but also in α-brass specimens which had the advantage of revealing the operative slip systems in the *interior*, if sectioned, electrolytically polished and etched. In general, they found two or three operative slip systems in each grain, but up to six systems were observed; this implies that slip occurs also on non-octahedral systems as there can only be four sets of visible slip markings on {111} planes. These authors detected slip also on {100} and {110} planes. It should perhaps be emphasized that observations on the free surface frequently did not correspond with those made on sections, which revealed that often more slip systems were operating in the more restrictive conditions away from the surface of the specimen.

In some cases, inhomogeneities in one grain propagated across the boundary into a neighbouring grain and, moreover, slip lines were observed which were continuous across the boundary. Ogilvie[15] has analysed these occurrences, and shown that slip crossed the boundary only when it was straight, and when the two crystals had a common ⟨110⟩, ⟨112⟩ or ⟨113⟩ direction. One of these directions was found to be within 2° of the direction of intersection of the slip planes at the boundary. The fact that different slip systems operate in adjacent regions of the one grain in an aggregate leads to different types of lattice rotation, and so the original crystal becomes disoriented in a complex fashion. Some of these disorientations lead to deformation bands which have been revealed by crystallographic etching of deformed coarse-grained aluminium and other metals (see Chapter 8).

Another result of the localized differences in slip is that the grain size is a significant variable. With a sufficiently large grain size, the interior of a grain may escape the restraints imposed by the grain boundaries, but as the grain diameter is reduced, these restraints will reach to the centres of the grains, and multiple slip will be widespread. In terms of strength, the interior of a large grain will be softer than that of a small grain (Fig. 9.5),

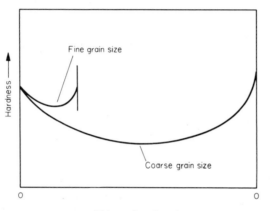

Fig 9.5   Effect of grain size on the deformation of grains. (After McLean, 1957, *Grain Boundaries in Metals*, O.U.P.)

so that during deformation the work hardening of a fine-grained aggregate will be greater than that of a coarse-grained aggregate.

9.4   Theories of the deformation of aggregates

A number of theories have been developed to predict the stress–strain curve of a polycrystalline aggregate from the behaviour of single crystals[49]. In Chapter 2 it has been shown for single crystals that

$$\tau = \frac{L}{A} \sin \chi \cos \lambda \qquad\qquad \textbf{9.2}$$

L/A is the tensile stress σ, while τ is the resolved shear stress on the slip plane, so

$$\sigma = \tau m \qquad\qquad 9.3$$

where $m = (\sin \chi \cos \lambda)^{-1}$ is the orientation factor.

The earliest calculations of the stress–strain curves of polycrystals assumed that in each grain only one slip system operated, namely that with the highest resolved shear stress. The first step was to determine a mean value $\bar{m}$ for the orientation factor, which could then be used in an expression relating polycrystalline stress σ_p and strain ϵ_p.

Assuming a random aggregate of grains, Sachs[16] found that $\bar{m} = 2 \cdot 238$, and related the tensile yield stress σ_Y of the polycrystal to the critical shear stress thus:

$$\sigma_Y = \bar{m}\tau_0 \qquad\qquad 9.4$$

τ_0 = critical resolved shear stress for a single crystal.

Now it is a reasonable assumption that the critical shear stress τ_0 of a single crystal is a function of the shear strain α
so

$$\sigma_Y = \bar{m} \cdot f(\alpha)$$
$$= \bar{m} \cdot f(\bar{m} \cdot \epsilon) \qquad\qquad 9.5$$

where ϵ = polycrystal strain.

In this way Kochendorfer[17] predicted a stress–strain curve for an aggregate from that of a single crystal.

The above approach is limited as it assumes that each grain in an aggregate behaves as an unrestrained single crystal and deforms on a single slip system. This is clearly incompatible with the experimental evidence. Von Mises[18] first pointed out in 1928 that, to cause any desired change in shape of a body, there must be five independent strain components (assuming the volume to remain constant), which for a crystalline solid means five separate slip systems. Clearly more than five could be used but, energetically, the optimum conditions for flow are obtained with five systems. In a face-centred cubic metal, there are 12 crystallographically equivalent octahedral slip systems, so those chosen would be the ones on which the resolved shear stress is greater for a given region of the solid.

Taylor[19] applied this criterion to the deformation of polycrystalline aluminium, and assumed that the five slip systems which operate in each grain are those which require the least work. He further assumed that all grains undergo the same amount of deformation for a given overall deformation. He then approached the averaging problem of the orientation factor $\bar{m}$ when five slip systems operate. The work done in terms of the macroscopic stress σ and the strain $d\epsilon$ is equated to the work done by the several slip systems.

The critical shear stress for slip is τ_0 while $d\alpha$ is the increment of strain on a slip system, so

$$\sigma \, d\epsilon = \sum_1^n \tau_0 \, d\alpha \qquad\qquad 9.6$$

where $n =$ number of slip systems.

Taylor made the assumption that the critical shear stress τ_0 was identical for all slip systems, and that the work hardening did not vary from grain to grain so that $(\tau_0)_1 = (\tau_0)_n$. Thus,

$$\frac{\sigma}{\tau_0} = \frac{\sum d\alpha}{d\epsilon} = m \qquad\qquad 9.7$$

By a detailed numerical treatment Taylor found that the mean value $\bar{m}$ for randomly oriented face-centred cubic metals was 3·06 so that

$$\sigma = 3{\cdot}06\tau_0$$

Bishop and Hill[20] developed another mathematical method for determining $\bar{m}$ which does not involve a fixed set of operating slip systems for each grain, having the smallest shear sum. They did not require homogeneous strain, a defect which the Taylor theory possessed, but they had to make assumptions about the strain distributions in the aggregate. The theory is also more general in so far as it can be applied to a number of tests in addition to the tensile test—they obtain the complete yield surface for a quasi-homogeneous face-centred cubic polycrystalline solid. In the case of the tensile test they have confirmed that the approximate value for $\bar{m}$ is about 3·1.

When Taylor carried out his investigations on aluminium in 1938, the orientation dependence of shear stress–strain curves of high purity single crystals had been little investigated, and a parabolic curve was considered typical. Kocks[21] has pointed out that the most appropriate single crystal curves for calculating polycrystalline stress–strain curves should be those from crystals of 'hard' orientations, where slip on several systems occurs from the start of the deformation. He refers specifically to the [001] and [$\bar{1}$11] orientations, which give much steeper stress–strain curves than crystals with soft orientations. Kocks has calculated the polycrystalline stress–strain curve (Fig. 9.6) using the stress–strain curve for a [$\bar{1}$11] crystal of high purity aluminium and the Taylor formula. He found reasonable agreement with the experimentally determined curve, but correction for a small transitional region gave even better agreement. A similar calculation with an average single crystal curve gave poor agreement.

Thus it seems that the Taylor relationship, particularly in the general form of Bishop and Hill, is of wide validity; however, there are still some unexplained discrepancies. For example, it is very rare to observe the operation within a grain of five slip systems, two or three being the usual number; this may be partly explained by the fact that most observations have been made on the free surface, and not in the interior; nevertheless, when the interior is studied using special techniques, the observations still do not support the theoretical assumptions. Taylor's theory also predicted the way in which grains would rotate during deformation to produce a texture. His predicted textures do not agree with those determined experimentally by Barrett and Levenson,[22] and it seems that the deviations are due to the inhomogeneities which have been observed within the grains during the deformation. Several types of inhomogeneity which occur have been discussed in Chapter 8.

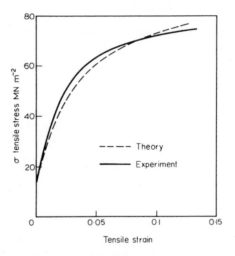

Fig. 9.6 Calculation of polycrystalline stress–strain curves. (After Kocks, 1958, *Acta metall.* **6**, 85.)

9.5 Grain boundary sliding at elevated temperatures

It has been known for many years that at temperatures above about $0.5\,T_M$ (T_M = melting point) grain boundary sliding takes place. The significance of this phenomenon as a contribution to creep strain was first emphasized by Hanson and Wheeler[23] in their experiments on the creep of aluminium, while the first systematic attempt to study the phenomenon using bicrystals of tin was made by King, Cahn and Chalmers,[24] who were able to measure the sliding along the common boundary by means of fiducial marks, and showed conclusively under these circumstances that the boundary was a zone of weakness rather than strength (in contrast to the role of boundaries during deformation at low temperatures). Further experiments have been carried out on aluminium, aluminium alloys[25,26] and copper, in which the strain due to boundary shear was plotted against time, to give curves which were usually not smooth, but with several cusps indicating marked changes in rate of shear. The beginning of the process often occurs after an incubation period and proceeds in a linear fashion. Such results do not lend themselves to a simple time law. The sliding process appeared to commence in a uniform fashion, but at a later stage the displacement varied from point to point along the boundary. Moreover, the experiments revealed that the shearing process became less limited to the boundary itself and spread over a narrow region on each side.

Activation energies for sliding in most cases correspond to that for volume self-diffusion, but some workers have found that the activation energies gradually reach these values as the disorientation between the crystals is increased. The role of impurities which form precipitates at the boundary have been shown to inhibit and finally eliminate shear, while soluble impurities seem either to have no effect, or raise the stress for sliding to take place.

9.6   Deformation of polycrystalline face-centred cubic metals

So far we have examined how individual crystals contribute to the deformation of polycrystalline aggregates, and the theories developed to calculate the stress–strain curves from the shear stress/shear strain curves of single crystals have been summarized. We must now see whether any of the features of the single crystal curves, which have been related to structural changes, can be recognized in polycrystalline curves. Moreover, it is necessary to consider those variables such as crystal structure, grain size and temperature which have a significant effect on the yield stress and the rate of strain hardening of polycrystalline aggregates.

9.6.1   *Stress–strain curves*

The typical stress–strain curve of a polycrystalline face-centred cubic metal is usually described as parabolic, and from time to time different relationships have been put forward to provide a general description. A typical equation is

$$\sigma = \sigma_0 + A\epsilon^n \qquad\qquad \textbf{9.8}$$

σ_0 = elastic limit, A is a constant. However, it seems that in some metals the index n varies with the amount of deformation. On the other hand, Jaoul[3] has divided the stress–strain curve into three stages (Fig. 9.7), which are found, for example, in 99·99 per cent aluminium deformed at 77 K.

Firstly, up to 1–2 per cent strain a parabolic relation holds

$$\sigma = \sigma_0 + A\epsilon^n \quad \text{up to } \epsilon = \epsilon_1 \text{ (Stage 1)} \qquad \textbf{9.9}$$

followed by a linear section of the curve where

$$\sigma = \sigma_0' + P\epsilon \quad \text{between } \epsilon_1 \text{ and } \epsilon_2 \text{ (Stage 2)} \qquad \textbf{9.10}$$

and finally a second parabolic section where

$$\sigma = \sigma_0'' + B\epsilon^m \quad \text{beyond } \epsilon_2 \text{ (Stage 3)} \qquad \textbf{9.11}$$

It is interesting to explore how far the stress–strain behaviour of face-centred cubic crystals is reflected in that of polycrystalline aggregates. In single crystals Stage 1 hardening represents essentially uninterrupted slip on the primary system; this stage would not be expected to be extensive

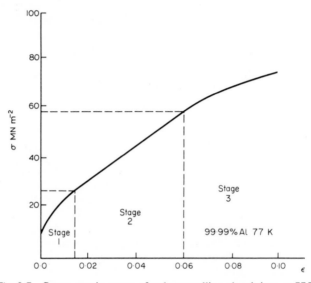

Fig. 9.7   Stress–strain curve of polycrystalline aluminium at 77 K

in polycrystalline aggregates because the grain boundary restraints lead to multiple slip, early in the stress–strain curve. On the other hand, Stage 2 linear hardening in single crystals seems to be closely paralleled by the linear hardening in polycrystalline stress–strain curves, and Stage 3 parabolic hardening is comparable in both single and polycrystalline specimens.

Jaoul[27] found that Stage 1 in polycrystalline stress–strain curves was usually limited to small strains up to 2 per cent elongation for pure metals, and stresses not greater than 50 MN m^{-2} (Table 9.1). The shape of the curve at this stage is fairly insensitive to temperature in the range 77–293 K, ϵ_1 and $\sigma_1 - \sigma_0$ having approximately the same values at each temperature. The index n is sensitive to grain size; for coarse-grained metals (1 mm

Table 9.1   Data on Stress-Strain Curves in Tension of Polycrystalline f.c.c. Metals (after Jaoul)

Metal	293 K							78 K	
	ϵ_1	σ_1	ϵ_2	σ_2	$\dfrac{d\sigma}{d\epsilon}$ (Stage 2)	E		ϵ_2	σ_2
	per cent	$MN\,m^{-2}$	per cent	$MN\,m^{-2}$	$MN\,m^{-2}$	$MN\,m^{-2}$		per cent	$MN\,m^{-2}$
Aluminium 99·99 per cent	1·6	23	1·6	23	800	79 000		6	60
Copper O.F.H.C.	1·5	50	8	120	1200	120 000		30	350
Silver	1·5	30	20	110	670	79 000		54	300

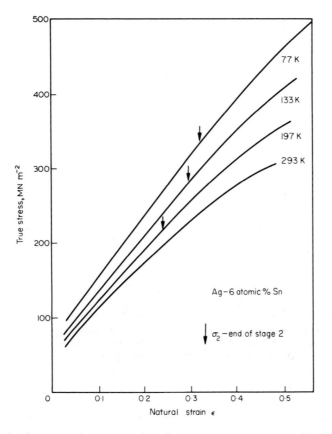

Fig. 9.8   Stress–strain curves of a silver–6 at. per cent tin solid solution at different temperatures. (After Hutchison)

diam.) it has a value of 0·7–0·8, but for finer grained materials (0·1 mm diam.) $n \simeq 0·5$ for 99·99 per cent aluminium. The constant A in eqn. **9.9** is approximately $E/400$ for very fine-grained material.

Stage 2 is barely detectable in polycrystalline aluminium at room temperature (Fig. 9.11), and the curve comprises essentially the two parabolic sections. This resembles the behaviour of aluminium single crystals at room temperature which exhibit Stage 3 hardening at very low strains. It will be remembered that in single crystals the extent of Stage 2 hardening is related to the ease with which dislocations can cross-slip to avoid obstacles in their paths; this leads to the prolongation of Stage 2 as the temperature of deformation is lowered, and to longer Stage 2 hardening in metals of low stacking fault energy. When these ideas are applied to polycrystalline Stage 2 hardening, the same general trends are noted. Fig. 9.8 contains data for a polycrystalline silver–6 atomic per cent tin solid solution deformed at 77 K, 133 K, 197 K and 293 K, which demonstrates the prolongation of Stage 2 hardening as the temperature falls.

9.6.2. *Role of stacking fault energy*

Data for aluminium, copper and silver are collected in Table 9.1, which reveals the effect of lowering the stacking fault energy. This can be seen by comparing the stress at which Stage 2 ends (σ_2) for aluminium at 77 K (60 MN m^{-2}) with that for copper (350 MN m^{-2}). Stage 2 hardening is clearly much more prolonged in the case of copper and silver. The effect of stacking fault energy on the stress–strain curves can be further examined by comparing the behaviour of a pure metal with that of several solid solutions with increasing solute concentration and decreasing stacking fault energy.[28] Figure 9.9a shows the stress–strain curves of silver and a series of silver–gallium alloys of the same grain size, deformed at 77 K. The stress δ_2 moves to higher values as the stacking fault energy of the solid solution decreases. The values of stacking fault energy for these alloys determined by the node method are given in Fig. 9.9b.[28]

Table 9.1 shows that the rate of work hardening in Stage 2 varies substantially from metal to metal; moreover, it does not seem to be independent of temperature, as in the case of the single crystal stress–strain curves. Some of this variation may well arise from changes in elastic moduli with temperature, and also possibly from uncontrolled differences in grain size. There is some indication that if the grain size is carefully controlled as, for example, in the experiments on silver–gallium alloys, the slope of Stage 2 does not vary markedly with composition (Fig. 9.9).

Jaoul found that if the modulus of elasticity E was taken into account, the slope of Stage 2 could be expressed thus:

$$\left(\frac{d\sigma}{d\epsilon}\right)_{\text{II}} = \rho \simeq \frac{E}{100} \qquad\qquad \textbf{9.12}$$

When σ/E is plotted against strain, curves for most face-centred cubic metals fall in a fairly narrow scatter band if the grain size is constant and the temperature is low.

Table 9.1 gives data relevant to the start of Stage 3 or parabolic hardening (σ_2, ϵ_2) in polycrystalline face-centred cubic metals. The index m (eqn. 9.11) has a value of less than 0·5. The results show that, as in the

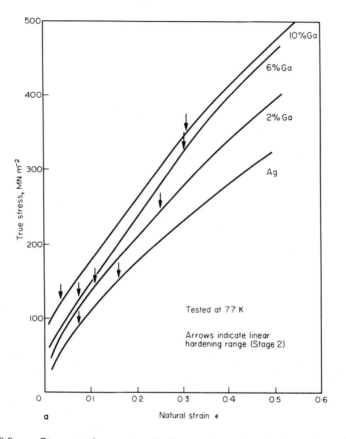

Fig. 9.9 **a**, Stress–strain curves of silver and silver–gallium solid solutions at 77 K (constant grain size). (After Hutchison)

case of single crystals, Stage 3 dominates the stress–strain curve at elevated temperatures. It represents the stage in which dislocations, previously blocked at obstacles on their slip planes, can cross-slip by the combined effect of stress and thermal activation, and thus allow plastic deformation to continue at a decreasing rate of work hardening.

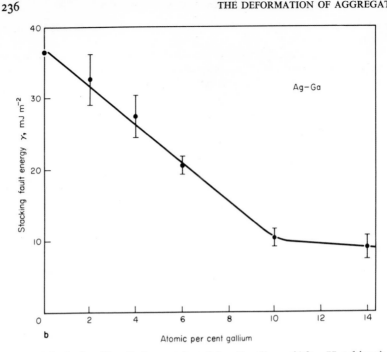

Fig. 9.9 **b,** Stacking fault energies of Ag–Ga alloys. (After Hutchison)

9.6.3. *Effect of grain size*

So far we have not considered the detailed effect of grain size. Even in the early stages of deformation, the grain boundaries limit dislocation movement, with the result that an initial parabolic hardening replaces the easy glide of single crystals. There is also the *complexity* factor which leads to multiple slip in most grains, which structural studies have revealed even at small strains. This is necessary to allow changes in shape while maintaining continuity across the boundary. A second aspect of the complexity factor is that the stress system even within one grain is very heterogeneous and dependent on the different behaviour of a number of adjacent grains. In general, these effects are likely to be more important at small grain sizes, while the coarser grain sizes will approach more closely the behaviour of single crystals.

While grain size can influence the whole of the stress–strain curve, the effect on the elastic limit is particularly marked. Figure 9.10 shows how the elastic limit of 99·99 per cent Al varies from about $1·5 \, \text{MN m}^{-2}$ to $3·5$ MN m^{-2} as the grain size changes from 2 to 50 grains/mm². In comparison, the critical shear stress for a single crystal is about $0·9 \, \text{MN m}^{-2}$. It should perhaps be emphasized that we are concerned here with the macroscopic elastic limit; favourably oriented grains will have undergone slight plastic deformation prior to this point in the stress–strain curve.

The dependence of the initial flow stress σ_y or a subsequent flow stress σ_f on grain size follows a relationship of the Hall–Petch type[31, 32]

$$\sigma_y = \sigma_i + k_y d^{-n} \qquad\qquad \textbf{9.13}$$

where d is the grain diameter, σ_i is a friction stress, and k_y is a constant associated with the propagation of the deformation across the boundaries. The index n, normally 0·5 in body-centred cubic metals, is not so well defined in face-centred cubic metals and alloys.

In Fig. 9.11 stress–strain curves at room temperature are plotted for polycrystalline aluminium (99·99 per cent) of a range of grain sizes, which reveal that the coarser grained specimens work harden substantially less

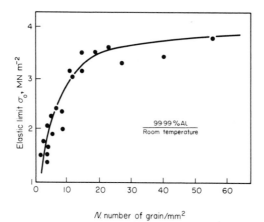

Fig. 9.10   Effect of grain size on the elastic limit of pure aluminium. (After Jaoul, 1964, *Étude de la Plasticité et Applications aux Métaux.* Dunod, Paris)

than the finer grained specimens, although the general form of the curves is similar. Aluminium at room temperature shows no linear Stage 2 hardening and gives essentially parabolic curves. In addition, the tensile stress–elongation curves are plotted for three single crystals, two with 'hard' orientations and one with a very 'soft' orientation. It is rather surprising to see that the curve of the crystal 5 oriented near [$\bar{1}$11] hardens even more rapidly than the polycrystalline specimens, probably because this particular orientation is particularly favourable for the formation of Lomer-Cottrell locks (Chapter 4). The crystal 6 is perhaps more typical of crystals deforming on several slip systems in so far as its stress–strain curve falls somewhat short of the coarsest polycrystalline specimen. On the other hand, the soft crystal 7 has a much lower work hardening capacity than any of the polycrystalline specimens.

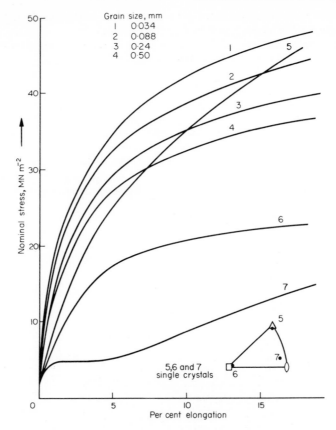

Fig. 9.11   Effect of grain size on stress–strain curves of pure aluminium

9.7   Deformation of body-centred cubic polycrystalline metals

Many polycrystalline body-centred cubic metals develop a sharp yield point if the grain size is fine, in contrast to face-centred cubic metals. The sharp yield point is known to be due to the presence of small concentrations of interstitial impurity atoms such as carbon, nitrogen and oxygen. These effects in iron and other body-centred cubic metals will be described as they represent normal behaviour, however, we must first consider shape of the stress–strain curves in the virtual absence of interstitial impurities.

9.7.1. *Stress–strain curves of pure body-centred cubic metals*

McLean[2] has pointed out that zone-refined iron work hardens little more than 99·99 per cent aluminium at room temperature. When the curves for various face-centred cubic and body-centred cubic pure metals

are compared, with corrections made for shear modulus and melting
point, most of the body-centred cubic curves fall below those of the face-
centred cubic metals (Fig. 9.12) and the rates of work hardening are substantially less. Like the face-centred metals, the stress–strain curves of

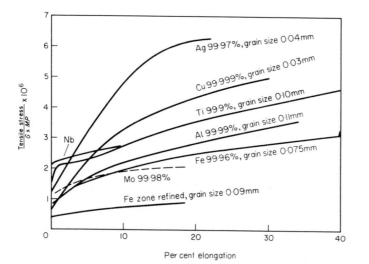

Fig. 9.12 Stress–strain curves of polycrystalline metals compensated for differences in melting point and elastic modulus. (After McLean, 1962, *Mechanical
Properties of Metals*, John Wiley, New York and London)

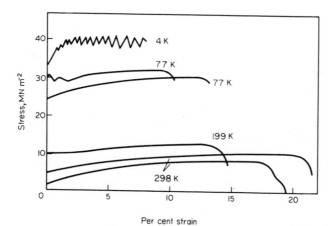

Fig. 9.13 Stress–strain curves of zone-purified iron at various temperatures.
(After Smith and Rutherford, 1957, *Trans, AIME*, **209**, 857)

body-centred metals are sensitive to grain size, the finer grained material
showing both a higher yield point and a greater rate of work hardening.
Single crystals curves usually cover much lower stress levels.

The yield stress of zone purified polycrystalline iron is very sensitive to
temperature (Fig. 9.13) in the range 4·2-298 K and it is this change which
dominates the stress–strain curves. The rates of work hardening are comparable at 298 K and 77 K, and are at a low level. This behaviour persists
to even lower temperatures but is hidden by the severe serrations which
develop in the stress–strain curves as twinning becomes a dominant form
of deformation around 4 K.

9.7.2. *Yield phenomena in polycrystalline iron*

In Chapter 6 it was pointed out that strong yield points are observed
in body-centred cubic crystals because dislocations are firmly locked by
interstitial atom impurities such as carbon, oxygen and nitrogen. In polycrystalline body-centred cubic metals phenomena associated with the yield
point are intensified, and while a sharp yield can be eliminated by the use
of zone purified metal, metals of normal purity levels usually exhibit yield
phenomena, under the appropriate testing conditions.

Taking the various phenomena in order of their occurrence, just prior
to the sharp upper yield point, some small plastic strain takes place. This
is referred to as pre-yield strain, and may vary between about 0·002 and
0·5 per cent.[29] The extent of the strain is time dependent but tends to reach
an equilibrium value within a few seconds.

The yield point in polycrystalline iron can be very marked, particularly
if attention is paid to axial loading of the specimen[30] and in some cases
yield drops of nearly 50 per cent of the applied stress have been observed
(Fig. 9.14). As in single crystals, the sharp yield point is followed by a
yield elongation zone in which plastic deformation is propagated through
the specimen by the passage of a front (or fronts) called Lüder bands, or
stretcher strains. If a wire specimen is used it is possible with a careful
testing procedure to pass one Lüders band from one end of the specimen
to the other. In these circumstances, this part of the stress–strain curve
occurs at constant applied stress. When the Lüders bands have covered the
whole specimen, the yield elongation zone or Lüders zone ends, and work
hardening is registered on the stress–strain curve, although it has been
taking place behind the Lüders front from the beginning of deformation.
As the upper yield point is vulnerable to testing variables such as size and
shape of specimen, axiality of loading, it is usual to consider as a more
reliable value the lower yield point which corresponds to the stress needed
to propagate a Lüders band through the specimen. Polycrystalline iron
exhibits strain ageing in the same way as single crystals. As the testing
temperature is raised the sharp yield point and yield elongation zone are
replaced by a serrated stress–strain curve, sometimes referred to as the
Portevin–Le Chatelier phenomenon. At the same time the rate of work

hardening becomes substantially greater than at lower temperatures. In this temperature range the carbon and nitrogen atoms diffuse sufficiently rapidly to keep up with the glide dislocations. They thus cause dislocation locking which the applied stress overcomes either by unlocking or by nucleation of new dislocations, and the cycle is repeated, in the extreme case over the whole strain range of the test.

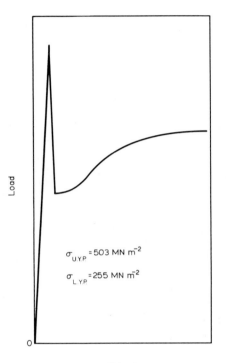

$\sigma_{U.Y.P.} = 503$ MN m^{-2}

$\sigma_{L.Y.P.} = 255$ MN m^{-2}

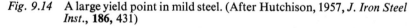

Fig. 9.14 A large yield point in mild steel. (After Hutchison, 1957, *J. Iron Steel Inst.*, **186**, 431)

9.7.3. *Effect of grain size*

In general, the grain size has a marked effect on the phenomena associated with the yield point. The influence is summarized in Fig. 9.15, which illustrates the changes in the stress–strain curves of polycrystalline iron over a wide range of grain sizes. In general, the whole curve is displaced to high stresses with decreasing grain size, as both the upper and lower yield points are raised.

The Lüders extension also increases as the grain size becomes finer. The Hall–Petch[31,32] relationship (eqn. **9.13**) has been found to apply to the

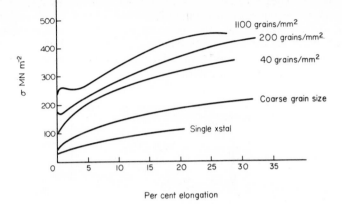

Fig. 9.15 Effect of grain size on the stress–strain curves of decarburized Armco iron. (After Jaoul, 1964, *Étude de la Plasticité et Applications aux Métaux.* Dunod, Paris)

dependence of the lower yield stress σ_{LY} on grain size d, giving a linear plot when σ_{LY} is plotted against $d^{-1/2}$. Some results for iron are shown in Fig. 9.16. The first terms on the right hand side of the equation, σ_i, is interpreted as a frictional stress, the value of which is independent of grain size (d) and is the intercept of the ordinate in Fig. 9.16. It represents the stress necessary to move unlocked dislocations on the slip planes of the crystals. The second term is a dislocation locking term which is more

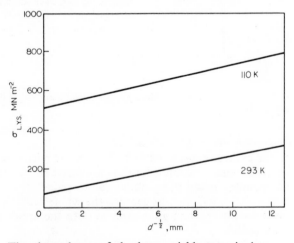

Fig. 9.16 The dependence of the lower yield stress in iron on grain size. (After Codd and Petch, 1960, *Phil. Mag.*, **5**, 30)

effective at fine grain sizes. The usual model used to derive such an equation assumes dislocations emanating from sources within the grains and piling up at the grain boundaries. Clearly the number of dislocations within a pile-up will be related directly to the glide path which is determined by the grain size. For larger grain sizes there will be larger pile-ups with higher stress concentrations at their heads, so the locking will be less effective, and the Lüders band will tend to propagate at smaller values of the lower yield stress.

As the grain size becomes larger not only are the upper and lower yield stresses lowered, but the yield drop tends to become less marked and eventually may disappear. Also the Lüders front is much less well defined until at very coarse grain sizes it is so diffuse that it is difficult to detect at all. In the extreme case of the single crystal, upper yield points are rarely observed in iron at room temperature without prior strain ageing.

9.7.4 *Effect of temperature*

Temperature has a striking effect on the stress–strain curves of polycrystalline iron, which normally exhibits a yield point at room temperature. As the temperature of deformation is lowered the upper yield point rises markedly, the yield drop and the yield elongation zone both become progressively larger. This is a general observation for body-centred cubic metals which has been confirmed also in molybdenum, niobium and tantalum.[33] At low temperatures, the dislocations pre-existing in the specimens are firmly locked by condensed atmospheres of interstitial atoms; a substantially larger stress is needed to generate new dislocations and to move them through the crystals, so the temperature dependence arises largely from the frictional stress. This can be established by examination of the effect of grain size on the lower yield stress over a wide temperature range. Figure 9.16 shows that the curves for 110 K and 293 K have approximately the same slope (k_y) which implies that the locking terms are similar; however, the intercepts on the ordinate are very different, indicating that the frictional stress σ_i increases substantially as the temperature is lowered.[34] There is also experimental evidence[35] that the frictional stress increases markedly with interstitial $(C + N)$ concentration in the range 0·001–0·03 wt per cent.

On the other hand, as the temperature of deformation is raised above room temperature, the upper and lower yield points and the yield elongation zone gradually disappear, to be replaced by curves showing serrations to a greater or lesser degree. Recent electron microscope observations of iron deformed in this temperature range have shown that the dislocation density is almost an order of magnitude greater than that in specimens deformed to the same strain at room temperature. This suggests that rather than repeated breaking away of dislocations from solute, there is permanent pinning of newly generated dislocations by the mobile carbon (and nitrogen) atoms, or more probably by carbide or nitride precipitates.

This role is so effective at these temperatures that new dislocations must be continually generated to continue the deformation, and so much higher dislocation densities result.

9.7.5. *Stress–strain curves of other body-centred cubic metals*

The metals, vanadium, chromium, niobium, molybdenum, tantalum and tungsten form an important group of body-centred cubic metals, all of which normally contain carbon, nitrogen or oxygen, and are capable of exhibiting yield phenomena. The yield stresses of all these metals are extremely sensitive to temperature below about $0.2\,T_M$. It is thus apparent that any theory developed for the strong temperature dependence of the yield stress of iron will find an application to other body-centred cubic metals. The stress–strain curves likewise show marked similarities, but differences arise from the degree to which brittle fracture intervenes in the plastic behaviour of each metal. Chromium, molybdenum and tungsten

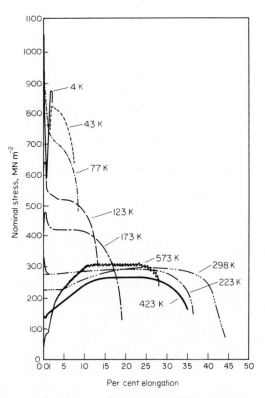

Fig. 9.17 Stress–strain curves of tantalum. (After Bechtold, 1955, *Acta. metall.*, **3**, 249)

are more prone to brittle fracture, whereas vanadium, niobium and tantalum show a wider range of plastic deformation behaviour.

Fig. 9.17 gives stress–strain curves of tantalum over the temperature range 4 K to 573 K in which is shown the typical development of a yield point followed at low temperatures by localized flow which leads to a falling stress and to fracture. In the intermediate temperature range, a yield elongation zone is formed, while at the highest temperatures, serrations or localized yielding occur throughout the stress–strain curve.

9.7.6. *Effect of interstitial impurities on mechanical properties*

In recent years the role of interstitial impurities such as hydrogen, carbon, nitrogen and oxygen on the mechanical properties of the Group 5A (vanadium, niobium, tantalum) and the 6A (chromium, molybdenum, tungsten) metals has been increasingly studied. Table 9.2 indicates

Table 9.2 Estimated interstitial atom concentration in solid solution in Group 5A and 6A metals after moderate cooling (p.p.m.)

| Interstitial element | Group 5A | | | Group 6A | | |
	V	Nb	Ta	Cr	Mo	W
Hydrogen	10000	9000	4000	0·1–1	0·1	not detectable
Carbon	1000	100	70	0·1–1	0·1–1	<0·1
Nitrogen	5000	300	1000	0·1	1	<0·1
Oxygen	3000	1000	200	0·1	1	1

that the two groups of metals have very different solubilities with respect to these four interstitial impurities, the solubility being substantially greater in the case of the Group 5A metals. The Group 6A metals will normally have second phase particles of oxide, carbide and nitride as well as a very dilute supersaturated solution of the interstitial impurities. Despite these differences, yield points can be obtained in the stress–strain curves of all these metals. This lends support to the view that the essential role of the interstitial element is to lock the dislocations firmly either by solute atom segregation or by forming a fine precipitate of oxide, carbide, etc., along the dislocation lines. Both mechanisms should be effective, but it would be expected that metals which have a higher interstitial solubility should be more likely to show strain ageing or serrated stress–strain curves at elevated temperatures. These metals have an adequate concentration of solute atoms which, when thermally activated, can diffuse to any unlocked dislocations, and thus cause strain ageing; this phenomenon is indeed observed in niobium and tantalum. On the other hand, a metal with a low interstitial concentration in solid solution and a dispersion of oxide, etc., cannot reprecipitate fine oxide particles on freshly generated dislocations.

The interstitial impurities have a profound effect on the brittle behaviour of all the body-centred cubic metals, including iron. This is discussed in Chapter 15.

9.7.7. *Theories of the yield point in polycrystalline body-centred cubic metals*

The body-centred cubic metals in contrast to the close-packed hexagonal and face-centred cubic metals exhibit a strong dependence of the yield stress on temperatures below about $T = 0 \cdot 2T_M$. The yield stress is also markedly dependent on strain rate. It is thus clear that to account for this behaviour a mechanism of locking or dragging dislocations is needed which is strongly dependent on temperature. Amongst the most significant mechanisms which have been proposed to explain the yield stress are:

1. Dislocation atmospheres of interstitial atoms.
2. Fine precipitates on dislocations.
3. Peierls–Nabarro force opposing the movement of dislocations (friction stress).

In Chapter 6 the solute atom–dislocation interaction has been discussed with reference to single crystals, together with the role of locked dislocations in causing the sharp drop in stress associated with the upper yield point. If no dislocations are free to move initially, new ones must be generated, and in the early stages of deformation they must move rapidly to achieve the imposed strain rate. This requires a high stress, but as the number of dislocations increases rapidly, their velocity need not be as high and the stress falls sharply.

In polycrystalline body-centred cubic metals the yield points are usually much more pronounced than in single crystals, because the grain boundaries stop mobile dislocations from travelling too far into unyielded material. In a single crystal it is only necessary to form one centre of yielding, whereas in a polycrystal yielding must occur separately in each grain. In the ideal case the upper yield point occurs at the stress when plastic yield crosses the first grain boundary, and if there is only one region in which this process commences, rather than several produced by local stress concentrations, there will be a very large yield point.

The Hall–Petch equation previously referred to defines σ_{LY}, the stress necessary to propagate the yield from grain to grain. The model due to Cottrell[36] is shown in Fig. 9.18, where some dislocations are released from a source S_1, and move along the slip plane to pile-up at the grain boundary. A stress concentration is thus set up.

Now let σ_n be the stress needed in the second grain to make the grain slip from a source S_2 at a distance l from the boundary. Once yielding occurs in a grain, σ_Y on a yielded slip band relaxes to σ_i, but at the grain boundary the resultant stress concentration causes a stress at S_2 of about $[(\sigma_y - \sigma_i)d]/4l$. Yielding occurs when this expression is equal to σ_n.

From this model the Hall–Petch equation, $\sigma_y = \sigma_i + k_y . d^{-1/2}$,[36] can then be derived and it follows that

$$k_y = 2\sigma_n . l^{1/2}   \textbf{9.14}$$

k_y in this analysis is a term which accounts for the unpinning of dislocations from their interstitial atmospheres and which, if the theory is correct, should show a marked temperature dependence.

There is some support for this view in so far as k_y increases as the carbon and nitrogen contents of iron are reduced;[37] however, a serious criticism is that k_y is almost independent of temperature and strain rate in the range

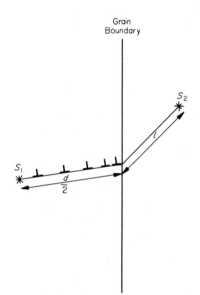

Fig. 9.18   Dislocation model for propagation of yielding. (After Cottrell, 1958, *Trans. AIME*, **212**, 192)

100–300 K. Moreover, k_y is nearly the same for the lower yield stress, and for flow stresses beyond the lower yield stress. This would not be likely if yielding occurred by the breaking away of locked dislocations. On the contrary, there is evidence that the pre-existing dislocations in the specimen remain firmly locked. Electron microscopic studies[38] have shown that in iron the dislocations are often locked by fine precipitates rather than a solute atom atmosphere, so that dislocations cannot break away according to the Cottrell model. The interstitial solubility data on refractory metals referred to earlier also indicates that, at least in molybdenum and tungsten, precipitates are more likely to lock dislocations than solute atoms. Another objection is that dislocation pile-ups are not very common in polycrystalline iron. On the other hand, in iron, Cottrell[39] considers that a whole range from weak to strong pinning can be obtained as the ageing treatment is intensified. Very light strain ageing in which only solute atoms segregate to dislocations would lead to a temperature dependence of k_y, while more pronounced ageing would result in formation of precipitate and a temperature insensitive k_y.

There are several alternative explanations of the role of grain size on to the yield point; k_y in iron appears to be insensitive not only to temperature but also to composition and heat treatment. It has been proposed[39] that it is a measure of the stress needed to create dislocation loops at grain boundaries. It is now necessary to seek another explanation for the marked temperature dependence of the yield stress. The experimental results quoted earlier indicate that this is likely to be a result of a strong temperature dependence of the Peierls force, which would give the same dependence in both single crystals and polycrystalline aggregates.

9.8   Deformation of polycrystalline solid solutions

We have examined the important variables in the strengthening of solid solution crystals, so it remains to determine whether such results are valid when applied to polycrystalline aggregates. Numerous experiments have indicated that those solutes which have the largest effect on the lattice parameter of the parent metal will likewise have the most pronounced effect on the strength. For example, Frye and Hume-Rothery[40] examined the hardening of polycrystalline silver by the addition of various solutes and found a roughly linear relationship between the hardness and the square of the change in lattice parameter.

The work of French and Hibbard[41] indicated that the change in yield stress per one atomic per cent solute increased linearly with the increase in lattice parameter for one atomic per cent solute. However, the relative valency of solvent and solute atoms also appears to be of significance from the work of Allen, Schofield and Tate,[42] who found that solid solutions of copper with zinc, gallium, germanium and arsenic had almost identical stress–strain curves (Fig. 9.19) when the electron atom ratio of each alloy was identical, the atomic concentrations being widely different.

Dorn and co-workers[43] attempted to deal with both the atomic size and valency differences by use of an empirical parameter, by the application of which to a number of aluminium alloys they were able to represent the data on one curve. Hibbard[44] concluded also that both valency and relative atomic size were significant factors. Very different stress–strain curves were obtained from a number of copper-base solid solutions with the same lattice parameter, but a better correlation was observed when alloys of the same e/a ratio were deformed.

Ainslie et al.[45] prepared a series of copper-base alloys with the same yield stress, and also found little correlation with the lattice parameter values but the e/a values were in most cases similar, thus confirming the earlier observations of Allen, Schofield and Tate.

More recently Fleischer[46] has calculated the effect of the change in shear modulus in the region of solute atoms and has been able to explain the hardening in copper base alloys in terms of a combination of the size effect and the modulus effect (Chapter 6). The modulus effect replaces the electron/atom ratio in the earlier analyses, and it is interesting to note that

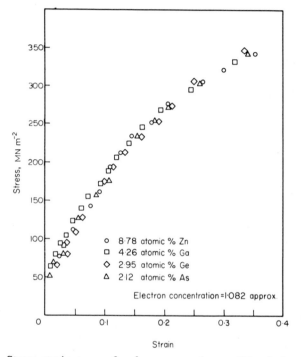

Fig. 9.19 Stress–strain curves for four copper-base solid solutions with the same e/a ratio. (After Allen, Schofield and Tate, 1951, *Nature (Lond.)*, **168**, 378)

the rate of change in modulus with atomic concentration of solute is greater as the valency of the solute increases (Fig. 9.20).

The effect of increasing the concentration of a given solute is to displace the whole stress–strain curve upwards, with the result that the curves are almost parallel to each other in many systems; however, some deviations from this behaviour exist. Comparisons should be made at a temperature at which Stage 2 deformation predominates and so any effect of stacking fault energy on the transition to Stage 3 is minimized. Bullen and Hutchison[47] examined polycrystalline copper-base solid solutions of silicon, aluminium, zinc and antimony; in each case the stress–strain curves for different concentrations of solute were practically parallel at 77 K. The hardening due to the solutes was referred to as an impedance of the form $I(1 + \epsilon)$, where I is an impedance factor which is proportional to the square root of the solute concentration and represents $(\sigma_{alloy} - \sigma_{metal})$ at a low strain, and ϵ is the strain in tension. This approach emphasizes the fact that the strengthening can be regarded as an impedance or frictional term which is superimposed on the hardening curve of pure copper, but it does not indicate the mechanism.

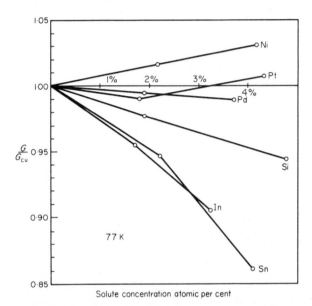

Fig. 9.20   Variation of shear modulus with solute concentration for Cu-base alloys. (After Fleischer, 1963, *Acta metall.*, **11**, 203)

Russell and Jaffrey[48] have compared the behaviour of copper with a 3 atomic per cent solid solution of tin in copper; while the flow stress and strain hardening in State 2 were independent of temperature in copper, the alloy exhibited a flow stress proportional to $T^{1/3}$ below a critical temperature when the solute atoms were assumed to be immobile. Above this temperature, the flow stress was proportional to T^{-2}, and the difference was attributed to the mobility of the solute atoms. On the other hand, the strain-hardening coefficient was independent of temperature. Recent work by Hutchison and Honeycombe[28] has underlined the effect of temperature on the flow stresses of solid solutions. Whereas the flow stress (after 0.01ϵ) of pure silver is very insensitive to temperature in the range 77–300 K (Fig. 9.21), the solid solutions show a progressively larger effect as the concentration increases. Figure 9.21 illustrates the magnitude of the strengthening for a number of different solute elements at a concentration of 6 at. per cent in silver. The results could be described in terms of a temperature-dependent and a temperature-independent contribution to the solution hardening, both related primarily to the valency of the solute, rather than the Fleischer parameter.

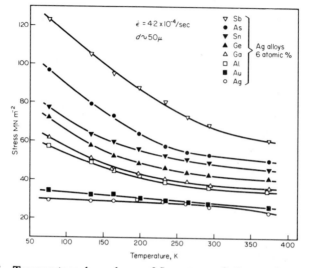

Fig. 9.21   Temperature dependence of flow stress of silver and several silver-base solid solutions. Grain size $\simeq 50\,\mu$. (After Hutchison and Honeycombe, 1967, *Metal Sc. J.* **1**, 70)

9.9   Dislocations in deformed polycrystalline aggregates

The dominant feature of thin-foil electron micrographs of deformed polycrystalline metals is the formation of a dislocation sub-structure within the grains. In metals of high stacking fault energy where cross-slip and climb of dislocations can occur readily, the as-deformed cell structure is often quite sharp after small deformations, but even metals of much lower stacking fault energy, e.g. silver, exhibit areas practically free from dislocations bounded by tangled walls of dislocations at moderate deformations. Only solid solutions of very low stacking fault energy reveal planar arrays of dissociated dislocations, for example, silver–10 atomic per cent gallium alloy ($\gamma < 10\ \mathrm{mJ\,m^{-2}}$) after 5 per cent deformation at 77 K, but after heavier deformation a dislocation sub-structure eventually develops so this can be regarded as typical of most polycrystalline deformed metals. The strain at which cell structures are first observed is a linear function of the solute concentration; the effect of lowering the deformation temperature is to raise the level of this relationship to higher strains. As the extent of deformation is increased, the cell size within the grains diminishes until a constant size is reached. Whilst it is difficult to assign precise values, the cell size is normally in the range 0·25–3 microns, but it is dependent on numerous variables, including the temperature of deformation, extent of deformation, composition of the alloy, etc. Some typical measurements are plotted in Fig. 9.22 for pure silver, and a 2 atomic per cent fallium alloy deformed at 77 K and 293 K.

The presence of dislocation sub-structures in deformed metals causes disorientations within the grains which are readily revealed by standard X-ray techniques. Similar effects occur in deformed single crystals, and

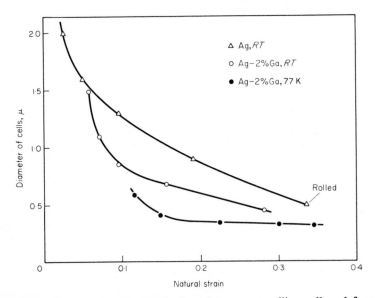

Fig. 9.22   Cell sizes in silver and silver–2 per cent gallium alloy deformed at 293 K and 77 K. (After Hutchison and Honeycombe, 1967, *Metal Sci. J.* **1**, 186)

lead to asterisms in X-ray Laue photographs. Disorientations can be as small as a few seconds of arc, but may be as high as several degrees in extreme cases. These localized disorientations are of importance during the annealing of deformed metals and will be referred to again in Chapter 11.

9.10   Deformation of alloys containing two phases

We have already considered the behaviour of single crystals containing a finely dispersed hard phase, as in precipitation-hardened or dispersion hardened systems. There are, however, many different types of distribution of two phases which occur in practice. For example, the two phases may both be ductile with the softer phase present either as a continuous network or as discrete particles, or as lamellae. On the other hand, the second phase can be very hard and with similar alternatives for its distribution.

Unkel[50] carried out the first systematic experiments on a number of two-phase alloys deformed by rolling, measuring the mean compression of a large number of grains of each phase. While the softer phase will always plastically deform first, Unkel pointed out that the relative rates of work hardening will determine the relative deformation of the phases.

9.10.1. *Two ductile phases*

A very approximate idea of the behaviour of such an alloy can be obtained by assuming that the strain is constant in all regions of each phase.[51] Thus the corresponding stresses σ_1 and σ_2 must differ, and the average stress σ can then be given as

$$\sigma = f_1\sigma_1 + f_2\sigma_2 \qquad \textbf{9.15}$$

where f_1 and f_2 are the volume fractions of the two phases. Fig. 9.23a expresses this criterion graphically, the dotted curve for 50 per cent of each phase having been calculated from the stress–strain curves of the pure phases. As expected, the flow stress at a given strain increases with the volume fraction of the harder phase. However, examination of a deformed duplex alloy quickly shows that the assumption that the strain is uniform within one phase is not correct. For example, in deformed 60/40 brass (α and β phases), it has been shown that the harder β phase suffers heavier strains at the α-β interfaces than in the interior of the β phase.[52]

Alternatively, it can be assumed that each phase experiences the same stress σ_1, and that this leads to different strains ϵ_1 and ϵ_2.

The average strain is then

$$\epsilon = f_1\epsilon_1 + f_2\epsilon_2 \qquad \textbf{9.16}$$

We now as before adopt two stress–strain curves for the pure phases and construct the curve for 50 per cent volume fraction of each phase (Fig. 9.23b). The flow stress is not now a linear function of f. However, this model is also unrealistic because there is assumed to be a strain discontinuity at the inter-phase boundaries, and again this can be shown by experiment not to be so. Observations on the tendency to recrystallize and the micro-hardness of the deformed phases in brass[52] and Al-Mg alloys[53] have demonstrated that the softer phase is much more heavily deformed in the early stages, but continuity of strain occurs across the boundary. in so far as the hard phase is more deformed in its vicinity and the soft phase less. While in the early stages of deformation, the soft phase deforms more, after further deformation the phases deform to practically the same extent, unless the volume fraction of the hard phase is less than about 0·3.

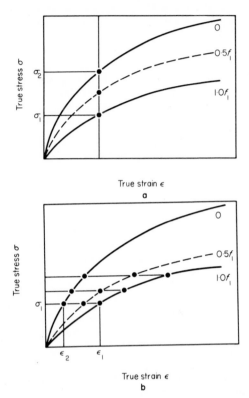

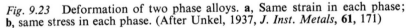

Fig. 9.23 Deformation of two phase alloys. **a**, Same strain in each phase; **b**, same stress in each phase. (After Unkel, 1937, *J. Inst. Metals*, **61**, 171)

9.10.2. *Ductile–brittle phases*

It is perhaps difficult to define a brittle phase, for many phases apparently brittle can be heavily deformed in favourable conditions, for example cementite in steels. The mode of distribution of the brittle phase is of great importance, for if it forms a continuous film around the grain boundaries disastrous brittleness occurs, for example bismuth in copper. On the other hand, if the brittle phase spheroidizes as a result of a high interfacial energy, the alloy is not only ductile but can have improved strength.

The strengthening effect of such dispersions, as distinct from the much finer dispersions encountered in aged alloys was first systematically studied by Gensamer,[54] who examined the role of the distribution of iron carbide in steels on their mechanical properties. For both ferrite–pearlite aggregates and tempered steels with spheroidized carbide particles, it was found that the yield stress was inversely proportional to the logarithm of the mean ferrite path (Fig. 9.24). This relationship breaks down only at very large mean ferrite paths.

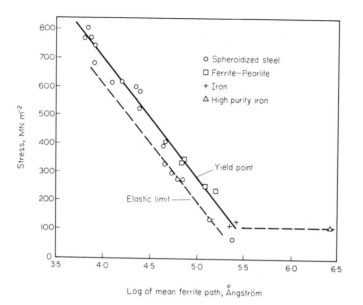

Stress, MN m^{-2}

o Spheroidized steel
□ Ferrite–Pearlite
+ Iron
△ High purity iron

Yield point

Elastic limit

Log of mean ferrite path, Ångström

Fig. 9.24 Effect of cementite dispersions on the yield strength of steels. (After Roberts *et al.*, 1954, *Relation of Properties to Microstructure*, American Society for Metals)

9.11   Whiskers, fibres and fibre reinforcement

Freshly drawn glass fibres possess very high strengths approaching the theoretical strength, but deterioration quickly sets in, as contamination occurs and surface cracks nucleate. Fine single crystal fibres and whiskers of metals and other crystalline solids prepared in a variety of ways, are in many cases extremely strong[4] because they can be made practically devoid of dislocations. Indeed, the achievement in this way of strengths approaching that of the theoretical strength are a spectacular vindication of the basic concepts of the dislocation theory.

In recent years, efforts have been made to utilize the high strengths of such fibres by embedding them in a more ductile matrix to provide new materials with valuable properties over a wide temperature range. The fibres are arranged axially along the rod, and do not prevent the initial deformation of the matrix, but by this means the load is transferred to the fibres which can be stressed almost to fracture.

9.11.1. *Growth of metal whiskers*

The growth of metal whiskers can be achieved in many ways and an extensive literature exists.[55,56,57]

There are four main methods of whisker growth:

1. from plated substrates,
2. from the vapour phase,
3. by reduction reactions,
4. by electrolysis and from solution.

1. Technological interest in metal whiskers arose from the observation some years ago that whiskers of cadmium grew spontaneously at room temperature on cadmium-plated condenser plates and eventually caused electrical short-circuits. Similar behaviour occurs with zinc and tin. The whiskers were found to be metallic single crystals up to 0·5 microns in diameter and up to 10 mm long. Zinc and cadmium usually grow along the hexagonal *c* axis. This type of whisker seems to grow from its base rather than the tip and in some cases growth is undoubtedly accelerated by the presence of impurities and/or internal stresses in the substrate. The temperature of growth can vary from as low as 233 K up to near the melting point. Oxide whiskers (e.g. Al_2O_3, Mo_2O_3, CuO, NiO) can be grown by heating the appropriate metal substrate in an oxidizing atmosphere.

2. Vapour phase methods are most successful with metals which possess high vapour pressures at their melting points, e.g. mercury, zinc and cadmium. The metal is condensed from the vapour phase often in the presence of an inert gas, and results in the formation of whiskers from 0·2 to 50 microns in diameter and up to several cm long. The method works with other metals such as silver, copper and iron, but the whiskers are extremely small. It is often assumed that growth from the vapour is associated with axial screw dislocations which accelerate the rate of growth, but it is by no means proved that these dislocations are present in a majority of the cases.

Graphite whiskers can be prepared using a d.c. arc in a pressure vessel containing argon at about $9·3 \, MN \, m^{-2}$.

3. Perhaps the most successful method of whisker preparation involves the high temperature reduction of salts in a hydrogen atmosphere. The whiskers grow profusely in and around the boat holding the salt. Most of the common metals including iron, copper, silver, nickel, cobalt, platinum and gold, also silicon and germanium, have been prepared in whisker form by this method.[56] For example, iron whiskers up to 10 cm long and varying in diameter from 0·5 microns to 1 mm have been grown by reduction of

ferrous chloride or ferrous bromide. Again in this method, screw dislocations are thought to be significant, but mechanisms are still in debate.

4. Electrolysis of aqueous solutions provides a simple method of preparation for some metals, e.g. silver from concentrated silver nitrate solution and lead from lead acetate solution.[56] Whiskers of metallic salts such
as KBr, KI, NaCl, have been grown simply from solution, and again the
axial screw dislocation has been used to explain the growth. Some NaCl
and KCl whiskers have been shown to possess such screw dislocations.

9.11.2. *Properties of metal whiskers*

A large number of whiskers of different metals, including iron, copper,
silver, zinc and also Al_2O_3 and SiO_2 have been deformed in tension and
have exhibited elastic strains between 2 and 5 per cent, i.e. greatly in
excess of normal materials in single and polycrystalline form. Typical
stress–strain curves for copper and iron whiskers are shown in Fig. 9.25,
in which deviations from elastic behaviour occur above 2 per cent elon-

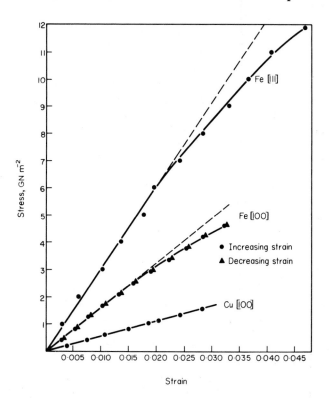

Fig. 9.25   Stress–strain curves of copper and iron whiskers. (After Brenner
1964, *Metall. Rev.*, **9**, 280)

gation in the case of iron. For iron whiskers with [111] orientation, the strength approached 12 GN m⁻². In some cases, the large elastic strains are followed by catastrophic failure, while in others a sharp yield point is exhibited followed by plastic deformation at constant stress and the propagation of slip bands along the whiskers. Typical elastic and mechanical properties of various whiskers are given in Table 9.3.

Table 9.3 Properties of whiskers

Material	Young's modulus, $GN\,m^{-2}$	Tensile strength, $GN\,m^{-2}$
Al_2O_3	525	15·2
Fe	193	12·4
SiC	689	20·7
SiO_2 (in air)	73	10·3
Asbestos	186	5·9
Soda glass	68	3·4

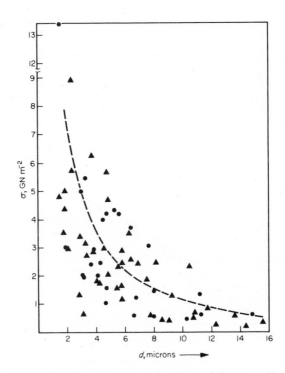

Fig. 9.26 Strength of iron whiskers as a function of diameter. (Brenner, 1964, *Metall. Rev.*, **9**, 280)

The strength of whiskers is a sensitive function of size, and to obtain a marked increase in strength over conventional material, the whiskers should have a diameter less than 10 microns. Figure 9.26 illustrates this trend for iron whiskers. The change in strength with size is related to the dislocation population in the whiskers, which becomes negligible for very small diameters of specimens. In some cases, e.g. zinc, the absence of dislocations has been established by transmission electron microscopy[58] or by the etch pit method in the case of iron.[59]

X-ray examination[63] of some whiskers, e.g. Al_2O_3, NaCl, has revealed axial screw dislocations by diffraction contrast, particularly in fine whiskers. In coarser specimens more complicated dislocation structures have been observed, while in other cases no dislocations have been detected.

Consequently, although growth on screw dislocations can be put forward as a reasonable mechanism in some cases, other mechanisms must also be capable of producing whisker growth.

9.11.3. *Fibre reinforcement*

Fibre reinforcement occurs when thin fibres are oriented in a ductile matrix.[60] Ideally, whiskers would be best suited to this application because of their strength; however, their small dimensions make them difficult to handle, so reinforcement is often done with fine drawn polycrystalline wires, which fall short of the whisker strengths. However, tensile strengths in the range 2·0–3·5 $GN\,m^{-2}$ are common. A number of investigations have been carried out on fine tungsten wires embedded in copper[61, 62] and silver, also stainless steel in aluminium which provide some basic information on mechanical properties. There are four stages in the stress–strain curves.[60]

1. Elastic deformation of fibres and matrix which occurs until just beyond the normal elastic limit of the matrix.
2. The matrix begins to deform plastically while the fibres still deform elastically.
3. Plastic deformation of both fibres and matrix.
4. Fracture of fibres followed by total fracture of the specimen.

Of the four stages, Stage 2 is the most significant and it accounts for a large proportion of the stress–strain curves (Fig. 9.27). In Stage 1, the deformation is totally elastic and Young's modulus E_c of the composite is

$$E_c = E_f V_f + E_m V_m \quad \text{(Stage 1)} \qquad \textbf{9.17}$$

where f and m refer to fibre and matrix respectively. V is the volume fraction.

However, in Stage 2, E_m is replaced by the slope of the stress–strain curve of the matrix at strain ϵ.

$$E_c = E_f V_f + \left(\frac{d\sigma_m}{d\epsilon_m}\right) . V_m \qquad\qquad \textbf{9.18}$$

As this is in the plastic region, the slope is much less than E_m, approx. $E_m/100$, so the second term is small, thus

$$E_c \simeq E_f . V_f \qquad\qquad \textbf{9.19}$$

The ultimate tensile strength σ_c of the composite is

$$\sigma_c = \sigma_f V_f + \sigma_m'(1 - V_f) \qquad\qquad \textbf{9.20}$$

where σ_f = U.T.S. of fibres, σ_m' = stress on matrix when fibres are at their maximum strain in tension.

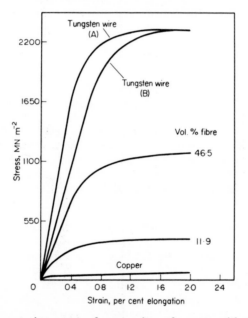

Fig. 9.27 Stress–strain curves of composites of copper with tungsten wires. (After McDanels *et al.*, 1960, *Metal Progr.*, **78**, 118)

Thus the U.T.S. of the composite is linearly related to the volume fraction of the fibres, a relationship supported by experimental results on copper strengthened with tungsten fibres (Fig. 9.28).

To obtain any benefit from the presence of fibres, the strength of the composite must be greater than the U.T.S. of the matrix σ_u, so

$$\sigma_c = \sigma_f V_f + \sigma_m' . (1 - V_f) > \sigma_u$$

from which we have a critical volume fraction of fibres V_{crit} which must be exceeded.

$$V_{crit} = \frac{\sigma_u - \sigma_m'}{\sigma_f - \sigma_m'} \qquad 9.21$$

$$\simeq \frac{\sigma_u}{\sigma_f}$$

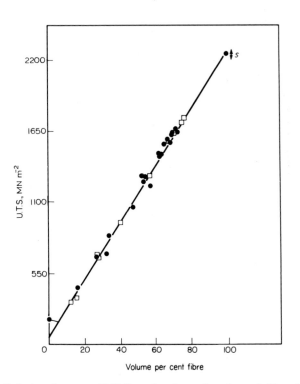

Fig 9.28 Relation between U.T.S. and volume fraction of fibre for copper strengthened with tungsten wires. (After McDanels *et al.*, 1960, *Metal Progr.*, **78**, 118)

Typical values of V_{crit} are given in Table 9.4, from which it is seen that for the stronger matrices, such as nickel and stainless steel, critical volume fractions of fibres vary from 8·4 to 53·4 in the range of attainable metal fibre strengths (0·7–3·5 GN m^{-2}).

So far it has been assumed that the fibres are continuous from one end of a specimen to the other; however, this is not always possible in practice, so the effect of length of the fibres should be considered. In general, if the fibres exceed a critical length l_c the behaviour is described by the above equations; however, the general equation is

$$\sigma_c = \sigma_f V_f \left[1 - (1 - \beta)\frac{l_c}{l} \right] + \sigma_m'(1 - V_f) \qquad \textbf{9.22}$$

where l is the actual fibre length and β is a constant $\simeq 0\cdot 5$.

The strengthening achieved by discontinuous fibres, and described by eqn. **9.22** is always less than that obtained with continuous fibres, but it has been shown that if $l_c/l > 5$, the difference is not greater than 20 per cent.

Table 9.4 V_{crit} values for Fibres in Metallic Matrices (after Kelly and Davies[60])

Matrix	*Matrix properties* MN m^{-2}		V_{crit}			
	σ_m'	σ_μ	$\sigma_f = 0\cdot69$ GN m^{-2}	$\sigma_f = 1\cdot72$ GN m^{-2}	$\sigma_f = 3\cdot45$ GN m^{-2}	$\sigma_f = 6\cdot90$ GN m^{-2}
Aluminium	27·6	82·7	8·34	3·25	1·61	0·80
Copper	41·4	207	25·52	9·83	4·86	2·41
Nickel	62·1	310	39·56	14·95	7·33	3·63
18/8 Stainless steel	172	448	53·41	17·80	8·42	4·10

From eqns. **19.21** and **19.22** the following relation is obtained, which enables the ratio of the two strengths to be determined for increasing values of l_c/l

$$\frac{\sigma_c \text{ (discontinuous)}}{\sigma_c \text{ (continuous)}} = 1 - \frac{1}{\dfrac{2l}{l_c}\left[1 + \dfrac{\sigma_m'}{\sigma_f}\left(\dfrac{1}{V_f} - 1 \right) \right]} \qquad \textbf{9.23}$$

$$= 1 - \frac{l}{2l_c} \quad \text{for } V_f = 1 \qquad \textbf{9.24}$$

In practice the ratio l_c/d, where d is the diameter of the fibre, is important and is defined as

$$\frac{l_c}{d} = \frac{\sigma_f}{2\tau} \qquad \textbf{9.25}$$

where τ is the yield stress in shear of the matrix.

This equation determines the length to diameter ratio above which the fibres will fracture, rather than the matrix deform around them. If σ_f is large, e.g. $6\cdot9$ GN m^{-2} and $\tau = 69$ MN m^{-2} then $l_c/d \simeq 50$ which is readily achieved with most fibres, in fact fibres with l_c/d ratios of 100–500 are not difficult to obtain.

Summarizing, for the greatest strengthening by fibre reinforcement, the fibres should have a high elastic modulus to achieve high strength at low strains, and a high tensile strength. Ideally, the fibres should occupy as large a volume fraction as possible, and they should be perfectly aligned in one direction without discontinuities; however, in practice considerable latitude exists for both these requirements.

General references

1. MCLEAN, D. (1962). *Mechanical Properties of Metals*. Wiley, New York and London.
2. MCLEAN, D. (1957). *Grain Boundaries in Metals*. Oxford University Press, London.
3. JAOUL, B. (1964). *Etude de la Plasticité et Application aux Métaux*. Dunod, Paris.
4. KELLY, A. (1973, 2nd ed.). *Strong Solids*. Oxford University Press, London.
5. TEGART, W. J. MCG. (1966). *Elements of Mechanical Metallurgy*. Mac-Millan, New York and London.
6. *Fibre Composite Materials*. (1965). American Society for Metals.

References

7. BRANDON, D. G., RALPH, B., RANGANATHAN, T. and WALD, M. (1964). *Acta metall.*, **12**, 813.
8. CHALMERS, B. (1948). *Proc. R. Soc.*, **A193**, 89.
9. CLARK, R. and CHALMERS, B. (1954). *Acta metall.*, **2**, 80.
10. GILMAN, J. J. (1953). *Acta metall.*, **1**, 426.
11. BOAS, W. and HARGREAVES, M. E. (1948). *Proc. R. Soc.*, **A193**, 89.
12. URIE, V. M. and WAIN, H. L. (1952). *J. Inst. Metals*, **81**, 153.
13. CARREKER, R. P. and HIBBARD, W. R. (1957). *Trans. AIME*, **9**, 1157.
14. BOAS, W. and OGILVIE, G. J. (1954). *Acta metall.*, **2**, 655.
15. OGILVIE, G. J. (1953). *J. Inst. Metals*, **81**, 491.
16. SACHS, G. (1928). *Z. d. Ver. deut. Ing.*, **72**, 734.
17. KOCHENDORFER, A. (1941). *Plastische Eigenschaften von Kristallen*. Springer Verlag, Berlin.
18. VON MISES, R. (1928). *Z. angew. Math. Mech.*, **8**, 161.
19. TAYLOR, G. I. (1928). *J. Inst. Metals*, **62**, 307.
20. BISHOP, J. F. W. and HILL, R. (1951). *Phil. Mag.*, **42**, 414, 1298.
21. KOCKS, U. F. (1958). *Acta metall.*, **6**, 85.
22. BARRETT, C. S. and LEVENSON, L. H. (1940). *Trans. AIME*, **137**, 112.
23. HANSON, D. and WHEELER, M. A. (1931). *J. Inst. Metals*, **45**, 229.
24. KING, R., CAHN, R. W. and CHALMERS, B. (1948). *Nature, Lond.*, **161**, 682.
25. RHINES, F. W., BOND, W. E. and KISSEL, M. A. (1956). *Trans. ASM*, **48**, 919.
26. WEINBERG, F. (1958). *Trans. AIME*, **212**, 808.
27. JAOUL, B. (1957). *J. Mech. Phys. Solids*, **5**, 95.
28. HUTCHISON, M. M. and HONEYCOMBE, R. W. K. (1967). *Metal Science Journal*, **1**, 70.
29. OWEN, W. S., COHEN, M. and AVERBACH, B. L. (1958). *Trans. ASM*, **50**, 517.
30. HUTCHISON, M. M. (1957). *J. Iron Steel Inst.*, **186**, 431.

31. HALL, E. O. (1951). *Proc. phys. Soc.*, **64B**, 747.
32. PETCH, N. J. (1953). *J. Iron Steel Inst.*, **173**, 25.
33. WESSELL, E. T. (1957). *Trans. AIME*, **209**, 930.
34. CONRAD, H. and SCHOECK, G. (1960). *Acta metall.*, **8**, 791.
35. HESLOP, J. and PETCH, N. J. (1956). *Phil. Mag.*, **1**, 866.
36. COTTRELL, A. H. (1958). *Trans. AIME*, **212**, 192.
37. ARMSTRONG, R. I., CODD, R. M., DOUTHWAITE, R. M. and PETCH, N. J. (1962). *Phil. Mag.*, **7**, 45.
38. LESLIE, W. C. (1961). *Acta metall.*, **9**, 1004.
39. COTTRELL, A. H. *The Relation between the Structure and Mechanical Properties of Metals* (1963). National Physical Laboratory Symposium No. 15 H.M.S.O.
40. FRYE, J. H. and HUME-ROTHERY, W. (1942). *Proc. R. Soc.*, **181A**, 1.
41. FRENCH, R. S. and HIBBARD, W. R. (1950). *Trans. AIME*, **188**, 53.
42. ALLEN, N. P., SCHOFIELD, T. H. and TATE, A. E. L. (1951). *Nature, Lond.*, **168**, 378.
43. DORN, J. E., PIETROKOWSKY, P. and TIETZ, T. E. (1950). *J. Metals AIME*, **188**, 933.
44. HIBBARD, W. R. (1958). *Trans. AIME*, **212**, 1.
45. AINSLIE, N. G., GUARD, R. W. and HIBBARD, W. R. (1959). *Trans. AIME*, **215**, 42, 1959.
46. FLEISCHER, R. L. (1963). *Acta metall.*, **11**, 203.
47. BULLEN, F. P. and HUTCHISON, M. M. (1963). *J. Aust. Inst. Metals*, **8**, 33.
48. RUSSELL, B. and JAFFREY, D. (1965). *Acta metall.*, **13**, 1.
49. DORN, J. E. and MOTE, J. D. (1963). The Plastic Behaviour of Polycrystalline Aggregates, in *Materials Science Research*, Vol. 1. Plenum Press, New York.
50. UNKEL, H. (1937). *J. Inst. Metals*, **61**, 171.
51. DORN, J. E. and STARR, C. D. (1954). *Relation of Properties to Microstructure.* American Society for Metals. p. 71.
52. HONEYCOMBE, R. W. K. and BOAS, W. (1948). *Aust. J. scient. Res.*, **1**, 70.
53. CLAREBROUGH, L. M. (1950). *Aust. J. scient. Res.*, **3**, 72.
54. GENSAMER, M. (1946). *Trans. ASM*, **36**, 30.
55. HARDY, K. H. (1956). *Prog. Metal Phys.*, **6**, 45.
56. BRENNER, S. S. (1963). *The Art and Growing of Crystals.* Edited by J. J. Gilman. John Wiley, New York and London. p. 30.
57. COLEMAN, R. V. (1963). *Metall. Rev.*, **9**, 261.
58. PRICE, P. B. (1961). *Proc. R. Soc.*, **A260**, 251.
59. COLEMAN, R. V. (1958). *J. appl. Phys.*, **29**, 1487.
60. KELLY, A. and DAVIES, G. J. (1965). *Metall. Rev.*, **10**, 1.
61. KELLY, A. and TYSON, W. R. (1964). *Proc. 2nd International Materials Symposium.* John Wiley, New York and London.
62. MCDANELS, D. L., JECH, R. W. and WEETON, J. W. (1960). *Metal Prog.*, **78**, 118.
63. WEBB, W. W. (1960). *J. appl. Phys.*, **31**, 194.
64. FORTES, M. A. and RALPH, B. (1967). *Acta Metall.*, **15**, 707.
65. MORGAN, R. and RALPH, B. (1967). *Acta Metall.*, **15**, 341.

Additional general references

1. ARSENAULT, R. J. (Ed.) (1975). *Treatise on Materials Science and Technology* Vol. 6, Plastic Deformation of Materials, Academic Press, New York.
2. HAASEN, P., GEROLD, V. and KOSTORZ, G. (Eds.) (1980). *Strength of Metals and Alloys*, Proc. Fifth International Conference on the Strength of Metals and Alloys, Aachen. Review: Deformation of Polycrystals, Mecking H. Pp. 1573–1594.
3. HANSEN, N., HORSEWELL, A., LEFFERS, T. and LILHOLT, H. (eds.) (1981). *Deformation of Polycrystals*: Mechanisms and Microstructures, 2nd Risø International Symposium on Metallurgy and Materials Science, Risø National Laboratory, Denmark.
4. HULL, D. (1981). *An Introduction to Composite Materials.* Cambridge University Press.

Chapter 10

Formation of Point Defects in Metals

10.1 Introduction

It has been shown that non-conservative movement of dislocations, in particular of jogs formed by dislocation intersections, can lead to the formation of point defects, either vacant lattice sites (vacancies) or interstitial atoms. In a sense, impurity atoms are point defects as well, but these have already been dealt with. Point defects can be produced not only by deformation, but by irradiation with atomic particles, and quenching from temperatures near the melting point retains an excess of vacancies.

The role of vacancies in diffusion is well known.[1] We shall consider later deformation processes which are diffusion controlled (e.g. creep at high temperatures), in which vacancy movement is fundamental to the process. In the recovery of metals from the effects of cold work, both interstitial atoms and vacancies are mobile at low temperatures while dislocations are not. Dislocation rearrangements occur at higher temperatures when they utilize vacancies to undergo the process of climb. Interstitial atoms are more mobile than vacancies, and tend thus to disappear more readily to sinks such as vacancies or grain boundaries, but they require more energy for their formation.

10.2 The Retention of Vacancies by Quenching

The equilibrium concentration C of point defects varies exponentially with temperature.

$$\frac{n_v}{n_a} = C = e^{-Q_F/RT} \qquad\qquad \textbf{10.1}$$

where Q_F is the energy required to form 1 mole of defects; n_v = no. of defects; n_a = no. of atoms.

Typical values for Q_F, the energy in eV/atom (to form one defect), and for the defect concentration C in copper are shown in Table 10.1.

Table 10.1   Point defects in copper

Defect	Q_F (eV)	Defect concentration C		
		300 K	800 K	1300 K
Vacancy	1	10^{-15}	10^{-6}	10^{-3}
Interstitial	4	10^{-67}	10^{-25}	10^{-15}

These data indicate that vacancies are more readily formed and can exist at high temperatures in reasonable concentrations, for example approximately one atomic position in ten in a given crystallographic direction will be occupied by a vacancy at 1300 K.

Vacancies move by atomic jumps, and the mobility like the concentration, is an exponential function of temperature, so that

$$\nu_v = A\,e^{-Q_M/RT} \qquad \textbf{10.2}$$

where ν_v = frequency of jumps of atoms into a vacancy; A = constant; Q_M = activation energy for movement of vacancies.

Now the rate of self-diffusion of a metal is the rate at which atoms make jumps into vacant lattice sites, and we are thus concerned with the number of jumps an atom makes with an equilibrium concentration of vacancies per second (ν_a).

$$\nu_a = \frac{n_v}{n_a} A\,e^{-Q_M/RT} \qquad \textbf{10.3}$$

where n_v = no. of vacancies; n_a = no. of atoms.

Substituting eqn. **10.1** in eqn. **10.3**:

$$\nu_a = A\,e^{-Q_M/RT} . e^{-Q_F/RT}$$

$$= A\,e^{-(Q_M+Q_F)/RT} \qquad \textbf{10.4}$$

So the rate at which diffusion occurs depends on two activation energies Q_F and Q_M, and the activation energy for diffusion Q_D is thus defined.

$$Q_D = Q_F + Q_M$$

By rapid quenching from near the melting point of a metal, it is possible to preserve a supersaturation of vacancies at room temperature. The subsequent movement and elimination of vacancies can then be followed by measuring a physical property such as electrical resistivity, or the stored energy in a metal determined by calorimetry at a series of temperatures, and a value for Q_M can be obtained. Likewise, if the quenching temperature is varied to give different vacancy concentrations, by measuring changes in physical properties, Q_F can be determined. A large number of experiments have been carried out with face-centred cubic metals such as copper, gold, silver and nickel. Typical experimentally determined results for Q_F, Q_M and Q_D, the activation energies expressed in eV/atom, the most usual form, are shown in Table 10.2.

Table 10.2 Activation energies for formation and movement of vacancies, and for self-diffusion

Metal	Q_F (eV)	Q_M (eV)	Q_D (eV)
Gold	0·95–1·05	0·60–0·70	1·81
Copper	1·0	0·70	1·8
Platinum	1·2–1·4	1·1	2·96
Aluminium	0·8–1·0	0·44	0·9–1·4

The high mobility of a supersaturation of vacancies is shown by the fact that the physical properties such as resistivity already show changes after resting at temperatures well below 273 K, but the removal of the excess defects is not complete for most common face-centred cubic metals until the range 473–673 K is reached.

10.3 Effect of quenching-in of vacancies on mechanical properties

Of particular relevance is the effect of a supersaturation of vacancies on the mechanical properties of metals, because this involves the interaction of vacancies with dislocations. The most basic experiments are those in which single crystals have been quenched from near the melting point and their stress/strain curves determined at room temperature. Maddin and Cottrell[6] were able to show that the critical shear stress for quenched aluminium crystals was raised by over 500 per cent above that of slowly cooled crystals (Fig. 10.1). The effect was confirmed by Kimura and Maddin[5] in copper single crystals; however, these experiments showed that the hardening was not a direct result of the quench, but occurred progressively on ageing at room temperature with an activation energy of 0·8 eV. The quench hardening is accompanied by a coarsening of the slip band structure, and an increased tendency for the primary slip system to overshoot the symmetrical position for duplex slip.

These results can be explained by assuming that the vacancies retained by quenching migrate to the dislocations and, either as isolated vacancies or groups of vacancies, pin the dislocations in a similar way to the pinning achieved by solute atoms or precipitate particles. The vacancies either singly or in groups can exist elastically attracted to the dislocations, or alternatively they can be annihilated at the dislocations with the result that jogs are created. It has been shown in Chapter 3 that jogs act as brakes on dislocation movement, so that this theory can be used to explain the increased flow stress observed. The occurrence of coarse slip suggests that the first glide dislocations to move form paths which are free of point defects, so that subsequent dislocations will tend to move along these planes of least resistance, with the result that coarse glide bands develop. The stress–strain curves of the quenched and aged crystals are, however, permanently raised, so some obstacles must remain in the crystals to resist the movement of subsequent dislocations.

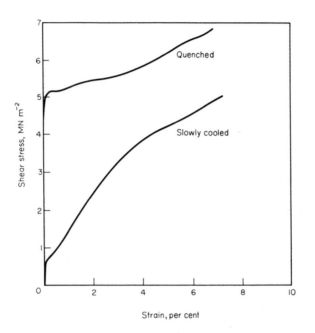

Fig. 10.1 Effect of quenching on the critical shear stress of aluminium crystals. (After Maddin and Cottrell, 1955, *Phil. Mag.*, **46**, 735)

10.4 Electron microscopy of quenched metals

There is now a considerable amount of structural evidence concerning point defects which has been obtained primarily from thin-foil electron microscopy. When a quenched foil of pure aluminium is examined, it is found to contain a high concentration of small dislocation loops.[7] A typical density is about $10^{15}/cm^3$ with diameters between 100 and 500 Å (Fig. 10.2). These loops form initially by the collection of vacancies into discs which then collapse to create sessile dislocation rings (Frank sessile)[8] of edge type, but with a Burgers vector normal to the close-packed planes. This mechanism results in a stacking fault bounded by the dislocation

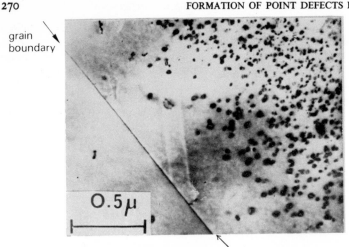

grain
boundary

Fig. 10.2 Dislocation loops in 99·995 per cent aluminium quenched from 650°C. Loops absent near the grain boundary. Thin-foil electron micrograph. (After Cotterill and Segall, 1963, *Phil. Mag.*, **8**, 1105)

loop, which should be revealed in the electron microscope by the usual light and dark striations. Such faulted loops have been observed in quenched zone-refined aluminium[9] (Fig. 10.3); they are much larger

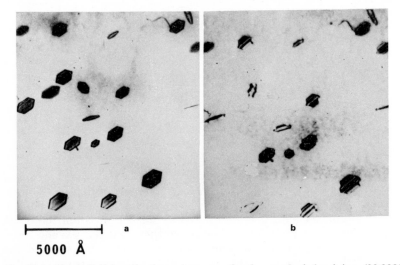

5000 Å

Fig. 10.3 Faulted dislocation loops in zone-refined, quenched aluminium (99·9999 per cent). **a** and **b** show successive changes due to beam heating during observation. Thin-foil electron micrograph. (After Cotterill and Segall)

(2000–3000 Å) than loops observed in less pure material. However, during observation many of the faulted loops transform to prismatic loops which do not possess fringe contrast (Fig. 10.3**b**).

The Burgers vector of a Frank partial dislocation is $\frac{1}{3}a[111]$, and is represented in the Thompson tetrahedron (Fig. 10.5) as αA, βB, etc. The removal of the stacking fault layer to produce the prismatic loops characteristic of quenched aluminium involves a reaction of the Frank partial dislocation with a Shockley partial (αB, αC, etc.) so that

$$\tfrac{1}{3}a[111] + \tfrac{1}{6}a[11\bar{2}] \rightarrow \tfrac{1}{2}a[110]$$

which in the Thompson tetrahedron notation is

$$A\alpha + \alpha B \rightarrow AB$$

The relationship of the dislocation loops to vacancies is supported by the observations that the loops are absent in the vicinity of grain boundaries (Fig. 10.2) and of dislocations. The grain boundaries are effective vacancy sinks which will reduce the supersaturation in a zone along each side up to 1 micron wide, while dislocations can also absorb a local excess of vacancies and become jogged. Field ion micrographic studies[42] have recently provided evidence of the relation between vacancies and dislocation loops.

It might be expected that Frank sessile dislocations would be stable in metals of lower stacking fault energy, where the removal of the stacking fault would not be energetically favoured. However, on quenching gold from 1183–1273 K, followed by ageing between 373 and 523 K, the defects are in the form of tetrahedra, not discs, and the faces of the tetrahedra are faulted (Fig. 10.4).[10] Depending on the orientation of the surface of the observed foil, the tetrahedra exhibit triangular, square or more complex profiles in thin-foil electron micrographs. Typical sizes are between about 200 and 750Å with a density of about 10^{14}/cm³. Assuming that these

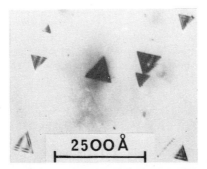

Fig. 10.4 Stacking fault tetrahedra in gold quenched from 1273 K. <112> orientation. Thin-foil electron micrograph. (After Cotterill)

defects also arise from collapse of vacancy aggregates, a vacancy concentration of 10^{-5} is sufficient to account for them. The origin of the tetrahedron is again a disc of vacancies on the close-packed plane which collapses to form a sessile loop, which reflects the symmetry of the close-packed plane, and so is triangular in shape bounded by the three close-packed directions. In the Thompson tetrahedron typical sides of the triangles are CD, DB and BC (Fig. 10.5a). The Burgers vector of this triangular Frank dislocation is αA (normal to the plane BCD). This dislocation now dissociates along each side of the triangle in the following way (Fig. 10.5b)

$$\tfrac{1}{3}a[111] \to \tfrac{1}{6}a[101] + \tfrac{1}{6}a[121]$$

$$\alpha A \to \alpha \beta + \beta A$$

The other two sides of the triangle dissociate in a similar fashion

$$\alpha A \to \alpha \gamma + \gamma A$$

$$\alpha A \to \alpha \delta + \delta A$$

Using Frank's rule, it can be easily shown that each of these reactions is energetically favourable. The $\tfrac{1}{6}a\langle 101 \rangle$ dislocations lie along the sides of the triangle, and are sessile dislocations of the stair-rod type (see Chapter 3), while the $\tfrac{1}{6}a\langle 121 \rangle$ dislocations are normal Shockley partials which can glide on the β, γ and δ planes respectively. In doing so they are separated from the $\tfrac{1}{6}a\langle 101 \rangle$ stair-rods by an area of stacking fault (Fig. 10.5) which gradually grows, until the Shockley partial dislocations meet. They meet along the other edges of the tetrahedron DA, BA and CA where they form three further stair-rod dislocations, which completes a stable tetrahedral array of dislocations separated by stacking fault.

The final combinations to produce the three stair-rod dislocations along DA, BA and CA are

$$\beta A + A\gamma \to \beta\gamma$$

$$\gamma A + A\delta \to \gamma\delta$$

$$\delta A + A\beta \to \delta\beta$$

which yields a stacking fault tetrahedron (Fig. 10.5c), all the edges of which are stair-rod dislocations with Burgers vectors of the type $\beta\gamma$, etc., or put in the alternative nomenclature

$$\tfrac{1}{6}a[121] + \tfrac{1}{6}a[\bar{1}\bar{1}\bar{2}] \to \tfrac{1}{6}a[01\bar{1}]$$

The overall result is that each $\tfrac{1}{3}a\langle 111 \rangle$ dislocation gives two $\tfrac{1}{6}a\langle 101 \rangle$ dislocations, which is an energetically favourable reaction.

Stacking fault tetrahedra have also been observed in quenched silver[11] but together with dislocation loops, while in copper[12] dislocation loops are the main structural feature. This suggests that the stacking fault energy increases in the order gold, silver, copper, but the behaviour is probably very sensitive in the presence of impurities. Loops have also been found in nickel,[13] which has a relatively high stacking fault energy.

The electron microscope has also been used to study the elimination of the defects on heating. For example, quenched loops gradually diminish in size and finally disappear in aluminium at around 473 K,[14] where the activation energy of 1·3 eV for the process suggests that the mechanism is dislocation climb, for the loops can shrink by emission of vacancies. The rate of disappearance is parabolic, becoming more rapid the further the loop shrinks. Measurements of the change of the electrical resistivity

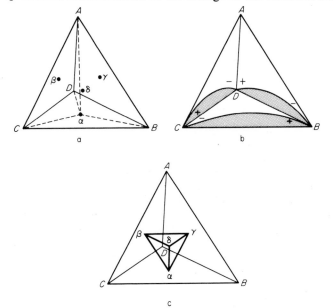

Fig. 10.5   Geometry of stacking fault tetrahedra. (After Silcox and Hirsch, 1959, *Phil. Mag.*, **4**, 72). **a.** Thompson tetrahedron $ABCD$; α, β, γ and δ are the mid-points of the faces; **b.** bowing out of partial dislocations βA, γA and δA in the slip planes ACD, ADB and ABC; **c.** relation between the defect tetrahedron $ABCD$ and the Burgers vector tetrahedron $\alpha\beta\gamma\delta$.

increment resulting from the quenching in of vacancies have shown that in aluminium the resistivity recovers in the same temperature range as that in which the loops disappear.[15] Actually the resistivity curves show two stages, the first occurring in a few minutes at room temperature, when it is assumed that vacancies are clustering and forming dislocation loops. The second stage occurs between 423 K and 473 K when the loops disappear. Tetrahedral defects are more difficult to remove by annealing. In gold they disappear in the temperature range 873–923 K, and this corresponds with a marked resistivity drop.[16] This contrasts with the behaviour of dislocation loops which would be expected to anneal at 623–673 K with an activation energy (for self-diffusion) of about 1·8 eV. The activation energy for removal of the tetrahedra appears to be closer to 5 eV.

In some alloys where the supersaturated state has been obtained by quenching, e.g. Al–4 per cent Cu,[17] Al–20 per cent Ag,[11] Al–Mg, many of the dislocations take a helical form (Fig. 10.6a). These structures arise from the interaction of quenched in vacancies with screw or mixed dislocations. Thus a vacancy or a group of vacancies is absorbed on to the dislocation; the dislocation line becomes curved (Fig. 10.6b) to form one turn of a

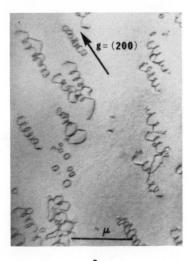

a

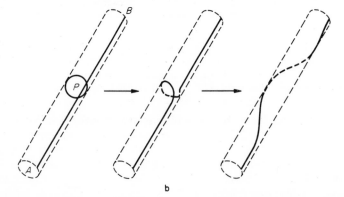

b

Fig. 10.6 **a**, Dislocation helices in an aluminium–magnesium alloy (after Thomas). Thin-foil electron micrograph; **b**, Adsorption of a vacancy *P* on a screw dislocation *AB* to form a helix. (After Cottrell, 1957, Symposium on Vacancies and other Point Defects in Metals and Alloys, *Institute of Metals Monograph No. 23*)

spiral. If other groups of vacancies are attracted at different points along the line, it will be transformed into a spiral or helical dislocation. Such dislocations have now been found in a number of alloys and other crystalline solids.[18] When the alloys are aged, the helical dislocations become the sites for nucleation of precipitates. In theory there is no reason why similar helices are not formed in pure metals, but it seems likely that they would tend to straighten out or break down into smaller loops. The presence of solute atoms with a tendency to segregate to dislocations slows down this process.

10.5   Formation of point defects during deformation

In Chapter 3 it was shown that a jog formed by intersection of two dislocations has a Burgers vector in a different direction to the rest of the dislocation, with the result that the jog will have to move non-conservatively to keep up. The jog tends to drag and, in moving, generates either a string of interstitials or vacancies. However, there is the possibility that the jog can move conservatively along the dislocation without forming point defects, if the dislocation is undissociated. On the other hand, if the stacking fault energy is sufficiently low to allow dissociation to partial dislocations, the jog connecting the two lengths of partial dislocation will be itself dissociated (Fig. 10.7a), and conservative movement along the dislocation line will be difficult. It is possible that the jog will contract (Fig. 10.7b) and then be able to move more readily along the dislocation. However, it is difficult to avoid the conclusion that at least in the case of metals of low stacking fault energy, point defects will be produced by the non-conservative movement of jogs.

An alternative mechanism is the interaction of a positive and negative edge dislocation on adjacent slip planes. As the two dislocations approach each other, the two half planes match up, but with a gap of one interplanar spacing which amounts to a row of vacancies. Such a mechanism is easily demonstrated in the bubble model. Alternatively a similar situation can exist when an edge dislocation bows around a screw dislocation.[19] On reforming the front beyond the screw dislocation, two parts of the moving dislocation A and B approach each other (Fig. 10.8), but because of the screw displacement are one interplanar spacing apart. Again a line of point defects will be created.

There has been some doubt as to the relative concentrations of interstitials and vacancies generated by plastic deformation, partly because interstitials are so readily eliminated even at low temperatures. Annealing experiments where the changes in electrical resistivity of a deformed metal are studied as it is raised from a very low temperature, on the whole indicate that there is a preponderance of defects with the activation energy of movement of vacancies rather than interstitials. For example, copper[20] deformed in liquid helium shows marked recovery at

Close—packed (III) planes

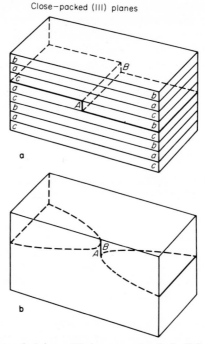

Fig. 10.7 **a,** An extended jog *AB* in an extended dislocation (f.c.c.); **b,** constriction of the jog *AB*. (After Balluffi *et al.*, 1963, *Recovery and Recrystallization*, AIME, Interscience, New York and London)

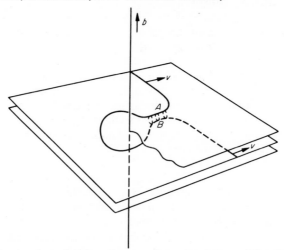

Fig. 10.8 Interaction of dislocations to give point defects. (After Balluffi *et al.*, 1963, *Recovery and Recrystallization* AIME, Interscience, New York and London)

300 K when vacancies are mobile, but none at 77 K when interstitials should move. More recently experiments on gold deformed at 77 K[21] have provided evidence for the formation of interstitials with a migration energy of about 0·15 eV but this represented only about 10–20 per cent of the total number of defects. The migration took place at 77 K and 90 K, and influenced the electrical resistivity. At room temperature, recovery of the resistance was much greater, and the migration energy was about 1 eV, which is appropriate for vacancy movement in gold.

10.6 Structural evidence for point defects in deformed metals

Some evidence for trails of interstitials and vacancies behind dislocation jogs has been found in copper-decorated silicon crystals[22] while similar effects have been observed by etching in sodium chloride[23] and lithium fluoride.[24] Perhaps the most striking results have been obtained by Price[25] from thin-foil electron microscopy studies of zinc and cadmium

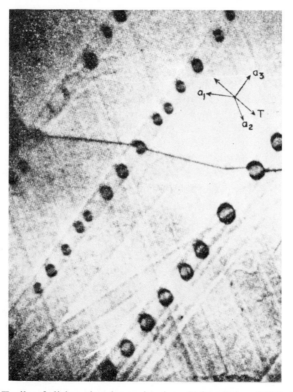

Fig. 10.9   Trails of dislocation loops in zinc. (After Price, 1960, *Phil. Mag.*, **5**, 873.) Thin foil electron micrograph × 27,000.

platelets. In these crystals the movement of edge dislocations is uniform over long distances, but screw dislocations form multiple jogs by cross-slip, which then drag behind the screw dislocations, with the result that eventually whole strings of dislocation loops are produced (Fig. 10.9). The evidence suggests that while the loops form by a dislocation inter-action, they disappear by the diffusion of point defects. MgO exhibits similar loops[26] but they have been identified as originating from vacancy discs. The evidence is growing that this type of dislocation–point defect interaction occurs fairly frequently.

There is also limited experimental evidence for the formation of tetra-hedral stacking fault defects in deformed metals and alloys of low stacking fault energy. Tetrahedra have been found in deformed silver[27] and in silver–gallium alloys[41].

10.7 Formation of point defects by irradiation

Fast-moving atomic particles provide a direct way of forming point defects in a crystalline solid. The simplest and least drastic method is to bombard the solid with high energy (~ 1 MeV) electrons and displace atoms from their equilibrium lattice positions into nearby interstitial sites, thus creating vacancy interstitial pairs. To form such a pair of defects in copper requires an energy of about 25 eV, and this necessitates an 0·5 MeV electron to provide sufficient energy by elastic collision. In copper bom-barded with 1·35 MeV electrons at 10 K, the subsequent electrical resisti-vity changes have been measured with increasing temperature. Between 37 and 60 K most of the radiation-induced increment in resistivity is lost by a process with a low activation energy (0·1 eV). This is interpreted as movement of interstitial atoms to recombine with the vacancies, leaving only a few vacancies to be subsequently moved at higher temperatures. This type of work suggests that several categories of defect can be distinguished.

1. Interstitial–vacancy pairs which are very unstable and tend to anneal out at very low temperatures.
2. Interstitials sufficiently far away from vacancies not to recombine rapidly. These defects tend to migrate to sinks which are mainly vacancies.
3. Isolated vacancies. Most of these defects become sinks for inter-stitials, while the others anneal out in the same way as in quenched metals, i.e. either by clustering or migrating to dislocations or grain boundaries.

Irradiation of crystalline solids in nuclear reactors has been widely in-vestigated in the last twenty years. The effective particles are fast neutrons with energies greater than 1 MeV. Collisions are infrequent in many metals, but each collision imparts about 5×10^4 eV to a struck atom. This atom is rapidly displaced, and in turn displaces many other atoms until many hundreds of atoms are involved. There is at the same time, a very short-lived 'thermal spike' or sudden release of thermal energy which is only of 10^{-11} sec duration, which accounts for most of the collision energy, only a small amount remaining stored in the damaged region. The exact structure of the damaged region is difficult to specify in detail, but cascades of vacancies are formed which tend to end in interstitial atoms (Fig. 10.10). For a metal such as copper the region affected by one atomic collision is about 100 Å diameter and after the event contains 100–500 interstitials and vacancies. However, neutron damage cannot be explained simply in terms of point defects alone.

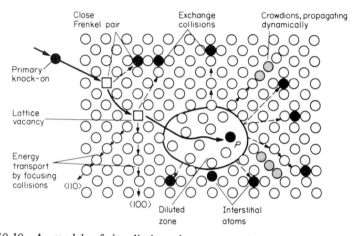

Fig. 10.10 A model of irradiation damage produced by one neutron collision. (After Seeger, 1962, *Radiation Damage in Solids.* International Atomic Energy Agency, Vienna)

In the case of the spike there are extended defects which are usually referred to as depleted or diluted zones which arise from long distance mass transport. This process occurs by a focusing effect where the energy is transmitted through the crystal lattice along linear chains of atoms and leads to the formation of *crowdions*, a defect involving the insertion of an atom in an atomic row to alter locally the inter-atomic spacing. Such reactions allow the formation of the depleted zones which some workers consider are the principle cause of radiation hardening.[29]

10.8   Structural studies on irradiated metals

Electron microscopy investigations have been carried out on annealed copper subjected to increasing neutron doses. At low levels a few dislocation loops with diameters up to 150 Å are seen, together with a high density of small dark spots.[30] As the neutron dose is increased, many large dislocation loops about 300 Å become visible. In the dose range 6×10^{17} neutrons/cm² to 10^{20} neutrons/cm², the density of defects is in the range 10^{15}–10^{16}/cm³. Similar results have been obtained on neutron irradiated gold, iron[43] and platinum,[31] and on copper irradiated with α-particles. When specimens are examined in a hot stage in the electron microscope, the defects disappear in the temperature range 573–673 K which is the range in which recovery from irradiation damage is known to take place.

A closer examination of the small dark spots about 25 Å diam. or less observed in a number of irradiated metals indicates that they are probably depleted zones rather than small dislocation loops.[32] They form readily in copper below 223 K as well as at normal radiation temperatures, and the evidence suggests that they form in situ rather than by diffusion. Furthermore, their concentration increases proportionately with the dose, unlike the larger dislocation loops which have also been observed. On annealing at fairly low temperatures (~ 623 K for copper and silver), more of the small defects appear presumably by growth of previously unresolved defects. Subsequently, at higher temperatures some of the spots grow at the expense of others.

The exact nature of the dislocation loops which are the more obvious feature of the microstructure is still under discussion. Evidence has been found for the existence of both interstitial and vacancy-type dislocation loops, but some experiments indicate that the majority are of the vacancy-type.[31]

Recent experiments on molybdenum[38,39,40] show that small clusters and loops are formed on neutron irradiation at 333 K (Fig. 10.11a). However, on annealing above 1073 K large loops develop (Fig. 10.11b) which are all of interstitial type, the dislocations having the Burgers vector $\frac{1}{2}a\langle 111 \rangle$. The defect structures become more coarsely distributed as the purity of the metal increases, and are thus visible at lower neutron doses.

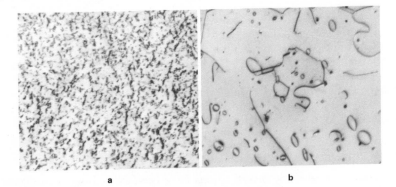

a   b

Fig. 10.11 **a**, Molybdenum irradiated to a fast neutron dose of 5×10^{19} cm⁻² at 333 K; **b**, as **a**, after annealing at 1203 K. (After Eyre). Thin foil electron micrographs, $\times 22,000$.

10.9   Effect of Irradiation on Mechanical properties

10.9.1.   *Single crystals*

Neutron irradiation has a marked effect on the mechanical properties of crystals. For example, the critical shear stress of copper crystals is raised from $0.5 \, MN \, m^{-2}$ up to $15 \, MN \, m^{-2}$ by a dose of 4.4×10^{18} neutrons/cm^2.[33] Moreover, a yield point is developed and the whole level of the stress–strain curve is substantially raised (Fig. 10.12). The slip lines undergo the same change as in the quenched crystals, becoming coarse and widely spaced. Blewitt and co-workers[33] found that the critical shear stress of

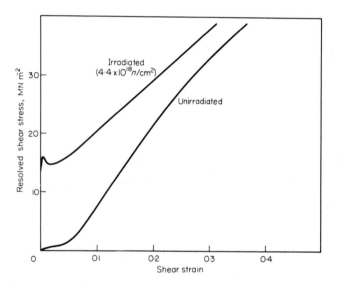

Fig. 10.12   The effect of neutron irradiation (4.4×10^{18} neutron/cm^2) on the stress–strain curve of a copper crystal. (After Makin, 1956, Progress in Nuclear Energy, *Metallurgy and Fuels*, Pergamon Press, Oxford)

copper crystals increased with the cube root of the neutron dose (Fig. 10.13). Most crystals showed a yield point and the yield elongation zone increased with the dose as well as with decreasing temperature of test. They found that with a dose of 10^{20} neutrons/cm^2, τ_0 was raised to $78.5 \, MN \, m^{-2}$. Diehl[34] examined the plastic behaviour of copper and nickel crystals up to doses of 1×10^{18} neutrons/cm^2, and found that τ_0 increased as the square root of the dose up to 2×10^{17} neutrons/cm^2, but at higher doses, saturation effects became apparent. Up to doses of 1×10^{18} neutrons/cm^2, the shear stress–strain curves show the usual three stages of hardening, although the yield elongation zone tends to obscure the true Stage 1. At high shear strains in Stage 3 the curves (including the unirradiated specimens) tend to converge. However, after heavier doses the curves of heavily irradiated material remain substantially above the level of the curves of the lightly irradiated material.

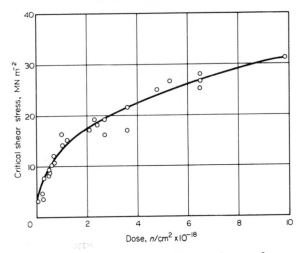

Fig. 10.13 Relation between neutron dose and τ_0 of copper crystals measured at room temperature. (After Blewitt *et al.*, 1960, *J. nucl. Mat.*, **2**, 277)

10.9.2. *Polycrystalline aggregates*

Turning to polycrystalline aggregates, there is much experimental evidence to show that substantial increases in strength occur as a result of irradiation with neutrons, and that the magnitude of the effects is a sensitive function of neutron dose, and the temperature at which the irradiation is carried out. The effect on the initial flow stress is the most marked (Fig. 10.14) and frequently leads to the development of a sharp yield point. For example, aluminium, copper and other face-centred cubic metals develop a sharp yield point after irradiation, but in contrast interstitial impurity induced yield points in steel and molybdenum are frequently eliminated. The ultimate tensile strength is also raised, but not to the same degree, consequently the yield stress/U.T.S. ratio is frequently adversely affected. The rise in yield stress and U.T.S. tends to follow a parabolic law with respect to the irradiation dose, saturation occurring in the region of 10^{20}–10^{21} neutrons/cm^2, but it should be emphasized that the exact values of strength attained in a particular metal or alloy will depend very much on the irradiation procedure, in particular the temperature reached by the sample. In general, the lower the irradiation temperature, the larger the strengthening effect observed.

In the body-centred cubic metals such as iron, molybdenum, etc., which

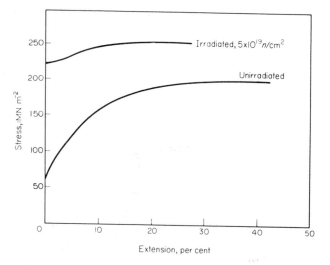

Fig. 10.14 Stress–strain curves of irradiated and unirradiated polycrystalline copper. (After Makin)

are prone to brittle fracture below a critical temperature (Chapter 15) irradiation with neutrons raises the ductile-brittle temperature.[35,36] This is a matter of considerable practical concern, because steel plays an important structural role in reactor design. The extent of the effect for a 0·2 per cent C steel is shown in Fig. 10.15, where a neutron dose of $10^{20}/cm^2$ raises the ductile brittle transition temperature from 243 K to

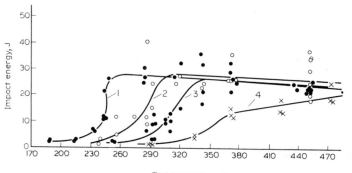

Fig. 10.15 The effect of neutron irradiation between 433 and 473 K on the ductile-brittle transition of a 0·2 C steel. 1, unirradiated; 2, 2·7 × 10^{18} neutron/cm²; 3, 1 × 10^{19} neutron/cm²; 4, 1 × 10^{20}; neutron/cm². (After Pravdyuk et al., 1962, Properties of Reactor Materials and the Effects of Radiation Damage, ed. D. J. Littler. Butterworths, London)

about 353 K.* Similar effects have been observed in molybdenum, doses of 10^{20} neutrons/cm^2 making the metal quite brittle at room temperature.[36]

The irradiation of uranium results in more complex phenomena which in part arise from the anisotropy of physical properties in uranium crystals (Chapter 12), and partly from the products of fission processes. Under irradiation, alpha uranium crystals undergo severe distortion, and in polycrystalline aggregates possessing preferred directions of crystal growth, marked dimensional changes can occur which are related to neutron dose.

$$l = l_0 \exp (Gf) \qquad\qquad\qquad \textbf{10.5}$$

where l and l_0 are the instantaneous and original lengths of a rod; $f =$ fraction of atoms which have undergone fission ('Burn-up'); $G =$ growth coefficient.

Changes in length of uranium rods occur also as a result of temperature cycling (Chapter 12) which is inevitable in a reactor so the combination of these phenomena can lead to serious design problems. Growth due to neutron irradiation has not yet been satisfactorily explained but the theories fall into two groups. One group attempts to explain the dimensional changes in terms of anisotropic movement of point defects produced by the irradiation, while other theories use the anisotropic nature of deformation by slip or twinning to account for the observed changes in shape.

The products of fission comprise a wide range of elements of intermediate atomic number which include the inert gases zenon and krypton. The uranium becomes supersaturated with respect to these gases, which then form very fine gas bubbles.[37] If the irradiation is carried out at a high temperature (above 873 K) the gas bubbles become coarse, and marked volume changes take place which are referred to as *swelling*. Neutrons interact with other elements forming inert gases, for example boron-containing steels are damaged by the evolution of helium, also a matter of practical concern, because boron has a high neutron capture cross-section and is added to steel to produce suitable material for reactor control rods.

General references

1. *Vacancies and Other Point Defects in Metals*, 1958. Institute of Metals, Monograph No. 23, London.
2. LITTLER, D. J. (Ed.) (1962). *Properties of Reactor Materials and the Effects of Radiation Damage*. Butterworth, London.
3. DAMASK, A. C. and DIENES, G. J. (1963). *Point Defects in Metals*, Gordon & Breach, New York, London.
4. *International Conference on Crystal Lattice Defects, Tokyo* (1963). Published by the Physical Society of Japan.
5. COTTERILL, R. M. J., DOYAMA, M., JACKSON, J. J. and MASHII, M. (Eds.) (1965). *Lattice Defects in Quenched Metals*. Academic Press, New York and London.

* Refer to Chapter 15.

References

6. MADDIN, R. and COTTRELL, A. H. (1955). *Phil. Mag.*, **46**, 735.
7. HIRSCH, P. B., SILCOX, J., SMALLMAN, R. E. and WESTMACOTT, H. K. (1958). *Phil. Mag.*, **3**, 897.
8. FRANK, F. C. (1950). *Plastic Deformation of Crystalline Solids.* Carnegie Institute of Technology, O.N.R. p. 150.
9. COTTERILL, R. M. J. and SEGALL, R. L. (1963). *Phil. Mag.*, **8**, 1105.
10. SILCOX, J. and HIRSCH, P. B. (1959). *Phil. Mag.*, **4**, 72.
11. SMALLMAN, R. E., WESTMACOTT, K. R. and COILEY, J. K. (1959). *J. Inst. Metals*, **88**, 127.
12. HIRSCH, P. B. and SILCOX, J. (1958). *Growth and Perfection of Crystals.* John Wiley, New York and London. p. 262.
13. KUHLMANN-WILSDORF, D., MADDIN, R. and KIMURA, H. (1958). *Z. Metallk.*, **49**, 584.
14. SILCOX, J. and WHELAN, M. J. (1960). *Phil. Mag.*, **5**, 1.
15. PANSERI, E. and FEDERIGHI, T. (1958). *Phil. Mag.*, **3**, 1223.
16. COTTERILL, R. M. J. (1961). *Phil. Mag.*, **6**, 1351.
17. THOMAS, G. and WHELAN, M. J. (1959). *Phil. Mag.*, **4**, 511.
18. AMELINCKX, S. (1957). *The Direct Observation of Dislocations.* John Wiley, New York and London.
19. MOTT, N. F. (1957). *Dislocations and Mechanical Properties of Crystals.* John Wiley, New York and London. p. 458.
20. BLEWITT, T. K., COLTMANN, R. R. and REDMAN, J. K. (1955). Conference on Defects in Crystalline Solids, Physical Society. London. p. 369.
21. TAKAMURA, J., FURAKAWA, K., MIURA, T. and SHINGU, P. H. (1963). Conference on Crystal Defects. *J. phys. Soc., Japan*, **18**, Supplement III, 7.
22. DASH, W. C. (1958). *J. appl. Phys.*, **29**, 705.
23. DAVIDGE, R. W. and WHITWORTH, R. W. (1961). *Phil. Mag.*, **6**, 217.
24. GILMAN, J. J. and JOHNSTON, W. G. (1960). *J. appl. Phys.*, **31**, 687.
25. PRICE, P. B. (1963). *Electron Microscopy and the Strength of Crystals*, ed. G. Thomas and J. Washburn. Interscience, New York and London.
26. GROVES, G. W. and KELLY, A. (1962). *J. appl. Phys.*, **33**, 456.
27. LORETTO, M. H., CLAREBROUGH, L. M. and SEGALL, R. L. (1965). *Phil. Mag.*, **11**, 459.
28. DENNEY, J. (1956). *Bull. Amer. phys. Soc.*, **1** (11), 335.
29. SEEGER, A. (Vienna, 1962). *Radiation Damage in Solids.* International Atomic Energy Agency. p. 105.
30. SILCOX, J. and HIRSCH, P. B. (1959). *Phil. Mag.*, **4**, 1356.
31. RUEDL, E. and AMELINCKX, S. (1963). Proc. International Conf. on Crystal Lattice Defects, 1962. *J. phys. Soc. Japan*, **18**, 195.
32. MAKIN, M. J., WHAPHAM, A. D. and MINTER, F. J. (1961). *Phil. Mag.*, **6**, 465; *ibid.*, **7**, 285 (1962).
33. BLEWITT, T. H., COLTMANN, R. R., JAMISON, R. E. and REDMAN, J. K. (1960). *J. nucl. Mater.*, **2**, 277.
34. DIEHL, J. (Vienna, 1962). *Radiation Damage in Solids*, **1**, 129. International Atomic Energy Agency.
35. WILSON, J. C. and BILLINGTON, D. S. (1956). *J. Metals*, **8**, 665.

36. BRUCH, C. A., MCHUGH, W. E. and HOCKENBURG, R. W. (1955). *J. Metals*, **7**, 281 (1956). *Ibid.*, **8**, 1362.
37. GREENWOOD, G. W., FOREMAN, A. J. E. and RIMMER, D. E. (1959). *J. nucl. Mat.*, **1**, 305.
38. DOWNEY, M. E. and EYRE, B. L. (1965). *Phil. Mag.*, **11**, 53.
39. MEAKIN, J. D. and GREENFIELD, J. G. (1965). *Phil. Mag.*, **11**, 277.
40. EYRE, B. L. and DOWNEY, M. E. (1967). *Metal Science Journal*, **1**, 5.
41. HUTCHISON, M. M. and HONEYCOMBE, R. W. K. (1967). *Metal Science Journal*, **1**, 186.
42. FORTES, M. A. and RALPH, B. (1966). *Phil. Mag.*, **14**, 189.
43. BRYNER, J. S. (1966). Acta metall., **14**, 323.

Additional general references

1. HIRSCH, P. B. (Ed.) (1975). *The Physics of Metals* Vol. 2, Defects, Cambridge University Press.
2. SMALLMAN, R. E. and HARRIS, J. E. (Eds.) (1977). *Vacancies '76*, Proc. of Conference on 'Point Defects Behaviour and Diffusional Processes', The Metals Society, London.
3. NABARRO, F. R. N. (Ed.) (1979). *Dislocations in Solids* Vol. 4, Chapter by Balluffi, R. W. and Granato, A. V. on Dislocations, Vacancies and Interstitials, North-Holland Amsterdam.

Chapter 11

Annealing of Deformed Metals

11.1 Introduction

Cold worked metals have dislocation populations between 10^9 and $10^{12}/cm^2$; point defects are also present, and the physical and mechanical properties are very different from those of annealed metals. The various processes by which the annealed state is reached from the deformed state are included under the general terms recovery and recrystallization. The term recovery has been used to cover processes which do not result in the replacement of the deformed grains by new grains, but nevertheless lead to structural changes on a fine scale within the existing grains. On the other hand, recrystallization is easier to define because it normally results in the absorption of the old grains by new equi-axed strain-free grains. Recovery and recrystallization are not, however, single processes, but include a number of different phenomena, some of which overlap the two broad processes. The driving force of both these phenomena is the reduction in *strain energy* achieved by the removal of excess point defects and dislocations. This is in contrast to normal grain growth which occurs in annealed metals, and where the driving force is the reduction in interfacial, or more specifically, grain boundary energy.

The subject can be approached by examining changes in physical and mechanical properties as a deformed metal is heated; these properties reflect the structural changes which include the movement, redistribution and annihilation of point defects and dislocations, and the replacement of the deformed grains by recrystallized grains. Alternatively, structural studies can be made using X-ray diffraction, optical and electron microscopy. While many of the structural changes occur at elevated temperatures, point defects can move well below room temperature even in metals with moderately high melting points, such as copper and iron, so a complete survey of recovery processes must include reference to structural changes at low temperatures.

We shall follow the same approach as that previously adopted to examine deformation processes by firstly looking at the behaviour of single crystals.

11.2 Recovery in single crystals

Single crystals subjected to Stage 1 hardening, and in particular crystals of zinc and cadmium which have deformed on a unique slip system, are capable of recovering completely their mechanical properties at room temperature. The classical work of Haase and Schmid[5] on zinc crystals demonstrated that even after substantial plastic strain, a crystal rested for 24 hours at room temperature underwent complete recovery, attaining the flow stress of the undeformed crystal. This process could be repeated several times. Zinc strained in simple shear[6] has been shown to recover completely after resting at room temperature. Successive shear stress/shear strain curves interrupted by recovery anneals were found to coincide within experimental error. The kinetics of the recovery process in zinc crystals is shown in Fig. 11.1 over a range of temperature. The recovery

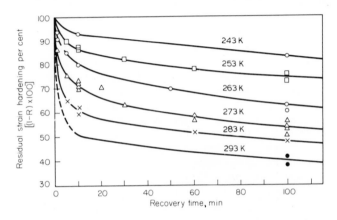

Fig. 11.1 Recovery of deformed zinc crystals at different temperatures. (After Drouard *et al.*, 1953, *Trans. AIME*, **197**, 1226)

of yield stress R, defined as $R = (\sigma_m - \sigma)/(\sigma_m - \sigma_0)$—where σ_m = flow stress of hardened crystal; σ = recovered flow stress; σ_0 = initial flow stress— was found to vary almost linearly with $\ln \theta$ where

$$\ln \theta = \ln t - \frac{Q}{RT} \qquad \text{(temperature-compensated time)} \qquad \textbf{11.1}$$

Fig. 11.2 shows the near linear plot obtained. An activation energy of 83.7 kJ mol^{-1} was obtained for the process, which corresponds approximately to the activation energy for diffusion of vacancies in zinc. Cottrell and Aytekin[7] obtained rather higher activation energies for the recovery of zinc, and found that the energy decreased with strain.

Several investigations on single crystals of zinc[7] and aluminium[8] have shown that the increment x of a physical or mechanical property above that of the undeformed metal is related logarithmically to the time of recovery t, so that

$$x = b - a \ln t \qquad\qquad \textbf{11.2}$$

where a and b are constants

$$\frac{dx}{dt} = -\frac{a}{t}$$

The rate of change of the property is inversely proportional to the time, being most rapid at the beginning of the process (Fig. 11.1). This behaviour has been confirmed for several different physical and mechanical properties, but particularly for the yield stress and the electrical resistivity.

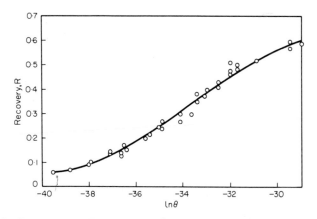

Fig. 11.2   Recovery as a function of $\ln \theta = \ln t - Q/RT$. (After Drouard *et al.*)

The complete recovery exhibited by zinc crystals is not normally found in crystals where slip on several systems is possible, e.g. aluminium. Even after light deformations, the properties do not return to the values typical of the annealed metals during recovery anneals. If deformation is restricted to Stage 1 of the stress–strain curve, then fairly complete recovery is possible but, even at this early stage, slip on other systems occurs. Consequently, irreversible dislocation reactions can take place which lead to the formation of incipient networks which are stable to high temperatures, and raise the flow stress above that of an undeformed crystal.

The completely recoverable deformation in zinc and cadmium crystals can be shown to produce no large scale structural changes such as the formation of sub-grains. X-ray reflections from cadmium crystals after 100–200 per cent elongation in tension are still quite sharp,[9] and on annealing complete recovery occurs without any recrystallization. Aluminium crystals on the other hand show spotty asterisms (streaks) in

X-ray Laue patterns which sharpen on annealing, but remain spotty. These effects are primarily due to the development of local inhomogeneities such as kink bands which are regions of lattice curvature in which secondary slip is likely to occur (see Chapter 8). Such structures cannot be completely removed by normal recovery processes involving migration and climb of dislocations. Aluminium crystals lightly deformed in tension will recrystallize[10] if annealed at a sufficiently high temperature, in contrast to the behaviour of cadmium or zinc crystals which will recrystallize only if sharply bent or twinned.

11.3 Polygonization

It has been pointed out that a cadmium crystal can be elongated considerably without causing any noticeable streaking in X-ray Laue patterns. On the other hand, if a crystal is bent around a large radius (e.g. 10 cm), asterisms are introduced and are difficult to eliminate by annealing near the melting point. Instead, the continuous streaks break down into well-defined intensity maxima (Fig. 11.3). Cahn[11] studied this phenomenon in zinc and aluminium, and proposed that the intensity maxima arose from small cells or polygons formed by dislocation rearrangement which was called *polygonization*. A bent crystal (Fig. 11.4a) has an excess of edge

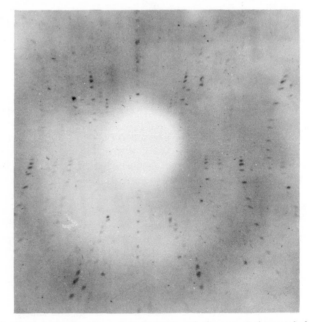

Fig. 11.3   Polygonization in a sharply bent cadmium crystal annealed at 300°C. X-ray Laue pattern. (Honeycombe)

dislocations of one sign in it. On annealing these dislocations migrate along the slip plane then climb to form walls at right angles to the slip plane (Fig. 11.4b). The structure produced is thus identical to a small angle boundary composed entirely of edge dislocations, that is a tilt boundary (p. 59). So the migration and climb of the dislocations has produced a disorientation θ which is inversely proportional to the dislocation spacing D.

$$\frac{b}{D} = 2 \sin \frac{\theta}{2} \simeq \theta$$

The driving force for this re-arrangement of dislocations is the reduction in strain energy achieved by a linear array. This means that isolated dislocations are elastically attracted to the array, and in this way the disorientation gradually increases. Cahn was able to reveal the polygonization

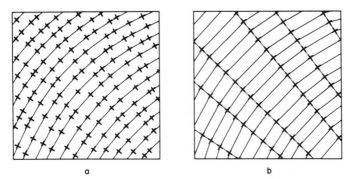

a   b

Fig. 11.4 Dislocation model of polygonization. (After Cahn, 1949, *J. Inst. Metals*, **76**, 121)

boundaries by selective etching of the dislocations, and found that the boundaries were in fact normal to the active slip plane. The regions between the sub-boundaries are called *sub-grains* or sometimes *cells*.

Subsequently these dislocation sub-boundaries have been etched up in various crystals, metallic and non-metallic. The phenomenon is particularly well demonstrated in silicon iron[12] (Fig. 11.5). Vogel,[13] using germanium, was able to prove that the disorientations calculated from the number of dislocation etch pits per unit length of the boundary corresponded with that determined by an X-ray diffraction method. Etch pit techniques have also been very successful with α-brass[14] and rock salt.[15]

In the general case, a polygonization boundary is not a simple tilt boundary, but can be composed of a grid of both screw and edge dislocations, so that the axis of misorientation need not be in the plane of the boundary. As the process is controlled by the rate at which dislocations climb, it will only occur readily at temperatures when vacancies are quite mobile, i.e. at $\frac{1}{2}T_M$ or higher. This means a temperature of 373–473 K for aluminium but substantially higher for metals such as iron, so marked formation of sub-boundaries occurs rapidly only in the higher part of the recovery range. It should also be pointed out that screw dislocations do not need to climb to join sub-boundary arrays; they can move conservatively but this may well be on alternative slip planes of higher energy,

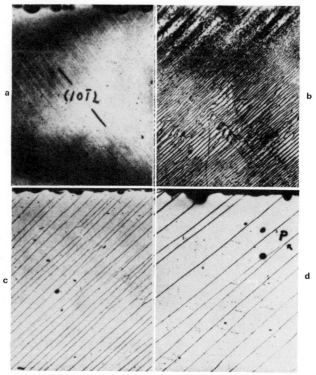

Fig. 11.5 Polygonization boundaries in deformed and annealed silicon iron crystals. **a**, Slip lines in as-bent crystal; **b**, deformed and annealed 10 min at 1223 K; **c**, deformed and annealed 24 hr at 1178 K; **d**, deformed and annealed 4 hr at 1573 K. Optical micrograph × 150. (After Dunn and Daniels, 1951, *Trans. AIME*, **191**, 147)

where thermal activation is helpful. Another factor which influences the rate at which polygonization takes place is the stacking fault energy. Metals of low stacking fault energy such as copper,[16] silver, lead and γ-iron polygonize much less readily than aluminium or nickel when the melting points are taken into account. This arises from the difficulty with which climb takes place when dislocations are widely dissociated (low stacking fault energy); before climb can occur over a length of dislocation line, the partial dislocations must re-associate. This is difficult in the absence of stress, but under creep conditions at high temperatures, climb can occur in metals of low stacking fault energy. However, more often the strain is removed by recrystallization.

The purity of the material is another important variable. Solute atoms can slow down the rate of climb of dislocations by segregating to jogs where vacancies are most likely to attach themselves. Moreover, once the sub-grain networks are formed, the segregation of solute atoms will limit the rate of growth of individual sub-grains. This process takes place by a coalescence of two sub-boundaries meeting usually at a Y junction, which gradually moves along creating a single boundary with a disorientation equal to the total across the two original boundaries.[17] This process is essentially a form of grain boundary migration and the driving force is the reduction in sub-boundary energy, but as this is not large the process occurs slowly.

11.4   Changes in physical properties during recovery

11.4.1.   *Stored energy*

When a metal is cold worked, most of the energy is expended in changing the shape and in generating heat; however, a small proportion up to about 5 per cent remains stored in the metal. The stored energy is mainly in the form of elastic energy in the strain fields of dislocations and point defects which are present in concentrations orders of magnitude higher than in annealed metals. The distribution of dislocations is, in addition, quite different to that in the annealed state, e.g., there are pile-ups against obstacles, tangles and diffuse sub-boundary arrays. During the processes of recovery and recrystallization the point defects and dislocations gradually approach the much lower concentrations typical of annealed metals, and considerable re-arrangements occur with the result that as the defects disappear, energy is released in the form of heat. Calorimetric studies can determine the rate of release of energy, and the amounts available at various temperatures, thus providing a sensitive means of studying the movement and disappearance of lattice defects.

Various calorimetric methods have been used,[1] but we shall consider firstly results from differential calorimetry, where two specimens, one deformed and the other annealed, are slowly heated at a constant rate. The two specimens are heated in separate furnaces and the power necessary to maintain the same rate of rise of temperature in each specimen is measured. The power needed for the deformed specimen will be less because it will gradually release its stored energy of cold work in the form of heat. Thus the power difference ΔP is proportional to the released energy, and is plotted against temperature to provide an overall view of the processes occurring during annealing.[18] A typical result for deformed polycrystalline copper of commercial purity is shown in Fig. 11.6.[19] The material had been deformed to 33 per cent elongation in tension prior to

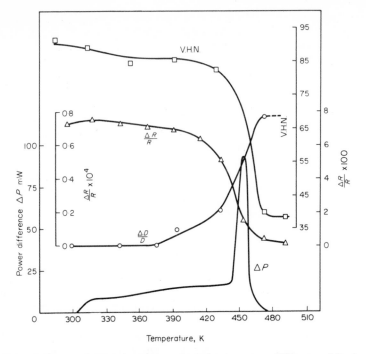

Fig. 11.6 Release of stored energy, plotted as power difference Δ*P*, from commercial copper deformed 33 per cent elongation in tension at room temperature. Changes in hardness (V.H.N.), resistivity (*R*) and density (*D*) are also shown. (After Clarebrough and Hargreaves, 1963, *Recovery and Recrystallization of Metals* (AIME). Edited by L. Himmel, Interscience, New York and London)

the experiment. Release of energy starts at about 343 K and occurs steadily until a sharp peak is reached just below 673 K, which can be shown by metallographic and X-ray methods to correspond to the growth of a new grain structure, i.e. recrystallization. The whole of the energy released below 623 K is due to recovery, and results from processes taking place in the original deformed grains. The same figure also shows the changes in hardness, density and electrical resistivity which all alter markedly in the recrystallization range, but also undergo smaller but significant changes at lower temperatures.

At this stage we shall concentrate on the processes occurring prior to recrystallization. Electron microscope examination of copper recovered in the range 373–623 K indicates that the as-deformed dislocation density has changed very little.* This does not exclude the possibility of dislocation

---

* However, substantial dislocation re-arrangements can occur during thin foil preparation, so such findings must be accepted with some reserve.

climb, but it suggests that dislocation annihilation may not be a significant factor in this temperature range for commercial copper. As the purity of the metal is increased, the recovery stage in the release of energy above room temperature becomes less significant, as shown by Fig. 11.7, where the behaviour of, A, 99·967 per cent and, B, 99·988 per cent copper is contrasted.

Very pure (99·999 per cent) copper has been examined using a technique in which the energy is released isothermally [20] at temperatures between 373 and 473 K. These experiments show two stages in the release of energy

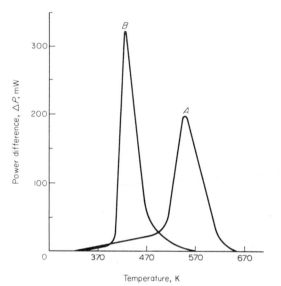

Fig. 11.7   Release of stored energy as a function of temperature for copper of two purities deformed to fracture in torsion. **A**, 99·967 per cent Cu; **B**, 99·988 per cent Cu. (After Clarebrough *et al.*, 1952, *Proc. R. Soc.*, **A215**, 507)

(Fig. 11.8), the first where the rate of release is most rapid at the beginning of the process but decreases quickly with time in a way comparable to the change of other physical properties during a recovery anneal. The second release of energy leads to a very pronounced peak, which accounts for 90 per cent of the stored energy and is associated with recrystallization. So with purer metal, the energy release associated with recovery above room temperature becomes less significant. These results suggest that much of the stored energy due to recovery is being lost either before it can be measured, or even during the deformation process. Furthermore, the results concerning the movement of point defects discussed in Chapter 10 indicate that recovery processes must be expected at very low temperatures, so that while room temperature is a convenient experimental point, it is a quite arbitrary dividing line for the release of the energy stored during plastic deformation.

A number of investigations have now been carried out with metals deformed at 77 K and lower, which conclusively indicate that a large amount of stored energy can be released below room temperature. Polycrystalline copper 99·999 per cent pure deformed in compression at 77 K liberates energy in two main stages, one between 77 and 200 K and the other

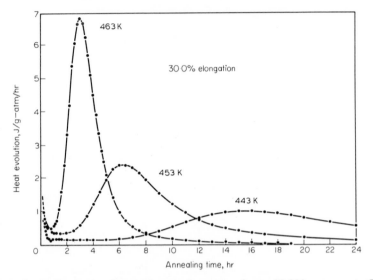

Fig. 11.8   Isothermal release of stored energy from 99·999 per cent Cu deformed 30 per cent elongation in tension, at room temperature. (After Gordon, 1955, *Trans. AIME*, **203**, 1043)

between 200 and 370 K.[21] Figure 11.9 shows a typical set of results for three degrees of deformation at 88 K, where the two main stages can be distinguished. More recent work has resolved these stages into further sub-stages, but this complication will be ignored at present. The amount of energy released clearly depends on the extent of the deformation, but typically after 66 per cent deformation about 42 J/g atom is evolved below room temperature,[22] and a similar amount of energy is evolved in the second stage above room temperature. The almost equal distribution between sub-zero and elevated temperatures indicates the importance of the low temperature recovery processes.

It is clear that the magnitude of the stored energy is by no means constant. Apart from the obvious variables such as the extent of deformation, mode of deformation and the composition, the temperature of deformation is particularly significant.

So far, we have referred primarily to measurements on metals of fairly low stacking fault energy which polygonize only with difficulty. Nickel which possesses a relatively high stacking fault energy, has also been extensively studied.[23, 24] One of the main differences from copper is that there is a well-defined second peak of energy release above room temperature which is associated not with recrystallization but with a recovery process. The process has an activation energy of about 1 eV. As the purity of the nickel is increased two peaks are resolved in the released energy

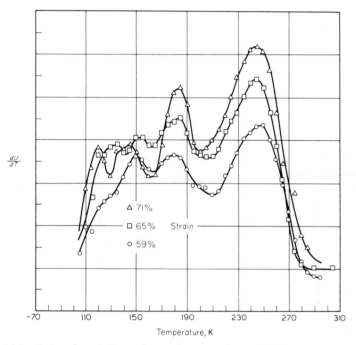

Fig. 11.9 Rate of evolution of stored energy from 99·999 per cent copper deformed at 88 K, as a function of temperature and deformation. (After Henderson and Koehler, 1956, *Phys. Rev.*, **104**, 626)

curve prior to the recrystallization peak. When the deformation is carried out at 77 K,[22] similar results to those of copper are found, viz. there is a release of energy between 130 and 220 K and another between 220 and 270 K.

11.4.2. *Electrical resistivity*

Electrical resistivity measurements provide a convenient method of following recovery processes, which is as sensitive to movement of both point defects and dislocations as the stored energy methods. In Chapter 10 we have seen how interstitial and vacancy defects can be distinguished by measuring resistivity changes over a range of temperatures.

Molenaar and Aarts[25] deformed copper, silver and aluminium wires at 90 K; on annealing at 293 K they found that at least 30 per cent of the increase in resistivity due to cold work was removed, while there was no detectable change in the flow stress. This indicates that no dislocation movements had occurred, and that point defect movement is likely to cause recovery of resistivity at quite low temperatures. Ideally, the metal should be deformed at liquid helium temperature (4·2 K) and the resistivity measured as the specimen warms up. Experiments done in this way on copper crystals (99·999 per cent)[26] have shown that practically no change in resistivity occurred up to 77 K, but holding for a long time at this temperature results in a small degree of recovery.[27] As the temperature is raised above 77 K a wide range of behaviour is observed, with the number of steps in the curve varying from two to as many as five, as the experimental conditions are varied. As might be expected, the amount of deformation prior to recovery is an important variable. If one selects the most prominent stages in the recovery of the resistivity, activation energies for copper are found in the ranges 0·2–0·3 and 0·7–0·9 eV.[28] Fig. 11.10 shows results for copper which illustrate two well-defined stages which lead to activation energies of this order. Adopting the simplest possible explanation, these can be attributed to the movement of interstitials and vacancies respectively, but it seems likely that the ultimate explanation will have to include other point defect configurations, such as

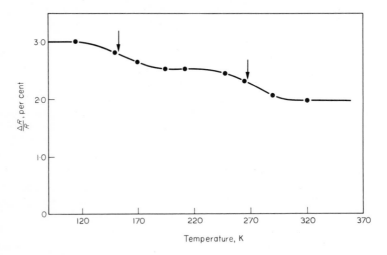

Fig. 11.10 Recovery of electrical resistivity in O.F.H.C. copper deformed 8 per cent at 90 K. (After Manintveld, 1952, *Nature*, **169**, 623)

di-vacancies, which may well be contributing to the behaviour below room temperature.

The resistivity continues to change at temperatures above room temperature as indicated in Fig. 11.6, together with the changes in stored energy, but these resistivity changes are of a smaller order except the final change which accompanies the recrystallization peak. Prior to recrystallization, aggregates of point defects are probably being removed by the formation of dislocation loops or tetrahedral defects which eventually disappear. These changes in resistivity above room temperature are also associated with changes in mechanical properties[29] which distinguishes them from the low temperature resistivity stages.

11.4.3. *Density*

Precision density measurements which have been made on nickel and copper during recovery (see Fig. 11.6) show that the density rises gradually. Only small changes occur which are difficult to measure with accuracy, the major change taking place during recrystallization. In nickel two steps in the density–temperature curve occur which correspond roughly to the two stored energy peaks for recovery.[30] Recently the methods used have been improved to give a higher degree of accuracy,[31] but it is unlikely that the method will become as widely used as stored energy and electrical resistivity measurements.

11.5   Recrystallization

Unlike recovery, recrystallization causes the mechanical and physical properties of the deformed metal to return completely to those of the annealed state. The mechanical properties such as hardness, yield strength and ultimate tensile strength change slowly during the recovery period, but during recrystallization these properties alter drastically over a very small temperature range (Fig. 11.6). Likewise, the elongation sharply increases to become typical of the annealed material. Physical properties such as electrical resistivity and density, while they sometimes undergo appreciable changes during recovery, also change sharply during recrystallization.

The phenomenon of recrystallization is a nucleation and growth process, in so far as unstrained nuclei begin to grow in the deformed metal when the temperature is high enough, and gradually absorb the whole of the deformed matrix. Figure 11.11 shows this process during the annealing of lightly deformed silicon ferrite. The first really extensive kinetic study was made by Anderson and Mehl,[32] who determined by microscopic means the fraction of the specimen which had recrystallized as a function of time at different temperatures to obtain basic data about the process. Figure 11.12 gives a typical set of results for aluminium deformed to three strains and annealed at 623 K. The curves show the following significant features which should be contrasted with the behaviour during recovery.

1. There is an incubation period.
2. Recrystallization starts slowly and gradually reaches a maximum rate.
3. The rate of transformation slows down again near complete recrystallization.

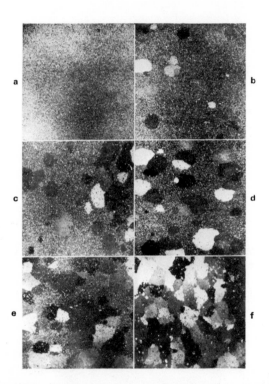

Fig. 11.11   Nucleation and growth of recrystallized grains in silicon ferrite after 4 per cent deformation. Annealing temperature 1043 K. × 5. **a,** 15 min; **b,** 70 min; **c,** 100 min; **d,** 120 min; **e,** 150 min; **f,** 190 min. (After Stanley and Mehl, 1942, *Trans. AIME*, **150**, 260)

Isothermal transformation curves are a convenient way of illustrating the role of the important variables of the process, particularly if taken in conjunction with the results on the release of stored energy during recrystallization

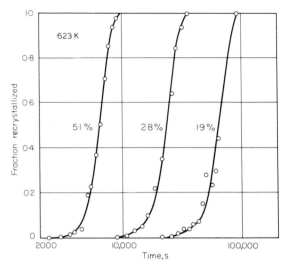

Fig. 11.12 Effect of deformation on the recrystallization kinetics of aluminium at 623 K. (After Anderson and Mehl, 1945, *Trans. AIME,* **161,** 140)

11.6 Variables in Recrystallization

11.6.1. *Strain*

In general, as the degree of cold work is increased the metal recrystallizes more readily, i.e. at a lower temperature or after a shorter time at a fixed temperature (Fig. 11.12). At smaller strains, fewer nuclei are created per unit volume and, moreover, the incubation time for formation of nuclei is longer. The smaller number of nuclei leads to a coarser recrystallized grain size. If the amount of strain is reduced, the *critical strain* is reached, that is the strain just necessary to initiate recrystallization, when the grain size can be extremely coarse after transformation is complete. This effect can be strikingly demonstrated in tapered tensile specimens where the strain can vary from zero to 20 per cent by using the appropriate dimensions. On annealing, coarser grains are formed in the regions of low strain and the tendency of the recrystallized grain size to diminish progressively with strain is revealed (Fig. 11.13). In pure metals such as aluminium and iron the critical strain is between 1 and 5 per cent elongation in tension, but it is dependent on the actual annealing temperature (Fig. 11.13), and for a given temperature tends to be raised by alloying. Critical grain growth can be carried out in a controlled fashion to produce single crystals by the strain anneal method (see Chapter 2). On the other hand, in practice the phenomenon is often troublesome during the annealing of worked material where large variations in strain have occurred. Also in extrusion at elevated temperatures when recrystallization occurs

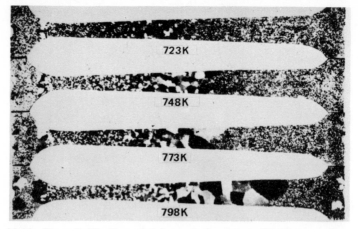

Fig. 11.13   Recrystallization of deformed aluminium (99·99 per cent) tapered specimens with strain gradients of 0 to 20 per cent elongation. Effect of annealing temperature. (After Eborall)

during the working process, strain gradients can lead to regions of extremely coarse grain size which impair the mechanical properties of the product.

As the dislocation population of a deformed metal increases with strain, it is obvious that the stored energy will likewise increase. Gordon,[20] working with high purity copper (99·999 per cent), has confirmed this, and has shown that the proportion of the energy released during recovery decreases as the strain increases (Table 11.1). This trend continues up to much heavier strains until stored energies of 50 J/g atom are released on annealing after true strains of 1·25 in the case of 99·98 per cent copper.

Table 11.1   Stored energy measurements on 99·999 per cent Cu

| Elongation per cent | Energy released, J/g atom | | $\dfrac{U_r}{U}$ |
	During recovery U_r	During recrystallization U	
10·8	1·13	10·5	0·10
17·7	0·79	15·1	0·05
30·0	1·05	20·9	0·05
39·5	0·79	26·0	0·03

11.6.2. *Temperature*

For a particular prior deformation, recrystallization will occur more rapidly as the temperature is raised (Fig. 11.8). It is clearly a thermally activated process where the time t, for a given percentage transformation, e.g. 50 per cent, is related experimentally to the temperature thus

$$\frac{1}{t} = A\,e^{-Q_R/RT} \qquad\qquad 11.3$$

where Q_R is the activation energy for recrystallization. Q_R can be determined readily if several experiments are carried out over a range of temperature, but it is not a unique value independent of metallurgical variables such as strain or purity. However, over a limited range for a particular material, time and temperature are interchangeable.

Isothermal experiments at several temperatures on the release of stored energy illustrate the effect of temperature on recrystallization.[20] As the temperature is raised, the peak of energy release moves to shorter times (Fig. 11.8) and the peaks are much sharper, although the total amount of energy released does not change. The critical strain is also a useful criterion for the commencement of recrystallization. As the temperature of annealing is raised, the critical strain is reduced (Fig. 11.13).

11.6.3. *Purity*

In general, the purer a metal, the lower the temperature at which recrystallization commences, other things being equal. This arises partly from the tendency of impurity atoms to segregate to interfaces such as grain boundaries which are preferential sites for the formation of recrystallization nuclei. For example, commercial copper begins to recrystallize in the range 473–773 K after heavy cold work; after similar deformation, high purity copper will recrystallize at room temperature. The release of stored energy from copper of two different purities has been measured after deformation in torsion to the same degree.[33] Within experimental error the same amount of energy was released in each case, but the evolution took place in different ways (Fig. 11.7). The purer material (curve B) showed a higher and sharper peak corresponding to the release of energy during recrystallization, while the metal of lower purity released a substantial amount of its energy *prior* to recrystallization. This behaviour has also been observed with nickel of different purities. Moreover, the recrystallization temperature of the copper was raised by 150 K by the presence of impurities (mainly phosphorus). Similar results were obtained with arsenical copper, where the amount of energy released was found to be higher than in the pure copper. This may not represent a fundamental difference, because in the case of the purer copper some energy is released during deformation, or before it can be measured. The role of phosphorus and arsenic in raising the recrystallization temperature, and allowing release of more energy during recovery is associated with the segregation of the atoms of these elements to dislocations by strain ageing. The dislocations are then less likely to climb at low annealing temperatures and help form recrystallization nuclei.

11.6.4. *Grain size*

Recrystallization occurs more readily in fine-grained metals because there is a larger grain boundary area per unit volume than in coarse-grained material. Recrystallization nuclei normally occur preferentially at grain boundaries so the latter influence the kinetics of the process. Stored energy results for 99·98 per cent copper of different grain sizes but identical deformation are shown in Fig. 11.14, where the coarser grained material has rather less stored energy and an appreciably higher recrystallization temperature.[34] As the prior strain is increased, the differences in stored energy become negligible, but a marked difference in recrystallization temperature persists. With copper of higher purity (99·999 per cent), the

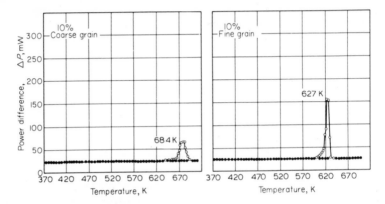

Fig. 11.14 Effect of grain size on the recrystallization of copper as shown by stored energy measurements. (After Clarebrough *et al.*, 1958, *Acta metall.*, **6**, 725)

trend is the same, and the stored energy is again higher in the fine-grained material.

11.7 Recrystallization kinetics

11.7.1. *Formal theory*

Recrystallization, like many phase changes, can be treated formally as a nucleation and growth process in which the fundamental quantities are N, the rate of nucleation, and G, the rate of growth of the recrystallization nuclei. One aim of recrystallization theory has been to establish a suitable model for the process, and then to derive a relationship for the fraction recrystallized in a given time in terms of N and G. Johnson and Mehl[35] derived a general relationship for phase changes occurring by nucleation and growth making the following assumptions:

1. That nucleation occurs randomly throughout the matrix.

2. That the rate of nucleation N expressed as the number of nuclei forming per second in unit untransformed volume is constant.
3. That the rate of growth G of the nuclei is constant.
4. That the growing nuclei are in the form of spheres which eventually impinge on each other.

The amount of the deformed matrix absorbed as a function of time can then be determined by a formal theoretical treatment[35] which gives an expression:

$$f(t) = 1 - e^{-(\pi/3)NG^3t^4} \qquad\qquad \textbf{11.4}$$

where $f(t)$ is the fraction transformed in time t.

This equation gives curves of approximately the form shown in Fig. 11.15. The exact form changes markedly for different values of N and

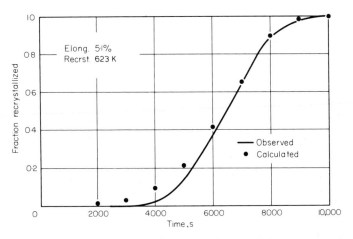

Fig. 11.15   Calculated recrystallization curve for aluminium compared with experimental results. (After Johnson and Mehl, 1939, *Trans. AIME*, **135**, 416)

G. A family of such curves can be reduced to a single curve by taking the term $\sqrt[4]{NG^3} \times t$ as abscissa. The amount of transformation thus depends on the value of this expression rather than separate values of N, G and t; there are many pairs of values of N and G which will give $\sqrt[4]{NG^3}$ a constant value.

Avrami[36] took the formal theory a stage further by examining the change in rate of nucleation N with time. It was assumed that prior to transformation, there are $\dot{N}$ preferred sites for nucleation, each of which has a nucleation frequency γ. These are gradually used up during recrystallization so that the nucleation rate N decreases exponentially thus

$$N = \dot{N}\gamma \, e^{-\gamma t}$$

whereas in the Johnson-Mehl treatment $N = \dot{N}\gamma$. When this expression is introduced into the theory, two expressions for the fraction recrystallized are obtained depending on the magnitude of γt.

For γt large

$$f(t) = 1 - e^{-kG^3 N t^3} \tag{11.5}$$

but when $\gamma t \to 0$ an equation almost the same as the Johnson–Mehl equation is obtained, viz.

$$f(t) = 1 - e^{(-kG^3 N t^4)/4} \tag{11.6}$$

Avrami proposed a general form of the formal equation for recrystallization as follows

$$f(t) = 1 - e^{-\beta t^k} \tag{11.7}$$

where β is a constant and $3 \leqslant k \leqslant 4$.

This equation applies to three-dimensional recrystallization, where recrystallized grains are much smaller than all dimensions of the specimen. However, if $2 \leqslant k \leqslant 3$, the equation is then valid for two-dimensional recrystallization, i.e. nucleation in thin sheets, when the new grains soon occupy the whole thickness of the specimen.

11.7.2. Experimental determination of N and G

A number of workers have measured N and G during recrystallization, the most extensive studies being those of Mehl and co-workers on aluminium and silicon iron.[32, 37] They used quantitative metallography on lightly deformed, fine-grained specimens in which on annealing the recrystallization nuclei could be easily observed and measured. To determine N the rate of nucleation, the number of nuclei in a set of specimens is determined and converted to the number of nuclei n per unit volume of untransformed material. N is then found directly from the slope of the curve of n against t. It is found that N varied exponentially with time

$$N = a \cdot e^{bt} \tag{11.8}$$

This sharp increase in N with time is a contrast to the decrease implicit in the formal Avrami theory.

The rate of growth G is determined by measuring the diameter D of the largest recrystallized grains in a series of specimens annealed for increasing times. When D is plotted against t, good straight lines are obtained, showing that G is independent of time.

With experimentally determined values of N and G, it is then possible to calculate the fraction of the material recrystallized as a function of time using the following expression derived from the formal theory for two-dimensional recrystallization, assuming that N varies exponentially with time.[37]

$$f(t) = 1 - e^{-\left\{\frac{2\pi G^2 a}{b^2}\left(\frac{e^{bt}}{b} - \frac{bt^2}{2} - t - \frac{1}{b}\right)\right\}} \tag{11.9}$$

where a and b are the constants in eqn. **11.8** and the equation is of the general form of eqn. **11.7** above. The curve thus determined is very similar to the experimental ones obtained on aluminium and silicon ferrite (Fig. 11.15).

The validity of an equation of the type **11.7** above has been confirmed more recently on zone refined aluminium with 0·0034 weight per cent copper, where $\log_e (1/1 - f(t))$ is plotted against $\log_e t$ giving good straight lines over a limited temperature range (Fig. 11.16).[38] In this case $k \simeq 2$, but

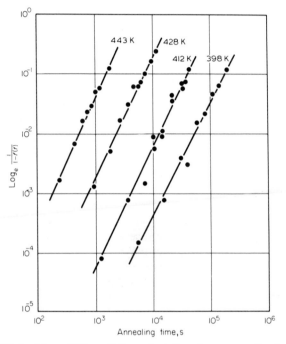

Fig. 11.16   Plot of $\log_e (1/1 - f(t))$ against time for zone-refined aluminium containing 0·0034 per cent Cu. Deformed 40 per cent by rolling at 273K. (After Vandermeer and Gordon, 1959, *Trans. AIME*, **215**, 577)

other investigations show wide variations. When the copper is removed from the aluminium, the kinetics change and a linear plot is obtained only at the earlier stages of the transformation. Later, the reaction is retarded by a marked reduction in the rate of growth of the new grains. This may arise from competition provided by polygonization in the deformed regions yet to be absorbed by the new grains.

The rates of nucleation and growth are strongly dependent on temperature, and are represented by equations

$$N = A\, e^{-Q_N/RT} \quad \text{and} \quad G = B\, e^{-Q_G/RT}$$

where Q_N and Q_G are the activation energies for nucleation and for growth respectively. A and B are constants. These equations, together with those above, lead to a temperature dependence of the rate of recrystallization as follows:

$$\frac{1}{f(t)} = A' \, e^{-Q_R/RT} \qquad\qquad \textbf{11.10}$$

where Q_R is now an activation energy for the whole process and A' a constant.

This activation energy is constant for transformation in a set of uniformly deformed specimens, but it varies markedly with strain. For example, with 99·999 per cent copper, Q_R is about 130 kJ/g atom after 40 per cent elongation but about 146 kJ/g atom after 10 per cent elongation.[20] The activation energy also varies with composition, reflecting the fact that impure metals recrystallize much less easily than pure metals. For example, Q_R increases from 63 kJ/g atom for zone purified aluminium to about 126 kJ/g atom when 0·007 atomic per cent copper is in solid solution.

11.8   The origin of recrystallization nuclei

Microscopic examination reveals that favoured sites for recrystallization nuclei include grain boundaries (Fig. 11.17),[41] phase interfaces,[42] twins,[43] deformation bands[44,45] and the surface of the material. These sites can often be characterized as regions of heavy distortion or high dislocation density, but it is equally true that they are frequently sites in the vicinity of marked changes in orientation. Some years ago Cahn[45] suggested that the nuclei were in fact sub-grains formed by polygonization, which accounted for the fact that nuclei occurred in the most heavily strained regions, but were themselves relatively undistorted, i.e. they possessed a low dislocation density. A quantitative theory[45] was put forward using this type of model. Detailed microscopic observations over the last twenty years have confirmed that sub-grain formation and subsequent replacement of deformed grains by recrystallization are closely related.

The advent of thin-foil electron microscopy has allowed the detailed study of dislocation arrangements and sub-boundaries before and during recrystallization, with the result that earlier theories of the origin of recrystallization nuclei have been confirmed and extended.

Bailey and Hirsch[39,40] have followed the recrystallization of polycrystalline copper, silver and nickel foils after heavy deformation which produces within the grains a well-defined cell structure, the cell boundaries comprising complex dislocation arrays. Typically the disorientations across the cell walls are about ten degrees after 90 per cent reduction by rolling. Recrystallization on the electron microscope scale is very inhomogeneous; however, numerous observations indicate that the process starts

from bulges in grain boundaries produced by migration. These then form
the recrystallization nuclei, but in order to grow there must be a substantial

grain
boundary

Fig. 11.17   Growth of recrystallized grains from, **a**, grain boundary; **b**, sub-
grains at boundary and within the grain. Aluminium polished and anodized.
Micrographs × 200. (After Crompton)

orientation difference between the sub-grain and other neighbouring, still deformed, material. Grain boundaries provide this large orientation change, so that larger sub-grains at boundaries can grow by crossing the boundary into the adjacent grain (Fig. 11.17b). A similar if less pronounced situation occurs at kink bands or other types of deformation band, so that sub-grains formed in these regions are favourably oriented for subsequent growth.[11,40]

Recent studies on aluminium[66] have confirmed that sub-grains provide the nuclei for recrystallized grains, and that those which act in this way have the largest disorientations with their neighbours. This work has shown that cell coalescence occurs apparently without extensive migration of the cell walls. Adjustments of cell boundaries are more likely in metals of high stacking fault energy such as aluminium, in contrast to the situation in copper and silver where little dislocation re-arrangement is detected prior to nucleation.

The importance of the nature of the dislocation distributions in deformed metals on the recrystallization behaviour has been emphasized by work on single crystal sheets of silicon iron of particular orientations.[44,47] The presence of a $\{100\}$ $\langle 110 \rangle$ orientation in the plane of the strip enables a silicon iron crystal to be deformed heavily without much orientation scatter,[48] in which circumstances recrystallization takes place with difficulty. Similar results have been obtained with copper and aluminium crystals of the $\{110\}\langle 112 \rangle$ type of orientation.[49] Hu[44] contrasted the recrystallization behaviour of the $\{100\}\langle 110 \rangle$ orientation in silicon iron with a $\{100\}\langle 001 \rangle$ orientation which recrystallizes readily on annealing after deformation. Examination of the as-deformed structures revealed deformation bands in the $\{100\}\langle 001 \rangle$ orientation which were absent in the other orientation. These bands were sharply defined and readily revealed by etching; at the band boundaries groups of smaller disorientations referred to as micro bands were observed about 1–3 microns in width. The presence of these disorientations is reflected in the deformation textures which are much less sharp than those from deformed $\{001\}\langle 110 \rangle$ crystals which do not develop deformation bands. Essentially it is these localized disorientations which lead to the greater ease with which recrystallization occurs. Electron diffraction photographs have shown that the disorientations in the microbands can be as much as 30°, whereas the disorientations within the larger bands are at the most a few degrees.

Thin-foil electron micrographs of Ag, Cu and Ni (Fig. 11.18) show that sub-grains grow preferentially in the microband regions by a coalescence process during which some sub-boundaries disappear without migration; a similar phenomenon has been observed in aluminium.[46,50] It has been proposed that the coalescence is achieved by rotation of one or more adjacent sub-grains until the disorientation is removed.[51] Subsequently the large coalesced sub-grains become recrystallization nuclei. It thus seems that electron microscope studies have confirmed earlier views that nuclei

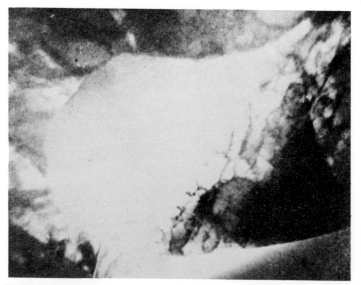

Fig. 11.18 Recrystallization nucleus in deformed nickel. Thin-foil electron micrograph × 6000. (After Bailey and Hirsch, 1962, *Proc. R. Soc.*, **A267**, 11)

occur by growth of sub-grains in deformation bands and at grain boundaries, although more detailed information is now available on the growth of the sub-grains and the dislocation distributions in the vicinity of nuclei.

11.9   Recent theoretical developments

The electron microscopic investigations referred to above have led to the development of a theory predicting the degree of recrystallization not in terms of N and G but in terms of the changes in surface energy and strain energy which occur during growth of nuclei. Bailey and Hirsch[39, 40] have derived an expression for the rate of growth of recrystallized grains using a model where a length of boundary $2L$ bulges out to form a spherical cap of radius R then migrates (Fig. 11.19). The driving force is assumed to be

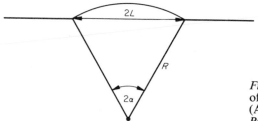

Fig. 11.19 Model for growth of a recrystallization nucleus. (After Bailey and Hirsch, 1962, *Proc. R. Soc.*, **A267**, 11)

the difference in strain energy (i.e. dislocation density) across the boundary.

The rate of growth of the recrystallized grain dV/dt can then be expressed as follows

$$\frac{dV}{dt} = Abf\left(E - \gamma\frac{dA}{dV}\right) \qquad\qquad \textbf{11.11}$$

where A is the surface area of recrystallized grain; $E =$ stored energy difference across boundary; $\gamma =$ surface energy; $b^3 =$ volume of one atom and f is a frequency factor defined thus

$$f = \left(\frac{vy}{RT}\right).e^{-\Delta F_A/RT} \qquad\qquad \textbf{11.12}$$

where ΔF_A is the free energy difference/atom in the initial and activated states; $v =$ jump frequency; $y =$ constant.

For growth to take place $dV/dt > 0$, thus

$$\left(E - \gamma\frac{dA}{dV}\right) > 0$$

and it can then be shown that $L > 2\gamma/E$. This indicates that there is a minimum length of boundary which can grow. In silver deformed moderately in tension $L \geqslant 10^{-4}$ cm, which is about the size observed in the electron microscope.

Finally the rate of recrystallization can be expressed in terms of X, the fraction of recrystallized material,

$$\ln\left(\frac{1}{1-X}\right) = N\left(\frac{\pi}{b}\right)\left(\frac{2\gamma}{E}\right)\left(\frac{x^3 + 3x}{\beta^3}\right) \qquad\qquad \textbf{11.13}$$

where $N =$ no. of migrating regions per unit volume; $x = \tan\alpha/2$; $\beta = 2/LE$.

11.10   Relationship between Recovery and Recrystallization

In the literature there is confusion over whether a prior recovery anneal makes a metal less likely to recrystallize or not. Part of the ambiguity undoubtedly arises from the fact that metals of low stacking fault energy do not undergo large dislocation rearrangements during recovery, whereas metals of high stacking fault energy such as aluminium or α-iron readily form sharply defined sub-structures which may compete with recrystallization in reducing the stored energy of the worked metal. The large variations in stored energies not only between different metals, but between the recovery and recrystallization stages in a given metal leave much scope for individual behaviour during recrystallization.

The influence of recovery on recrystallization is best illustrated by work on the annealing of zone-purified iron [52, 53] ($\sim$99·998 per cent purity) and less pure irons. Zone-refined iron, provided it is deformed not more than about 25 per cent in tension, will not recrystallize even at the highest temperatures in the α range; instead, it polygonizes very readily within the original grains, the process beginning as low as 473 K. This behaviour accounts for the difficulty in preparing large grains or single crystals of high purity iron by the strain anneal method. The difficulty can be surmounted by diffusing into the iron a small concentration of carbon prior to straining. The role of carbon is to pin the dislocations and prevent them from climbing readily to form a stable polygonized structure of sufficiently low energy to inhibit nucleation and growth. After heavier deformations (80–90 per cent reduction by rolling), zone purified iron recrystallizes readily after about 1 hour at 573 K. Thin-foil electron microscopy of heavily deformed material shows that no polygonization occurs prior to recrystallization; the dislocation tangles appear to undergo no detectable rearrangements.

The addition of impurities, e.g. carbon, alters this behaviour markedly. Firstly, even lightly deformed, less-pure iron recrystallizes, and when deformed heavily recrystallizes at least 523–573 K above similarly treated zone-purified iron. Moreover, there is distinct evidence of dislocation rearrangement or polygonization prior to recrystallization. For example, Armco iron after 96 per cent reduction followed by annealing at 723 K for 2 hours, develops a sharp sub-grain structure about 1 micron in diameter; recrystallization commences at 823 K. Similar behaviour has been observed in aluminium. Zone-refined aluminium does recrystallize, but without first forming sharply defined subgrains.[54] On the other hand, less pure aluminium first forms a very clear-cut sub-grain structure prior to recrystallization.

Evidence for the significant influence of recovery on recrystallization also comes from stored energy measurements and kinetic studies. The effect of impurities in copper is not only to displace the recrystallization energy peak to higher temperatures, but to release more energy in the recovery stage so that proportionally less is liberated during recrystallization. Similar results have been obtained with nickel of 99·85 and 99·6 per cent purity.[23] Experiments on the isothermal release of stored energy from dilute aluminium–copper alloys[38] have shown that recovery and recrystallization overlap, and consequently the driving force for recrystallization is reduced. Some curves plotting rate of release of energy against time appear to show typical recovery kinetics (Fig. 11.20), but this is a result of the two phenomena overlapping. The retarding effect of recovery is revealed in the kinetics when an attempt is made to fit the results to the relationship for the volume fraction recrystallized (X_v) in time t.

$$X_v = 1 - e^{-Bt^2} \qquad\qquad \textbf{11.14}$$

Normally a plot of $1/1 - X_v$ versus time of annealing gives a straight line for each annealing temperature; however, in this work deviation from linearity occurred. The deviations were greater for *lower* temperatures and *less pure* material, conditions which tend to favour sub-grain formation rather than immediate recrystallization. The impurities will tend to prevent excessive boundary movement needed for the growth of recrystallization nuclei, so opportunity is given for the more limited dislocation migration to form fairly stable sub-boundaries.

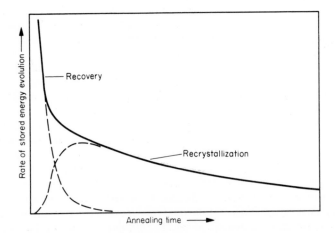

Fig. 11.20   Effect of overlapping recovery and recrystallization on release of stored energy (schematic). (After Vandermeer and Gordon, 1963, *Recovery and Recrystallization of Metals* (AIME). Edited by L. Himmel, Interscience, New York and London)

11.11   Grain boundary movement during recrystallization

We have dealt with the question of the origin of recrystallization nuclei but have not yet considered in detail the subsequent growth of the new crystals. The rate of growth appears to be quite sensitive to the presence of small concentrations of impurities, and to the structure of the matrix in which the grain is growing, i.e. the dislocation arrangements and densities. Furthermore, the orientation relationship between the growing grain and the matrix is a significant factor in determining growth rates.

The driving force for grain growth in a deformed matrix is the strain energy difference, i.e. the difference in dislocation density within the growing grain and the matrix, and typically in this type of growth the boundary migrates *away* from its centre of curvature. This is in contrast to the growth of grains in a fully annealed metal (Section 11.12, Fig. 11.23), which grow towards their centres of curvature.

11.11.1. *Effect of orientation*

It has been known for a long time that the mobility of grain boundaries during recrystallization is strongly dependent on orientation.[55] Grains closely oriented to each other have stable boundaries, and so also do grains near the twin relationship (60° rotation about ⟨111⟩ for face-centred cubic crystals). In single crystals grown by strain anneal methods small residual grains[56] which persistently remain on the surface frequently possess either of these orientations.

In the case of face-centred cubic metals, such as aluminium and copper, the most mobile boundaries appear to be those which are rotated by 30–40 degrees about a ⟨111⟩ axis. So if a metal is deformed locally (by indentation or a scratch) to create on annealing a large number of nuclei, those which tend to grow and eliminate the other nuclei, possess this type of orientation relationship with the matrix. One further point becomes evident from such experiments; the grain boundary mobility is dependent also on the orientation of the boundary itself with respect to the two crystals.

Low angle boundaries move slowly, but even when the disorientation reaches 10–15° the rate of movement is often an order of magnitude less than that of high angle (30–50°) boundaries.[57,58] On the other hand, very low angle boundaries (< 1°) of simple type, e.g. tilt, can be more mobile the smaller the angle across the boundary. Experiments have been done on zinc crystals[59] and on ice,[60] but they differ from normal experiments in so far as a stress was applied at temperature, and the movement of the boundaries was therefore stress induced.

The relative mobility of high angle boundaries has been examined in pure aluminium as a function of rotation about a common ⟨111⟩ axis.[61] The migration rate rose sharply to a maximum at a disorientation of about 40°. Other work on movement of the boundary in lead crystals[62] has shown that with zone-purified material little difference in mobility occurred as the disorientation angle about a ⟨111⟩ axis varied from 23 to 40°, or when the orientation across 'random' boundaries was changed in the range 15–60°. However, as impurities were added to the lead, the rate of movement of the boundaries became much more sensitive to the nature of the boundary (Fig. 11.21), and very low concentrations of impurity had a marked effect on the rate of migration.

A few experiments have been done on zone-purified lead bicrystals[62] with a 'coincidence' boundary, i.e. a boundary of a special orientation where a high proportion of atoms (e.g. 1 in 5) on one grain surface are correctly oriented for the adjacent grain as well. The migration rate of such a boundary, e.g. 38° tilt about a ⟨100⟩ axis, was found to be much faster than that of a non-coincidence boundary, e.g. 41° tilt about a ⟨100⟩ axis.

This is explained in terms of the lower energy needed to move atoms across a 'semi-coherent' boundary, i.e. one with good atomic fit, than one with bad atomic fit.

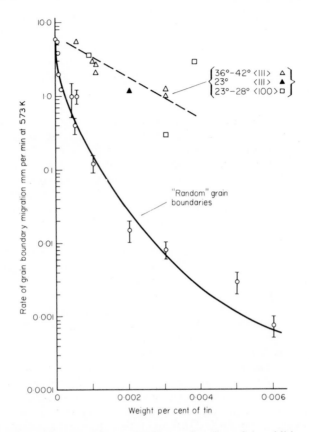

Fig. 11.21 Boundary migration at 573 K as a function of tin additions to zone-refind lead. Results for random boundaries and special coincidence boundaries. (After Aust and Rutter, 1959, *Trans. AIME*, **215**, 820)

11.11.2. *Impurities*

It is well established that impurities in small concentrations in solid solution markedly retard recrystallization. For example, 0·005 atomic per cent tellurium in copper raises the recrystallization temperature by over 473 K,[63] and similar effects are well known in lead.[64] More recently, stored energy measurements have emphasized the role of impurities, but it is difficult to decide whether their influence is on the rate of nucleation, the rate of growth of the nuclei, or on both these variables.

It is only recently with the availability of highly pure zone-refined metals that any really quantitative experiments could be done to determine the true role of impurities in recrystallization. Work on zone refined tin and lead[65] has shown that extremely small traces of impurity have a marked effect on grain growth during recrystallization. Zone purified aluminium (99·9999 per cent) produced by 12 zone passes exhibits a rate of growth of boundaries two orders of magnitude faster than material subjected only to 4 passes, when a recrystallization technique is used which moves one boundary into a heavily polygonized matrix. The activation energy for growth Q_G changes from 54 kJ/g atom to 350 kJ/g atom in this range of purity. This large effect of soluble impurities has been also found in work on zone refined lead with additions of tin, silver and gold, where a similar technique involving the migration of an isolated boundary was used.[66] The speed of migration is reduced by nearly two orders of magnitude when one solute atom of gold or silver in 10^6 is added (Fig. 11.22), and in the same composition range the activation

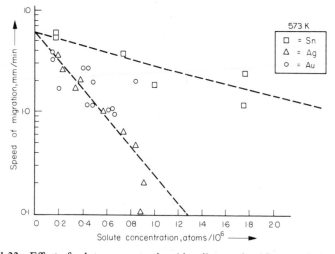

Fig. 11.22   Effect of solute concentration (tin, silver and gold) on grain boundary migration in zone-refined lead at 573 K. (After Rutter and Aust, 1960, *Trans. AIME*, **218**, 682)

energy for growth changes from 21 kJ/g atom for pure lead to 125 kJ/g atom for the lead with silver or gold added.

The drastic effect of such small impurity concentrations can only be explained by a mechanism involving absorption of solute atoms at the moving interface. The solute atoms then exert a drag on the boundary which moves at the rate at which the solute atmosphere can migrate.[67] Thus two temperature dependent processes are taking place; firstly the solute atoms are migrating towards the boundary, and then there is atom transfer across the boundary to cause it to move.

The movement of a 'pure' boundary, i.e. one in which solute atoms are absent, requires an activation energy much lower than that for volume diffusion (e.g. 25 kJ/g atom for lead) so the rate controlling process must be an easier one, the most likely being grain boundary diffusion. The process of growth is envisaged as an interchange of vacancies and atoms across the boundary, and the activation energy for growth should be equal to that for migration of vacancies. Most data are unsuitable to check this point because of the presence of impurities, but work on zone refined aluminium has given Q_G nearly the same as Q_M, the activation energy for vacancy movement.[68] The mobility M of a pure grain boundary can then be expressed in the following form.

$$M = \frac{\alpha b}{NkT} e^{-Q_M/kT} \qquad\qquad 11.15$$

where N = no. of atoms/unit volume, b = distance moved per jump by atoms, k = Boltzmann constant, α = constant.

The process may take place by individual atoms jumping across the interface (single process mechanism), or by groups (group process mechanism). The exact mechanism chosen will of course influence the details, but not the form of the above relationship.

The reaction changes radically as soon as solute atoms are present. Lücke and Detert[67] put forward a theory which assumed that the solute atoms segregate to the boundaries and create a drag on movement. At high impurity concentrations (or low temperatures) the rate of migration would then be determined by the rate at which solute atoms could diffuse, so that Q_G should be equivalent to the activation energy for volume diffusion of the solute. One difficulty is that the results of Aust and Rutter[66] on lead show that silver and gold which diffuse more rapidly than tin, cause the boundary to move much less rapidly. Further advances have been made by Lücke and Stüwe[69] and by Cahn,[70] who developed a more rigorous theory in which two limiting conditions of drag by solute atoms are defined.

1. At high grain boundary velocities

$$P_i \simeq \frac{C_0 N_v}{kTV} \int_{-\infty}^{+\infty} \left(\frac{dE(x)}{dx}\right)^2 D(x)\,dx \qquad\qquad 11.16$$

where P_i = total force exerted by the impurity atoms on the boundary; D = diffusion coefficient of solute; x = distance of solute atom from grain boundary; C_0 = concentration of solute; V = velocity of migration; N_v = number of atoms per unit volume; $E(x)$ = interaction energy between a solute atom and the boundary.

2. At low boundary velocities

$$P_i \simeq 4N_v C_0 VkT \int_{-\infty}^{+\infty} \frac{\sinh^2 [E(x)/2kT]}{D(x)}\,dx \qquad\qquad 11.17$$

These relationships show that for high velocities the drag is inversely proportional to the velocity, but at low velocities a more slowly diffusing solute will produce a greater drag than a rapidly diffusing one. At high velocities the reverse will be the case. Under the latter conditions, increased drag will occur when the variables are adjusted to give greater opportunity for local changes in composition, e.g. decreasing the rate of migration or pulling in a faster moving impurity. It should also be noted that the theory predicts a drag regardless of the sign of $E(x)$, that is, in cases of both adsorption and desorption at the boundary.

11.12   Normal grain growth

Finally, in this review of phenomena which take place during recrystallization, we must consider normal grain growth, by which is meant the growth of grains in a fully recrystallized matrix, which leads usually to a gradual coarsening of the grain structure. Typically in such circumstances grain boundaries are not advancing into a deformed matrix but into annealed grains, and the migration occurs towards the centre of curvature of the moving boundary.[71] Two stages in the process are illustrated in Fig. 11.23a, where this principle applies; in contrast Fig. 11.23b shows grain boundary movement during recrystallization.

The driving force for grain growth is the reduction in the surface energy of the grain boundaries achieved by reducing the grain boundary area per unit volume.[72] An excellent analogy is the growth of grains in a soap froth where surface energy again controls the process.[73] In the soap froth all grain boundaries meet at 120° as the surface tensions are all equal and the triangle of forces law is obeyed. To meet this energetic requirement, grains develop curved sides, concave or convex, and growth always occurs towards the centres of curvatures of the boundaries so that a grain boundary which is convex *outwards* will tend to move *inwards* and vice versa.

To a first approximation in a metal the surface energies of the various boundaries are similar, so again the angle of intersection of grain boundaries in equilibrium tends to approach 120°. If we consider for simplicity an idealized two-dimensional array of grains, only a hexagon is able to have all angles at 120° with *straight* grain boundaries. In figures with more than six sides, the boundaries must be concave outwards (Fig. 11.24) to achieve included angles of 120°; on the other hand, grains with less than six sides are convex outwards to give 120° intersections. So when grain growth occurs with boundaries migrating towards their centres of curvature, grains with less than six sides will experience *inward* movement of boundaries and will tend to shrink, while grains with more than six sides will tend to have *outward* movement of grain boundaries and so will grow.

The driving force for grain boundary movement is surface tension, and in the case of the soap bubbles the adjustment of grain boundaries to the

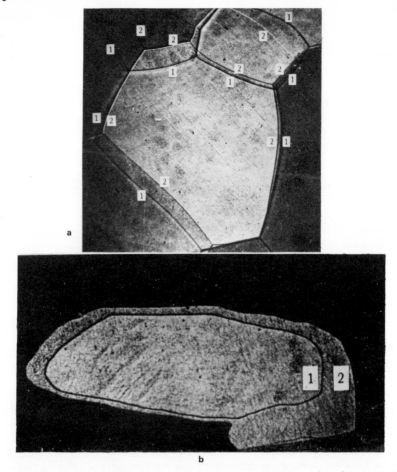

Fig. 11.23 Grain growth in, **a**, annealed aluminium, migration *towards* centres of curvature; **b**, recrystallization, migration away from centres of curvature. The specimens were anodized to show two stages of growth. × 75. (After Beck, 1952, *Metal Interfaces*, ASM)

surface tension requirements results in pressure differences ΔP within the bubbles where

$$\Delta P = \frac{2\gamma}{R} \qquad \textbf{11.18}$$

where γ = surface tension, and R = radius of curvature.

The pressure will be higher in the smaller grain with small radii of curvature so they will tend to lose pressure by diffusion of gas through the

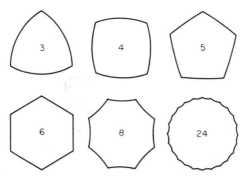

Fig. 11.24   Two-dimensional "grain" boundaries in equilibrium. (After Smith, 1952, *Metal Interfaces*, ASM)

grain boundary films. The adjacent grains will then expand and absorb the smaller grain either in stages or rapidly, depending on the number of grain boundaries the smaller grain possesses.

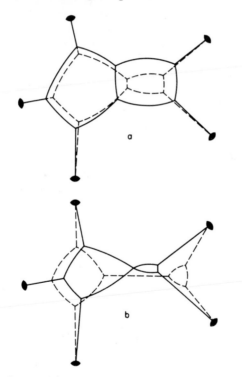

Fig. 11.25   Grain growth by elimination of grains with less than six sides. (After Smith, 1952, *Metal Interfaces*, ASM)

Grains with less than six sides tend to reduce their number of sides, 5 to 4 to 3, by a series of reactions illustrated in Fig. 11.25. Adjacent 5- and 4-sided grains shrink until an unstable situation is reached with four boundaries meeting at a point. This unstable configuration changes instantaneously to a situation where each of the original two grains has lost a side, and the three-sided grain then shrinks continuously until it disappears. This discussion is concerned only with the simple two-dimensional model. The more elegant three-dimensional treatment is found in reference 73.

In a metal, the grain boundary movement is envisaged as a transfer of atoms across the boundary from one grain to another under the stimulus of thermal activation. Atoms transferred to a *concave* surface of a crystal will be more surrounded by neighbours and, thus, less likely to move than atoms which transfer in the other direction to the convex surface of the adjacent crystal where they are less strongly bonded. The grain boundary will thus tend to move towards its centre of curvature. Examination of polished sections of polycrystalline metals shows that morphologically the situation is similar to the two-dimensional case discussed above; however, there will be some variation in the surface tensions of the boundaries with orientation, so exact 120° angular relationships cannot always be expected. Moreover, the sectioning process only gives a two-dimensional view of solid angular relationships, so to obtain true angular values a detailed statistical treatment must be used. However, even cursory examination of grain growth from metal surface observations confirms the general principles revealed by study of soap films.

11.13   Kinetics of grain growth

As a starting point it is a reasonable assumption that the rate of grain growth in a fully recrystallized metal should be proportional to the interfacial energy per unit volume. Beck et al.[74] found an equation of the form

$$\frac{\mathrm{d}D}{\mathrm{d}t} = \frac{A\gamma}{D} \qquad\qquad \textbf{11.19}$$

was valid, where γ = surface energy, D = average grain diameter, A = constant, t = time.

Integrating,

$$D^2 = D_0{}^2 + A\gamma t$$

where D_0 is the grain size at $t = 0$. If we consider the cause when there has been much grain growth, D_0 can be ignored and

$$D = (A\gamma t)^{1/2} \qquad\qquad \textbf{11.20}$$

so that the grain diameter during grain growth should be proportional to $t^{1/2}$. In practice the index (n) is rarely 0·5 but seems to vary in the range

0·1 to 0·5. Amongst the factors which contribute to these variations are composition, specimen shape, presence of inclusions in grain boundaries which tend to inhibit growth, etc., all of which seem to depress the value of the index below 0·5. Burke[75] found that in 70/30 α-brass of commercial purity the index was approximately 0·2, but in high purity material of the same composition the value approached 0·5.

Recent work[76] on zone refined aluminium and aluminium with small traces of impurities indicates that n is 0·33 for both cases. So while there are theoretical reasons for expecting pure metal to possess a value of 0·5 for n, very slight traces of impurity can have a large effect. This is in keeping with the theory of Lücke and Detert discussed in Section 11.11, where solute atoms are assumed to segregate to boundaries during grain growth associated with recrystallization. There seems little doubt that regardless of whether the boundary is moving because of differences in strain energy, or because of differences in surface energy, solute atoms play a significant part in determining the rate of movement even at extremely low solute concentrations.[77] Gordon[77] has suggested that $n = 0·5$ when the driving force is high, i.e. small grain size, and the migration proceeds without hindrance from solute atoms. At the other extreme with a low driving force, the migration rate is solely determined by solute atom movement, and n again is approximately 0·5. However, in between these two limiting conditions there is a transition zone where n is much less than 0·5 and if observations are made in this region difficulties arise. For example, experimentally determined activation energies will vary, and in any case will be of doubtful value.

General references

1. TICHENER, A. L. and BEVER, M. B. (1958). *The Stored Energy of Cold Work. Prog. Metal Phys.*, **7**, 247.
2. HIMMEL, L. (Ed.) (1963). *Recovery and Recrystallization of Metals.* Interscience, New York and London.
3. *Creep and Recovery* (1957). American Society for Metals.
4. Recrystallization, *Grain Growth and Textures* (1966). American Society for Metals.

References

5. HAASE, O. and SCHMID, E. (1925). *J. Phys.*, **33**, 413.
6. DROUARD, R., WASHBURN, J, and PARKER, E. R. (1953). *Trans. AIME*, **197**, 1226.
7. COTTRELL, A. H. and AYTEKIN, V. (1950). *J. Inst. Metals*, **77**, 389.
8. KUHLMANN, D., MASING, G. and RAFFELSIEFER, J. (1949). *Z. Metallk.* **40**, 241.
9. HONEYCOMBE, R. W. K. (1951–52). *J. Inst. Metals*, **80**, 45.
10. CRUSSARD, C., AUBERTIN, F., JAOUL, B. and WYON, G. (1950). *Prog. Metal Phys.*, **2**, 193.

11. CAHN, R. W. (1949). *J. Inst. Metals*, **76**, 121.
12. DUNN, C. G. and HIBBARD, W. R. (1955). *Acta metall.*, **3**, 409.
13. VOGEL, F. L. (1956). *Trans. AIME*, **206**, 946.
14. JACQUET, P. A. (1954). *Acta metall.*, **2**, 752.
15. AMELINCKX, S. (1954). *Acta metall.*, **2**, 848.
16. YOUNG, F. W. (1958). *J. appl. Phys.*, **29**, 760.
17. DUNN, C. G. and DANIELS, F. W. (1951). *Trans. AIME*, **191**, 147.
18. CLAREBROUGH, L. M., HARGREAVES, M. E., MICHELL, D. and WEST, G. W. (1952). *Proc. R. Soc.*, **A215**, 207.
19. CLAREBROUGH, L. M., HARGREAVES, M. E. and LORETTO, M. H. Reference 2, p. 63.
20. GORDON, P. (1955). *Trans. AIME*, **203**, 1043.
21. HENDERSON, J. W. and KOEHLER, J. S. (1956). *Phys. Rev.*, **104**, 626.
22. VAN DER BEUKEL, A. (1961). *Physica*, **27**, 603.
23. CLAREBROUGH, L. M., HARGREAVES, M. E., LORETTO, M. H. and WEST, G. W. (1960). *Acta metall.*, **8**, 797.
24. BELL, F. and KRISEMENT, O. (1962). *Acta metall.*, **10**, 80.
25. MOLENAAR, J. and AARTS, W. H. (1950). *Nature, Lond.*, **166**, 690.
26. BLEWITT, T. H., COLTMAN, R. R. and REDMAN, J. K. (1955). Report on a Conference on Defects in Crystalline Solids. Physical Society, London, p. 369.
27. MEECHIN, C. J. and SOSIN, A. (1958). *J. appl. Phys.*, **29**, 738.
28. BALUFFI, R. W., KOEHLER, J. S. and SIMMONS, R. O. Reference 2. p, 1.
29. KAMEL, R. and ATTIA, E. A. (1961). *Acta metall.*, **9**, 1047.
30. CLAREBROUGH, L. M., HARGREAVES, M. E. and WEST, G. W., *Phil. Mag.*, **1**, 528, 1956.
31. KUHLMANN-WILSDORF, D. and SIZAKI, K. (1963). Proc. Internt. Conf. on Crystal Lattice Defects, Kyoto, p. 54.
32. ANDERSON, W. A. and MEHL, R. F. (1945). *Trans. AIME*, **161**, 140.
33. CLAREBROUGH, L. M., HARGREAVES, M. E. and WEST, G. W. (1955). *Proc. R. Soc.* **A232**, 252.
34. CLAREBROUGH, L. M., HARGREAVES, M. E. and LORETTO, M. H. (1958). *Acta metall.*, **6**, 725.
35. JOHNSON, W. A. and MEHL, R. F. (1939). *Trans. AIME*, **135**, 416.
36. AVRAMI, M. (1939). *J. chem. Phys.*, **7**, 1103; **9**, 177 (1941).
37. STANLEY, J. K. and MEHL, R. F. (1942). *Trans. AIME*, **150**, 260.
38. VANDERMEER, R. A. and GORDON, P. Reference 2, p. 211.
39. BAILEY, J. E. and HIRSCH, P. B. (1962). *Proc. R. Soc.*, **A267**, 11.
40. BAILEY, J. E. (1963). *Electron Microscopy and the Strength of Crystals*. Ed. by G. THOMAS and J. WASHBURN. John Wiley, New York and London.
41. BECK, P. A. (1952). Chapter in *Metal Interfaces*. A.S.M., p. 208.
42. HONEYCOMBE, R. W. K. and BOAS, W. (1948). *Aust. J. scient. Res.*, **1** (1), 70.
43. BRINSON, G. and HARGREAVES, M. E. (1958–59). *J. Inst. Metals*, **87**, 112
44. HSUN HU (1963). Reference 2, p. 311.
45. CAHN, R. W. (1950). *Proc. phys. Soc.*, **A364**, 323.
46. WEISSMANN, S., IMURA, T. and HOSOKAWA, N. Reference 2, p. 241.
47. HSUN HU (1962). *Trans. AIME*, **224**, 75.
48. KOH, P. K. and DUNN, C. G. (1955). *Trans. AIME*, **203**, 401.
49. LUTTS, A. H. and BECK, P. A. (1954). *Trans. AIME*, **200**, 257.

50. FUGITA, H. (1961). *J. phys. Soc. Japan*, **16**, 397.
51. LI, J. C. M. (1962). *J. appl. Phys.*, **33**, 2958.
52. *Nouvelles Propriétes Physiques et Chemique des Métaux de très haute pureté*, CNRS, Paris, 1960.
53. TALBOT, J. Reference 2, p. 269.
54. DIMITROV, O. (1959). *Comp. Rend.*, **249**, 265.
55. BECK, P. A., SPERRY, P. R. and HSUN HU (1950). *J. appl. Phys.*, **21**, 420.
56. LACOMBE, P. and BERGHEZAN, A. (1949). *Métaux et Corrosion*, **25**.
57. RUTTER, J. W. and AUST, K. T. (1960). *Trans. AIME*, **218**, 682.
58. WALTER, J. L. and DUNN, C. G. (1960). *Trans. AIME*, **218**, 914.
59. BAINBRIDGE, D. W., LI, C. H. and EDWARDS, C. H. (1954). *Acta metall.*, **2**, 322.
60. HIGASHI, A. and SIKAI, N. (1961). *J. phys. Soc. Japan*, **16**, 2359.
61. LIEBMANN, B. and LÜCKE, K. (1956). *Trans. AIME*, **206**, 1413.
62. AUST, K. T. and RUTTER, J. W. Reference 2, p. 131.
63. SMART, J. S. and SMITH, A. A. (1943). *Trans. AIME*, **152**, 103.
64. BECK, P. A. (1940). *Trans. AIME*, **137**, 222.
65. HOLMES, E. L. and WINEGARD, W. C. (1959–60). *J. Inst. Metals*, **88**, 468.
66. AUST, K. T. and RUTTER, J. W. (1959). *Trans. AIME*, **215**, 119.
67. LÜCKE, K. and DETERT, K. (1957). *Acta metall.*, **5**, 628.
68. DIMITROV, O. In reference 52.
69. LÜCKE. K. and STÜWE, R. P. Reference 2, p. 171.
70. CAHN, J. W. (1962). *Acta metall.*, **10**, 789.
71. HARKER, D. and PARKER, E. R. (1945). *Trans. ASM*, **34**, 156.
72. EWING, J. A. and ROSENHAIN, W. (1900). *Proc. R. Soc.*, **A67**, 112.
73. SMITH, C. S. (1948). *Trans. AIME*, **175**, 15.
74. BECK, P. A., KREMER, J. C., DEMER, L. J. and HOLZWORTH, M. L. (1948) *Trans. AIME*, **175**, 372.
75. BURKE, J. E. (1949). *Trans. AIME*, **180**, 73.
76. GORDON, P. and EL-BASSYOUNI, J. A. (1965). *Trans. AIME*, **233**, 391.
77. GORDON, P. (1963). *Trans. AIME*, **227**, 699.

Additional general references

1. CAHN, R. W. and HAASEN P. (Eds.) (1983). *Physical Metallurgy*, 3rd edition, Chapter by Cahn on recrystallization, North-Holland, Amsterdam.
2. HAESSNER, F. (Ed.) (1971). *Recrystallization of Metallic Materials*. University of Stuttgart.
3. COTTERILL, P. and MOULD, P. R. (1976). *Recrystallization and Grain Growth in Metals* Surrey University Press.
4. HANSEN, N., JONES, A. R. and LEFFERS, T. (Eds.) (1980). *Recrystallization and Grain Growth of Multi-Phase and Particle Containing Materials*, Proc. 1st Risø International Symposium on Metallurgy, Risø National Laboratory Denmark.

Chapter 12

Anisotropy in Polycrystalline Metals

12.1 Introduction

Single crystal studies have revealed that the mechanical properties are markedly anisotropic even in cubic crystals, and in materials of non-cubic structure this anisotropy can be much more pronounced. It should now be clear that anisotropy of the plastic properties is on the one hand an inevitable result of the behaviour of dislocations in the crystals, the directional properties of which arise primarily from the anisotropy of the elastic properties. On the other hand the macroscopic restraints imposed on deformation by shear in tensile tests and other more complex deformation processes play a significant part. At first sight it might be thought that effects of anisotropy can be easily eliminated in a polycrystalline aggregate by having a random array of crystals. However, the formation of an annealed polycrystalline aggregate by first working an ingot, then recrystallizing either at various stages of the deformation, or after deformation is complete, does not usually lead to an array of randomly oriented grains. Most deformation processes tend to induce anisotropy by virtue of the intrinsic crystallographic nature of deformation by slip and twinning, and this is further intensified on recrystallization. Non-random arrays of grains are referred to as preferred orientations or textures. We shall consider briefly in this chapter the types of preferred orientation which occur in polycrystalline aggregates in both the deformed and annealed states. Many physical properties vary in magnitude with direction in the crystals, for example the magnetic properties and the thermal properties which will be briefly considered. Textures have a number of important technological applications, e.g. magnetic alloys, texture strengthening.

12.2 Development of preferred orientations (textures)

Most methods of deforming a polycrystalline metal result in the predominance of a particular crystallographic direction or plane parallel to the direction of deformation in a majority of individual crystals comprising the aggregate,[1,2,4] In a tensile test on a single crystal the slip direction rotates until it approaches the axis of tension, and in compression the slip direction rotates until it is in the plane of compression. Similarly, rotations occur in the individual grains of aggregates which suffer more complicated stressing even in simple processes such as tension; this leads to the development of preferred orientations or textures, the nature of which depends on the particular deformation process used, and also on the crystal structure of the metal which determines the crystallography of the operative slip systems. There are several main types of deformation texture which are determined primarily by the deformation process employed, and the characteristic stress distribution each process imposes on the metal. Some examples of ideal textures are given in Table 12.1. The textures are most

Table 12.1   Some common textures produced by cold work

Crystal structure	Mode of working		Texture
f.c.c.	Wire drawing and extrusion	$\langle 111 \rangle$ $\langle 100 \rangle$	Parallel to wire axis
b.c.c.	Wire drawing and extrusion	$\langle 110 \rangle$	Parallel to wire axis
c.p.h.	Wire drawing and extrusion	$\langle 10\bar{1}0 \rangle$	Parallel to wire axis
f.c.c.	Rolling	$\{110\}$	Parallel to rolling plane
		$\langle 112 \rangle$	Parallel to rolling direction
b.c.c.	Rolling	$\{001\}$	Parallel to rolling plane
		$\langle 110 \rangle$	Parallel to rolling direction
c.p.h.	Rolling	$\{0001\}$	Parallel to rolling plane
		$\langle 11\bar{2}0 \rangle$	Parallel to rolling direction

Note. These textures are idealized. In practice, considerable scatter occurs, and very frequently other textures are present.

conveniently represented on a stereographic projection where the centre of the projection represents a significant direction such as the wire axis or the normal to the sheet in rolling.

12.2.1. *Fibre textures*

This type of texture is characteristic of uniaxial deformation processes such as wire-drawing, swaging, extrusion, etc., during the course of which the grains are elongated in the direction of working (Fig. 12.1b and d), and the preferred orientation is revealed by the arcing of rings in X-ray powder photographs (Fig. 12.1a and c). The fibre texture is characterized by a crystallographic direction of low indices parallel to the axis, for example body-centred cubic metals deformed by wire drawing have a fibre texture with the $\langle 110 \rangle$ direction parallel to the wire axis.[2,4] The perfection of the texture increases with the degree of deformation, but it can vary appreci-

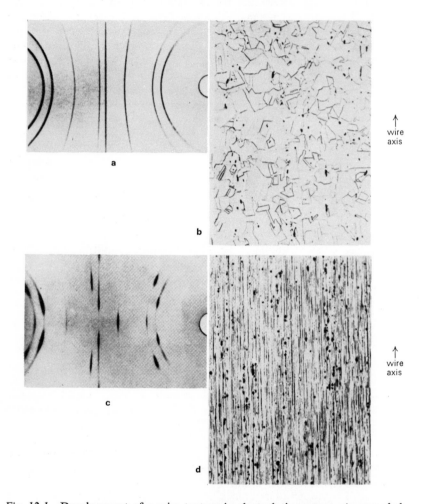

wire
axis

wire
axis

Fig. 12.1   Development of a wire texture in electrolytic copper. *As annealed:* **a**, powder photograph; **b**, micrograph. *After 95 per cent reduction by drawing:* **c**, powder photograph; **d**, micrograph. (After Wassermann and Grewen, 1962, *Texturen metallischer Werkstoffe*, Springer Verlag, Berlin) Micrographs × 225

ably from surface to core of the wire because of macroscopic inhomo-geneity in the deformation process. On the other hand, face-centred cubic metals can have a double fibre texture[2,4] with both $\langle 111 \rangle$ and $\langle 100 \rangle$ directions parallel to the wire axis. The $\langle 111 \rangle$ texture tends to predominate in metals of higher stacking fault energy, e.g. aluminium has a single $\langle 111 \rangle$ texture, whereas silver and brass with relatively low stacking fault energies have a predominantly $\langle 100 \rangle$ texture with some $\langle 111 \rangle$. With hexagonal metals such as magnesium, the basal planes rotate until they contain the wire axis which tends to coincide with the $\langle 10\bar{1}0 \rangle$ direction in the basal plane. Other hexagonal metals show more complex behaviour, which is no doubt influenced by the presence of twinning as a mode of deformation, and also slip on prismatic and pyramidal planes (see Chapter 4).

In general terms extruded and swaged metals, provided recrystallization does not occur, have fibre textures similar to those occurring as a result of wire drawing.

12.2.2. *Rolling textures*

A rolling texture not only has a specific crystallographic direction parallel to the direction of rolling, but also a plane of low indices in the rolling plane. Like the fibre textures, there is scatter about an ideal orientation which decreases with increasing amounts of deformation (Fig. 12.2). Furthermore, more than one marked texture can co-exist in a particular rolled metal as well as much smaller amounts of minor textural components.

Face-centred cubic metals and alloys have a fairly simple predominant rolling texture, viz. $\{110\} \langle 112 \rangle$ with a $\{110\}$ plane in the rolling plane and a $\langle 112 \rangle$ direction in the rolling direction. Many of the common face-centred cubic metals develop this texture on cold rolling, together with varying amounts of a second texture which in copper is irrational until heavy deformations when it approaches $\{112\}\langle 111 \rangle$. Several other minor textural components have been discovered which are probably significant in the growth on annealing of the various recrystallization textures.

With body-centred cubic metals, the predominant preferred orientation is $\{001\} \langle 110 \rangle$ with cube planes in the rolling plane, but again there are other texture elements such as $\{112\} \langle 110 \rangle$ and $\{111\} \langle 112 \rangle$. Hexagonal metals tend to have the basal plane parallel to the rolling plane and the close-packed direction $\langle 11\bar{2}0 \rangle$ parallel to the rolling direction, as might be expected, for the compressive stresses would tend to rotate the slip plane into the plane of the sheet. This texture is most commonly found in those metals with near-ideal axial ratios, such as magnesium and cobalt.

So far we have considered rolling in only one direction (straight rolling). When two directions at right angles are used alternately, the process is referred to as cross rolling. With body-centred cubic metals, cross rolling tends to intensify or sharpen the $\{001\} \langle 110 \rangle$ texture characteristic of straight rolled sheet. On the other hand, in some face-centred cubic metals, the effect of cross rolling has been to reduce the degree of anisotropy developed in straight rolling.

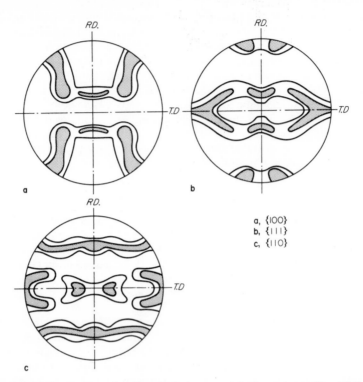

a, {100}
b, {111}
c, {110}

Fig. 12.2 Pole figures of cold-rolled commercial aluminium after 99·5 per cent reduction. **a**, {100}; **b**, {111}; **c**, {110}. *RD* = rolling direction, *TD* = transverse direction. (After von Göler and Sachs, 1927, *Zeit. Phys.*, **41**, 873)

12.2.3. *Role of stacking fault energy*

There is now substantial evidence[7,8,10] that stacking fault energy plays a part in determining deformation textures in face-centred cubic metals. Examination of the rolling texture of high purity copper shows that it does not conform to any simple texture and can best be described by the irrational texture {146} ⟨211⟩. On the other hand, pure silver during rolling develops a simple {110} ⟨112⟩ texture which has been described as a typical 'alloy' or 'brass' texture.[9] The same texture is developed when zinc is added to copper, and the transition is complete at 10 wt per cent zinc. Fig. 12.3 contrasts the {111} pole figures for the rolling textures of (a) copper, (b) α-brass. Recent stacking fault energy measurements indicate that silver has a substantially lower stacking fault energy (∼ 20 mJ m^{-2}) than copper (40–50 mJ m^{-2}), so it is not surprising that the silver texture is more like the α-brass texture.

Systematic experiments[10] have demonstrated that the {110}⟨112⟩ 'brass' texture can replace the 'copper' texture in a number of face-centred cubic metals, if solid solution alloying additions are made which lower the stacking fault energy sufficiently. For example, Smallman[10] has shown that copper–zinc, copper–aluminium and copper–germanium alloys

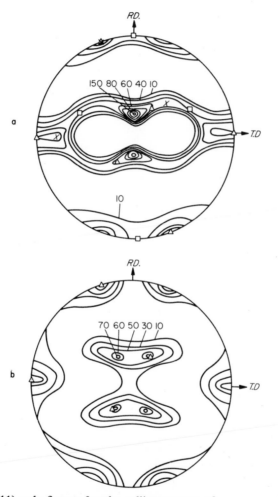

Fig. 12.3 {111} pole figures for the rolling texture of, **a,** copper; **b,** α-brass. *X* represents the ideal {146} ⟨211⟩ texture in copper, Δ represents the ideal {110} ⟨112⟩ texture in α-brass. The figures are relative intensities of pole distribution. (After Smallman and Green, 1964, *Acta metall.*, **12,** 145)

demonstrate the texture transition, and that the effectiveness of the solute is not related to the degree of atomic misfit, but to the stacking fault energy which is dependent more on the solute valency. So, for alloys of the same concentration, the order of effectiveness is germanium, aluminium, zinc. While the accurate measurement of stacking fault energies is still not possible, the transition in the texture seems to occur at an energy of about 30–40 mJ m^{-2}. The significance of stacking fault energy is further underlined in the behaviour of the whole range of gold–silver solid solutions in which the two atomic species have similar atomic size and the same valency. Pure gold with a higher stacking fault energy than silver exhibits the copper rolling texture, but as the silver concentration is increased, the $\{110\}\langle112\rangle$ texture gradually appears, until it is strong at 60 per cent silver and complete at 80 per cent silver, when the stacking fault energy is between 30 and 40 mJ m^{-2}. An example of the correlation between stacking fault energy and type of texture in copper–aluminium alloys is shown in Fig. 12.4.[67]

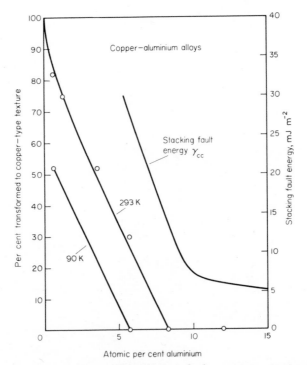

Fig. 12.4 Correlation between stacking fault energy, temperature of deformation and texture of copper–aluminium alloys. (After Christian and Swann, 1965, in *Concentrated Solid Solutions*, ed. T. B. Massalski. Gordon and Breach, New York, London, Paris)

The results described above were obtained after deformation at room temperature, but the texture developed is in fact temperature dependent. A rise in temperature of deformation favours the copper type texture rather than the brass type (Fig. 12.4), so that only a metal or alloy with very low stacking fault energy will retain the brass texture when the temperature is raised, for example, to $0.5T_M$. Another example is 18/8 austenitic steel, which has a fairly low stacking fault energy. As the rolling temperature is increased from 473 K to 1073 K, the $\{110\} \langle 112 \rangle$ brass texture is gradually replaced by the pure metal texture,[11] and it has been shown that the stacking fault frequency determined by X-ray analysis* decreases at the same time. So the effect can be attributable to an increase in stacking fault energy, although it is basically a result of a change in the deformation process as the temperature is raised.

A further factor which is particularly important in the development of the theory of texture change is that the brass texture $\{110\} \langle 112 \rangle$ is the texture which seems to form first at low deformations in all face-centred cubic metals, but as the deformation proceeds it is replaced by more complex textures in the case of metals with high stacking fault energies, and only those with low energies retain the originally developed texture in a sharper form. For example[8] commercial aluminium develops a $\{110\} \langle 112 \rangle$ texture after about 45 per cent reduction, but this is soon replaced by a texture approximating to $\{112\} \langle 111 \rangle$. On the other hand, silver retains the brass type texture $\{110\} \langle 112 \rangle$ up to 99 per cent reduction, but there is some evidence that beyond this deformation rotations away from the brass texture have commenced.

12.3   Theories of face-centred cubic rolling textures

We may regard the rolling process as two simpler deformation processes, compression normal to the rolling plane and tension in the rolling direction (plane strain). From single crystal studies we know that the axis of a grain deforming on a single slip system (111) [$\bar{1}$01] in tension will rotate towards the symmetry boundary of the basic stereographic triangle (Chapter 4) in the direction of the operative slip direction [$\bar{1}$01]. On reaching the boundary or a little beyond it, conjugate slip takes place on ($\bar{1}\bar{1}$1) [011], and this duplex slip results in the rotation of the tension axis to the stable end position [$\bar{1}$12]. Similarly, a compressive stress rotates the specimen axis until it eventually coincides with [101]. Calnan and Clews[12] regarded the rolling texture as comprising those orientations which simultaneously satisfied the tension and compression textures described above. As a result of a detailed analysis of grain rotations they predicted that the rolling texture of a face-centred cubic metal would be $\{110\} \langle 112 \rangle$ with several minor components. Calnan[13] attributed differences in the pure metal and

* The X-ray techniques are summarized by Warren.[66]

alloy textures to the fact that alloy crystals exhibit the phenomenon of overshooting of the symmetry boundary by the stress axis (Chapter 6), which has been explained on the basis that the latent slip system (conjugate system) hardens more than the primary system. However, overshooting of the symmetry boundary does not determine the final stable orientation produced by deformation, for, even in alloy crystals, overshooting is eventually eliminated and the $\{110\}\langle112\rangle$ end orientation attained. The fact remains that deviations from this texture occur at a late stage of the deformation, so there must be some type of preferential slip or another mode of deformation.

Single crystal studies have indicated that extensive slip on the cross-slip plane occurs at the end of Stage 2 and in Stage 3 of the deformation of face-centred cubic crystals. The transition to Stage 3 has been shown to be very dependent on stacking fault energy in so far as metals with low stacking fault energy cross-slip less readily, because the dislocations have to associate prior to moving on to the cross-slip plane (Chapter 4). Recent studies[8] on texture development in rolled face-centred cubic metals predict in the first instance the 'brass' or $\{110\}\langle112\rangle$ texture as a result of rotations of the tension and compressive axes caused by slip on the two systems with the highest resolved shear stresses under the bi-axial stress conditions† of rolling. However, when extensive cross-slip occurs, the orientation is shown to spread and to tend towards a new orientation $\{112\}\langle111\rangle$. This describes the behaviour of aluminium and, at heavier deformations, pure copper. Moreover, the tendency of metals to cross-slip increases markedly with the temperature of deformation, and it has been shown that the texture transition is likewise temperature dependent. At the higher rolling temperatures, only those alloys of low stacking fault energy, where cross-slip is very difficult, will retain the 'brass' texture. There is thus a good correlation between stacking fault energy, ease of cross-slip and the transition from the 'brass type' to the 'pure metal' rolling texture.

An alternative theory[14,15] for the development of the stable $\{110\}\langle112\rangle$ texture in metals and alloys of low stacking fault energy proposes that deformation twinning which is more prevalent in these alloys, is responsible for the observed major texture, and can also account for the weak $\{110\}$ $\langle001\rangle$ component. On the other hand, the pure metal texture is explained solely in terms of rotations due to slip, but slip on both $\{111\}$ and $\{100\}$ planes is required. The evidence for cubic slip in pure face-centred cubic metals is, however, limited, and cannot be considered to be substantiated as a major alternative slip process. On the other hand, there is considerable evidence for fine deformation twins in alloys of low stacking fault energy, and also in some pure metals (Chapter 8) even after tensile deformation of single crystals. The much heavier deformations involved in 80–99 per cent reduction in rolling must lead to extensive twinning in these

† A system of stress in which stresses, e.g. tensile and compressive, are operative in two directions normal to each other.

alloys which is difficult to detect but, nevertheless, must be an important variable in texture development.

It is probable that the texture transition in face-centred cubic wires of $\langle 100 \rangle$ to $\langle 111 \rangle$ as the stacking fault energy is raised, can be interpreted in a similar way to the rolling textures, involving the occurrence of cross-slip as the basic cause of the transition. Some work has also been done on body-centred cubic metal rolling textures, which without the influence of cross-slip would be expected to approximate to $\{112\} \langle 110 \rangle$; however, when cross-slip occurs, the stable orientation reached changes to $\{100\}$ $\langle 011 \rangle$.

12.4   Recrystallization textures

12.4.1. *Introduction*

In a deformed polycrystalline aggregate the grains usually exhibit preferred orientations or textures which are characteristic of the mode of deformation, e.g. rolling, drawing, etc. When such an aggregate is annealed, the recrystallized grains also possess a preferred orientation which in many cases is even stronger than the deformation texture.

The texture may arise from:

1. dislocation rearrangements (polygonization) in the deformed grains, or
2. recrystallization of the deformed grains (primary recrystallization), or
3. subsequent growth of selected grains in the recrystallized material. Often the grain growth process involves only a few grains and a coarse-grained structure results. This phenomenon has been called '*secondary recrystallization*', but is often referred to as *exaggerated grain growth* or *coarsening*.

There are many kinds of recrystallization textures even amongst metals of the same crystal structure, and often one metal will have different textures when the annealing conditions are changed, or it may have two textures co-existing in the one sample. In this account the underlying principles will be emphasized rather than an attempt made to give detailed descriptions and explanations of the behaviour of many metals after different annealing treatments.

12.4.2. *Origin of annealing textures*

In the past, a great deal of controversy centred around two theories of the origin of annealing textures: one based on the existence of *oriented nuclei* in the deformed metal, and the other on the growth of nuclei in specific crystallographic directions (*oriented growth*). Like many controversies, subsequent work has shown that both mechanisms probably play a part in the development of recrystallization textures. We now know that at least some recrystallization nuclei grow from sub-grains which are formed by deformation and are sharpened by subsequent annealing, and that the sub-grains which grow are likely to be in preferred sites where the driving force is large. However, it is also true that the grains which grow the fastest have optimum orientation relationships with the matrix, so that oriented growth is a significant process in the formation of annealing textures.

In general, the sharper the deformation texture, the better defined will be the recrystallization texture;[16] an extreme case is the annealing of a heavily deformed single crystal. Perhaps the best example of a sharp re-crystallization texture is the *cube* texture {100} ⟨001⟩ which is obtained in many face-centred cubic metals on recrystallization after the cold rolling of sheet material.[2] In its extreme form, the material resembles a single crystal, so closely oriented are the individual recrystallized grains to the ideal orientation (Fig. 12.5). The pole figure obtained from X-ray data

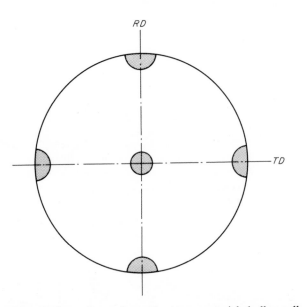

Fig. 12.5 {100} pole figure for an iron–50 wt per cent nickel alloy rolled heavily then annealed at 1373 K. Cube texture. (After Burgers and Snoek, 1935, *Z. Metallk.*, **27**, 158)

shows how sharp the distribution of orientations can be. Fig. 12.6 illustrates the effect of increasing prior deformation on copper (99·96 per cent) in the range 80–95 per cent reduction in thickness of the sheet. The percentage of cubically aligned grains formed on subsequent annealing rises from 5 per cent to over 95 per cent. The figure also shows that the perfection of the texture improves as the annealing temperature is raised. The cube texture has been observed in copper, gold and aluminium, also in iron–nickel alloys, but the addition of alloying elements to copper, e.g. zinc, tin, arsenic, suppresses the texture in favour of a more complex one, $\{113\} \langle 211 \rangle$.

The pure copper rolling deformation texture is irrational, but if the deformation is heavy it approaches the orientation $\{112\} \langle 111 \rangle$. On close examination this texture is found to be associated with traces of $\{100\} \langle 001 \rangle$. On annealing it is the $\{112\} \langle 111 \rangle$ texture which is host for the cube texture,

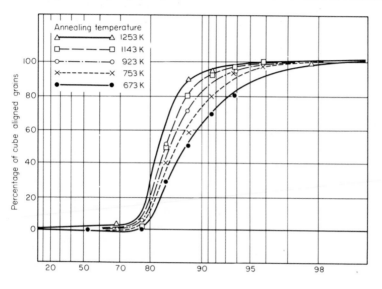

Fig. 12.6 Effect of increasing cold work and annealing temperature on the perfection of the cube texture in annealed copper. (After Baldwin, 1946, *Trans. AIME*, **166**, 591)

which is nucleated by the traces of cube texture in the as-deformed material. So the first requirement for the development of a strong annealing texture is a minor texture component accompanying the major deformation texture.

The criterion which decides which of the minor textures in the deformed specimens will grow to absorb the deformed grains during annealing is the mobility of the boundary between the major recrystallization orientation and the major deformation texture. The boundary mobility is a function of the energy; Read and Shockley[17] have shown for low angle boundaries that the energy is related to the disorientation thus:

$$E = E_0 \theta (A - \ln \theta)$$

where E = grain boundary energy, θ = disorientation, E_0 and A are constants.

This equation applies very well even up to large angles and predicts a maximum at $\theta = 30°$ which has been confirmed by experiments on silicon iron.* In face-centred cubic metals, the boundary of greatest mobility is that produced by a 30–50° rotation around a $\langle 111 \rangle$ axis,[18, 19] while it is 25–40° about a $\langle 110 \rangle$ axis in body-centred cubic metals, and in the close-packed hexagonal metals a 30° rotation about the [0001] axis.

When the orientation relationships between deformation and recrystallization textures are examined, a large number can be described by a rotation about a crystallographic axis: in the case of face-centred cubic metals this is normally 30–50 degrees about $\langle 111 \rangle$ or occasionally $\langle 100 \rangle$.[20] In a face-centred cubic metal such as copper there will be eight equivalent orientations formed by clockwise or anti-clockwise rotations about the four $\langle 111 \rangle$ axes, but the full range of orientations is not usually developed. However, Liu and Hibbard[20] deformed a single crystal of copper on a {110} plane in a $\langle 112 \rangle$ direction to obtain a sharp deformation texture; on annealing they obtained a recrystallization texture comprising the eight orientations theoretically predicted.

The cube texture in polycrystalline face-centred cubic metals has also been explained in terms of the orientation relationships for maximum grain boundary mobility. Beck and Hu have shown that this annealing texture has a favourable orientation relationship with all the four possible preferred orientations in the deformed matrix.[21] Consequently, the cube orientations will tend to eliminate competitive orientations as soon as they are in effective contact with the matrix. This process is assisted by a fine grain size and heavy deformation as well as high annealing temperatures. On the other hand, the texture will be suppressed by any factor which prevents the new grain growing to achieve effective contact with matrix grains, e.g. precipitation.

In metals such as silver[18] and copper,[22] the deformation texture on rolling can be changed from the brass type to the copper type by raising the temperature of deformation. On annealing each deformation texture then gives rise to its characteristic recrystallization texture, the brass type producing a {113} $\langle 211 \rangle$ type, while the copper type gives rise to the cube texture. In both cases the recrystallization texture can be derived from the deformation textures by rotation of between 25 and 40° about a $\langle 111 \rangle$ axis, consistent with the original observations of Beck. Similar behaviour has been observed in face-centred cubic alloys, e.g. 18/8 austenitic steel,[23] where an increase in the temperature of deformation similarly changes the deformation and recrystallization textures. Rolling at 473 K produces the brass rolling texture, and on annealing a texture similar, but not

* It should, however, be pointed out that the derivation is based on a dislocation model which does not give a true representation of a grain boundary above 10–15° disorientation.

identical to that obtained in brass. On the other hand, rolling at 1073 K gives a copper deformation texture which recrystallizes on annealing to a well-defined cube texture.

Some definitive studies of recrystallization textures in rolled body centred metals and alloys, for example iron–3 per cent silicon alloy, have been made with rolled single crystals of various initial orientations.[24] The usual preferred orientation in cold rolled polycrystalline iron and silicon–iron sheet is {001} ⟨110⟩ with some {111} ⟨112⟩ which on annealing goes to a {001} ⟨001⟩ or {110} ⟨001⟩ type texture.[2] If a single crystal is chosen to give a sharp {001} ⟨110⟩ deformation texture, this recrystallizes with difficulty to a complex structure in which, however, the new grains are related to the deformed orientation by a rotation of 25–30° about ⟨110⟩ axes. If on the other hand, orientations are used which give an alternative {111} ⟨112⟩ deformation texture, the recrystallization texture is predominantly a strong {110} ⟨001⟩ texture, which can also be formed from the deformation texture by rotation of 25–30° about an ⟨110⟩ axis.

In similar experiments on orientated crystals of iron–2 per cent aluminium alloy giving {100} ⟨110⟩ and {111} ⟨112⟩ deformation textures as-rolled, the primary recrystallized grains were aligned in a {113} ⟨332⟩ texture and a {110} ⟨001⟩ texture.[25] The relative proportions of the two deformation textures varied from surface to the middle of the rolled sheet, and as the {100} ⟨110⟩ deformation texture diminished, so the recrystallization texture {113} ⟨332⟩ became less pronounced in the annealed alloy. At the same time, the {110} ⟨001⟩ recrystallization texture became dominant when the {111} ⟨112⟩ texture was stronger in the deformed alloy. Such relationships are in keeping with the theory that grain boundaries of maximum mobility occur when the lattice disorientation across the boundary involves a large rotation (25–40°) about a ⟨110⟩ axis (for body-centred cubic metals).

Grain boundary mobility is reduced by the presence of fine dispersions of a second phase. It is known that the microstructure of recrystallized alloys can be influenced by the presence of fine precipitate,[26] and the rate of recrystallization is altered primarily by reducing the rate of growth G. That the texture is also altered has been confirmed by work on the recrystallization of cold-rolled iron containing 0·8 per cent copper.[27] In this material, if the copper is allowed to precipitate after deformation, the rolling texture is largely retained on subsequent annealing, preventing the formation of the usual annealing texture.

12.5   Secondary recrystallization (exaggerated grain growth)

When certain types of well-defined recrystallization texture, e.g. the cube texture, are annealed at a high temperature, a few grains grow preferentially, and absorb the matrix completely (Fig. 12.7). Examination of the orientations of coarse grains which grow in cube-textured copper and aluminium[28] reveals that they are related to the cubically-aligned grains by a 38° rotation in one sense about a $\langle 111 \rangle$ axis, or 22° in the opposite sense. Bowles and Boas[29] showed that if the primary texture was destroyed by cross rolling, secondary recrystallization does not occur. This phenomenon is another example of the application of the oriented-growth principle, where small grains widely disoriented from the main primary recrystallization texture are able to grow and absorb the whole sample by virtue of their energetically favoured boundaries. If the primary texture is reduced in perfection, the driving force for the growth of the secondary grains is less.

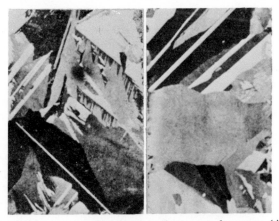

Fig. 12.7   Secondary recrystallization in cube-textured copper. (After Bowles and Boas, 1948, *J. Inst. Metals*, **74**, 501)

The development of the well-known $\{110\} \langle 001 \rangle$ texture by secondary recrystallization in silicon iron has been shown to depend very much on the presence of a uniform dispersion of precipitate which can maintain a small grain size in the primary matrix.[30] The nature of the precipitate seems unimportant in so far as manganese sulphide, silica, silicon nitride and a number of other nitrides and carbides[31] have been shown capable of encouraging secondary recrystallization to the $\{110\} \langle 001 \rangle$ texture by restraining general grain growth, and allowing a few favourably oriented secondary grains to grow. However, the size and distribution of the particles has been proved to be important;[32] in general, the finer and more concentrated the dispersion, the more complete is secondary recrystallization. The most effective method of achieving this is to utilize a second phase, which is precipitated from solid solution so that the dispersion an be easily controlled by varying the rate of cooling prior to working and annealing. A similar mechanism has been shown to function in an aluminium-manganese alloy with small concentrations of a manganese-rich phase.[33]

More recently, it has been found that orientation relationships between secondary and primary recrystallized grains are not always the controlling factor in secondary recrystallization. This is particularly true in 'two-dimensional' grain structures obtained in heavily rolled thin sheet, when on recrystallization the grain diameter approximates to the sheet thickness, and most grain boundaries intercept the sheet surfaces. In these circumstances the surface energy and its dependence on the orientations of the crystals, becomes an important variable. Walter and Dunn[34] have found in high purity iron–3 per cent silicon alloy sheet that the surface energy is lowest for the {100} orientation in normal annealing conditions, so that in secondary recrystallization, grains with this orientation relative to the surface tend to grow. However, grains with {110} planes in the surface also tend to grow, and subsequent investigation has shown that the preferred orientation depends on the atmosphere during annealing.[35] If the oxygen content is high, absorption of oxygen atoms causes the surface energy of the {100} planes to be the lowest, but if the oxygen content of the atmosphere is very low, the {110} planes have the lowest surface energy, and thin grains with this orientation to the surface will grow preferentially. So depending on the annealing conditions, the secondary recrystallization texture can be {100} ⟨001⟩, {110} ⟨001⟩ or a mixture of both. It is interesting to note that zone purified iron[35] exhibits similar behaviour to high purity iron–silicon alloys except that the primary recrystallization texture is different, viz. {111} ⟨110⟩, emphasizing that in these circumstances the reaction is not a simple oriented-growth process. This is also clear from the fact that, in non-oxidizing conditions, a secondary texture {110} ⟨001⟩ can grow in a primary texture of the same kind.

12.6   Anisotropy of mechanical properties

In Chapter 2, the anisotropy of mechanical properties of single crystals has been discussed and explained in terms of the crystallography of deformation. It follows that polycrystalline aggregates with textures should have different mechanical properties as the direction of measurement in the material changes, and that it should be true of both deformed and annealed metals alike provided a texture is present. Such anisotropy can either be useful or a serious disadvantage. For example, in the deep drawing of cups, ears frequently develop symmetrically around the rim (Fig. 12.8) with associated variations in thickness of the cup, as a result of the anisotropy of the mechanical properties. On the other hand, with the appropriate textures a sheet can be considerably stronger in useful directions, than material in which a texture has not been developed.

The directionality of mechanical properties in rolled, and in rolled and annealed sheet is a familiar aspect of the deformation of copper, copper-base alloys, aluminium, etc. Intensive study of deformation and annealing textures and the influence on them of metallurgical variables, such as the amount of reduction prior to annealing, the annealing temperature, etc., has enabled the production of sheet with the minimum amount of texture, which eliminates anisotropy of flow during drawing and the resulting 'earing' phenomenon. Texture control is now an important and fairly well understood aspect of mechanical metallurgy, although the number of possible variables can often make precise prediction of behaviour difficult.[1, 4]

Earing in copper and brass sheet has been extensively studied.[36, 37] In brass, four ears commonly occur at 45° to the rolling direction and are associated with the {110} ⟨112⟩ annealing texture,[38] which is similar to the rolling texture; however, a six-ear type can develop when this texture is changed by manipulation of some of the metallurgical variables. While the effects of the latter are undoubtedly complex, it is known that certain factors which sharpen the annealing texture, e.g. high final annealing temperature, heavy final rolling reduction, lead to more pronounced earing.

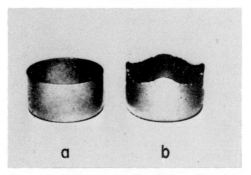

Fig. 12.8   Deep-drawn copper cups. **a**, Random grain orientation; **b**, single texture strip {100} ⟨001⟩. (After T. L. Richards, 1949, *Prog. in Met. Phys.*, **1**, 281. Pergamon Press, Oxford)

With copper the 45° earing is observed, but in addition four ears, two at 90° to the rolling direction and two parallel to it, are frequently obtained.

Some progress has been made in relating the earing behaviour of α-brass strip to the pole figures obtained by X-ray analysis,[39] in so far as the number of ears is found often to coincide with the number of areas of high density of normals to crystallographic planes occurring in pole figures. For example, the {111} slip planes tend to be parallel to the rolling direction, and at 90° to the rolling direction when earing is pronounced in the 45° direction. These latter positions would be the optimum ones for flow because a tensile stress acting in these directions would encounter the maximum number of {111} planes oriented to sustain the maximum resolved shear stress (45° to the stress axis).

The anisotropy of cube texture copper has been studied in detail.[40] Fig. 12.9 shows the marked variations in mechanical properties which have been found in this material. The ears occurring in the rolling direction and at 90° to it have been shown to be related to the magnitude of the tensile stresses in a radial direction during deep drawing. However, this simple approach is of doubtful validity, because the drawing process cannot be considered as a process in which radial tensile stresses predominate. Bourne and Hill[41] have shown that near the rim during drawing, a circumferential compression is the principal factor in determining the formation of ears. They measured the relative thicknesses and widths of tensile specimens at different elongations as a function of sheet direction, and found these strain ratios to be sensitive to orientation. From such data they were

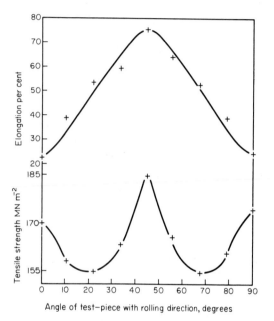

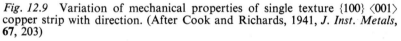

Fig. 12.9   Variation of mechanical properties of single texture {100} ⟨001⟩ copper strip with direction. (After Cook and Richards, 1941, *J. Inst. Metals*, **67**, 203)

able to deduce stationary values of the circumferential yield stress which they associated with the maximum and minimum tendencies to form ears during drawing.

More recently attention has turned to the anisotropy of mechanical properties *normal* to the plane of the sheet as compared with those in the plane of the sheet. This 'normal' directionality can be more pronounced than the 'planar' directionality, and can give rise to improved performance during drawing operations. This type of anisotropy is difficult to measure directly, but it is indirectly reflected in the behaviour of tensile specimens in the plane of the sheet.[5] These specimens exhibit different plastic strain across their width and thickness. This has been shown to be a consequence of difference in strength of material parallel and perpendicular to the plane of the sheet.[42] The ratio of the width to thickness strains, R, is found to be a constant independent of strain, and is a useful measure of directionality. It does, however, vary with the direction of the tensile axis in the rolling plane, but for textured material it is always greater than 1 (Fig. 12.10).

The fact that $R_0 \neq R_{45} \neq R_{90}$ is a result of planar anisotropy, while the slope of the curve deviating from 1 is due to 'normal' anisotropy. Studies of a number of steels for deep drawing have indicated that, when R is high,

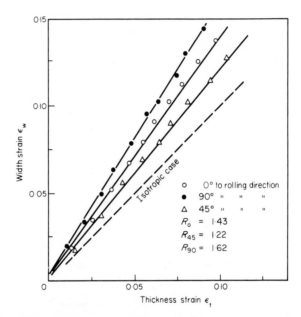

Fig. 12.10 Width strain versus thickness strain for mild steel tested in three directions. (After Whiteley, 1964, *Fundamentals of Deformation Processing*, Sagamore Army Materials Conference, Syracuse University Press, New York)

the deep drawing performance is good.[43] The textured material with a high R has an increased wall strength during drawing, and thus can support a higher punch load than isotropic material, with the result that a larger die blank can be drawn without wall failure by cracking. The superiority of aluminium killed steels over rimming steels in this work was found to be due to the stronger 'normal' directionality in the former material. There is evidence to suggest that this difference in degree of texture is due to the presence of aluminium nitride particles in the killed steels, which encourage texture sharpening by selective grain growth on annealing. The relationship between the ratio R and texture has been confirmed by detailed X-ray investigations of the preferred orientations which develop in cold rolled and annealed low carbon steels.[44]

The strain ratio concept has now been applied to a number of different materials and a general correlation of drawability with R has been obtained (Fig. 12.11) with a very high value for titanium sheet which, as titanium possesses an hexagonal structure, would be expected to develop marked anisotropy.

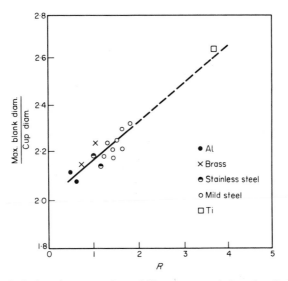

Fig. 12.11 Relation between drawability measured by the Swift cupping test, and the average strain ratio R. (After Whiteley, 1964, *Fundamentals of Deformation Processing*, Sagamore Army Materials Conference, Syracuse University Press, New York)

12.7   Texture strengthening

In recent years textures have been used to impart strengthening to metal sheets.[45] The best example is to be found in the highly anisotropic hexagonal metals such as titanium. The rolling texture of the hexagonal metals is ideally $(0001) \langle 10\bar{1}0 \rangle$: the metals magnesium, titanium and zirconium have a high proportion of the crystal grains lying with their basal planes in the plane of the sheet with varying degrees of scatter about the ideal texture. If we now consider the tensile deformation of a strip with the ideal texture, the three slip directions for basal slip are in the plane of the strip, moreover, these directions are also those used in prismatic slip $\{10\bar{1}0\}$ and pyramidal slip $\{10\bar{1}1\}$ (Fig. 4.29) which commonly occurs in titanium, zirconium and magnesium. Consequently, on tensile deformation, all slip displacements will tend to be in the *plane* of the strip, which will thus not thin during deformation. It has already been mentioned that thinning resistance in a tension test is measured by the ratio R of width strain to thickness strain. In a normal isotropic polycrystalline metal, i.e. one with no texture, $R = 1$, but in the case of the ideal hexagonal texture discussed, $R = \infty$; however, this value is substantially reduced by orientation spread in the texture. Typical values of R for hexagonal metals which form the appropriate texture are:

    α-titanium alloys   3–5
    zircalloy-2 (zirconium-base), up to 7·7 (heavily rolled)

With very strong textures, R is not very dependent on the direction in the rolling plane corresponding with the specimen axis. The mechanical properties of such oriented sheets are not outstanding in uniaxial tension: for example, the yield stress in tension would be fairly low if non-basal slip took place. However, with bi-axial tension, which can be considered as a hydrostatic tension plus a uniaxial compressive stress normal to the plane of the sheet (Fig. 12.12), the yield stress is reached only when the tensile

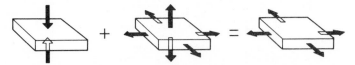

Fig. 12.12   Stresses for yielding under bi-axial tension. (After Hosford and Backofen, 1964, *Fundamentals of Deformation Processing*, Sagamore Army Materials Conference, Syracuse University Press, New York)

stress reaches a value equal to the compressive yield stress normal to the plane of the sheet. This compressive yield stress through the sheet will be approximately the same as the uniaxial tensile stress parallel to the sheet, in randomly oriented materials, so yielding will take place at the uniaxial tensile yield stress. However, in heavily textured material, the absence of suitably oriented slip systems which would permit thinning of the sheet leads to high compressive yield stresses, and thus substantial strengthening under conditions of biaxial tension. An extreme case of this has been reported for a single crystal of beryllium[46] compressed along the c axis at 77 K. No yielding was observed before the stress reached 3900 MN m^{-2}: it then fractured catastrophically. The strengthening by forming a texture which will not deform readily is referred to as 'texture hardening'. Such a phenomenon should be of value in many sheet metal applications, and in the manufacture of pressure vessels. It is also likely that sheet which is texture hardened will show greater resistance to denting. A recent study of titanium alloys has shown that the Ti–4 per cent Al alloy[47] can give R values up to 6 as a result of a strong basal plane rolling texture. This material is thus likely to be very useful from the textural hardening point of view.

As the basal rolling texture is usually preserved and even sharpened on recrystallization, the recrystallization textures must also be considered as potentially useful methods of textural hardening.

As cubic materials have a multiplicity of slip systems they do not show as marked anisotropy of mechanical properties as do the hexagonal metals. However, Hosford and Backofen,[45] using the original Taylor analysis which determines the properties of the polycrystalline aggregate from those of the single crystals (see Chapter 9), have shown that some texture hardening can be expected. The analysis has been used to demonstrate that the drawability of steel sheet is higher when the texture contains {111} components in the plane of the sheet, and reduced when {100} components are predominant.

12.8   Anisotropy of magnetic properties

The ferromagnetic properties of body-centred cubic α-iron are anisotropic. The saturation magnetization is the same for crystals of different orientations, but the rate of approach to saturation varies markedly with orientation.[48] Fig. 12.13 shows that the ⟨100⟩ orientation saturates more quickly than either the ⟨110⟩ or ⟨111⟩: the cube direction is thus referred to as the direction of easiest magnetization while the cube diagonal direction is the most difficult.

Consequently, it will be appreciated that a polycrystalline iron sheet with

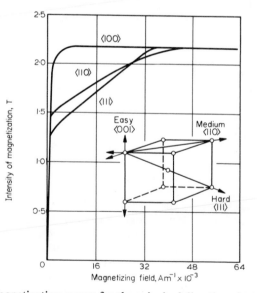

Fig. 12.13   Magnetization curves for the principal directions in an iron crystal. (After Honda and Kaya, 1926, *Science Reports, Tohoku University*, **15**, 721)

the appropriate texture can be magnetically superior to a randomly oriented sheet. The electrical industry uses iron–3·5 per cent silicon alloy sheet for transformer cores, and to achieve minimum hysteresis losses, ideally there should be a highly developed cube texture {100} $\langle 001 \rangle$, which has two easiest directions of magnetization $\langle 100 \rangle$ in the plane of the sheet. The next best is a {110} $\langle 001 \rangle$ or cube-on-edge texture which has only one easiest direction of magnetization in the plane of the sheet. In the 1930's the problem of producing a {110} $\langle 001 \rangle$ texture of a high degree of perfection by recrystallization of cold worked silicon–iron sheet was solved[49] by careful attention to metallurgical variables, such as the degree of reduction between intermediate anneals, and the annealing temperature. Subsequent investigation[50] has shown that the sharpest texture is obtained by secondary recrystallization from a primary recrystallized texture which need not be sharp. The sheet is cold rolled to a small thickness (0·2–0·5 mm), then annealed at 1073 K to allow primary recrystallization to take place, producing a fairly fine grain size (0·01–0·02 mm). If the grain boundaries of the primary texture are locked by fine precipitate, on raising the temperature to 1173–1273 K, a very coarse new texture appears which is {110} $\langle 001 \rangle$. This texture is not obtained in high purity silicon iron;[51] however, if nitrogen is introduced to form silicon nitride particles, the cube-on-edge texture appears. The role of these inclusions has been dealt with in Section 12.5. A typical example of the influence of inclusions (manganese sulphide) on texture, and consequently the anisotropy of magnetic properties is given in Fig. 12.14, where the much lower magnetic torque anisotropy of high purity silicon iron is compared with the strong magnetic anisotropy of similar material with a dispersion of manganese sulphide particles.[30]

In recent years a further technological step has been taken, in so far as cube texture {100} $\langle 001 \rangle$ silicon iron sheet can now be manufactured on a large scale with a high degree of texture perfection. Many of the essential practical details have not been revealed, but a general outline of the process can be given. Again the texture control is primarily achieved by use of small second phase particles. The growth of this texture was first reported in the literature in 1957.[52,53] The silicon iron sheet is best rolled to about 0·04 mm to obtain two-dimensional conditions for the growth of the final cube texture. Firstly, however, a fine grained primary recrystallization texture of the cube-on-edge {110} $\langle 001 \rangle$ type is obtained with a grain diameter several times the sheet thickness. This texture already possesses minor elements of {100} $\langle 001 \rangle$.[54] For successful subsequent growth of the cube texture, the matrix must be stabilized either by the edge effect when grain boundaries are locked on intersecting the surface, or by the presence of a dispersion of a finely divided phase such as manganese sulphide, silicon nitride, etc. The growth of {100} planes in the plane of the sheet as a coarse-grained secondary texture can then occur, provided the conditions are such that the {100} planes have the lowest surface energy as well as the correct orientation relationship with the primary texture for maximum growth rate. A high oxygen content in the annealing atmosphere will cause the {100} planes to have the lowest surface energy. The final secondary anneal is in the range 1323–1473 K, when the coarse texture which develops is aligned to within 5° of {100} ⟨001⟩. A similar texture has

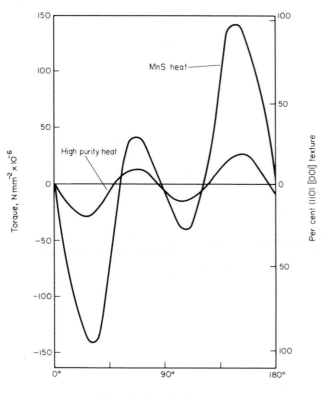

Fig. 12.14   Magnetic torque curves for a high purity iron–silicon alloy and a similar material with a dispersion of manganese sulphide. (After May and Turnbull, 1958, *Trans. AIME*, **212**, 769)

recently been obtained using a powder metallurgy technique in which carbonyl iron powder and silicon powder are sintered, then rolled.[55] The recrystallized grain size can be controlled by choice of powder size.

12.9   Anisotropy of thermal expansion

The thermal properties, thermal conductivity and expansion are isotropic in cubic metals, but they are highly anisotropic in many non-cubic metals. Some values for the coefficients of thermal expansion parallel and perpendicular to the c axis are given in Table 12.2.

Table 12.2. Coefficients of thermal expansion of some non-cubic metals

Metal	Structure	c/a	Coefficient of thermal expansion at 293 K, $\times 10^{-6}\,K^{-1}$	
			$\parallel$ *to c axis*	$\parallel$ *to a axis*
Cadmium	c.p.h.	1·886	52·6	21·4
Zinc	c.p.h.	1·856	63·9	14·1
Magnesium	c.p.h.	1·624	27·0	25·4
Zirconium	c.p.h.	1·589	6·96	5·65
Tin	tetragonal	0·546	30·5	15·5
Uranium (α)	orthorhombic	$a = 2\cdot85$ Å	$a = 33\cdot24$ $b = -6\cdot49$	
		$b = 5\cdot87$ Å	$c = 30\cdot36$	
		$c = 4\cdot96$ Å		

The anisotropy of thermal expansion causes stresses to be set up in non-cubic metals when they are subjected to temperature changes, because in a random aggregate of crystals the thermal expansion in a given direction changes as a boundary is crossed. Individual grains cannot conform to the changes in dimension imposed by the thermal expansion, as they are firmly joined to their neighbours; consequently, stresses are set up which, if the temperature fluctuation is great enough, lead to plastic deformation (Fig. 12.15). It has been shown by Boas and Honeycombe that polycrystalline specimens of cadmium, zinc and tin when repeatedly cycled

Fig. 12.15   Plastic deformation in polycrystalline cadmium after 10 thermal cycles between 293 and 423 K. $\times 80$

between 293 and 423 K undergo substantial plastic deformation,[56] whereas an isotropic metal such as lead remains undeformed. The deformation is clearly related to the extent of thermal anisotropy, for magnesium, which has only a small anisotropy of thermal expansion, deforms to a much smaller degree than zinc or cadmium. The latter metals show marked evidence of slip, twinning and grain boundary movement which are practically absent in magnesium under similar conditions. Several estimates have been made of the stress developed as a function of thermal expansion, either using a bi-crystal model[56] or a random aggregate.[57] An expression obtained using the latter model is

$$\sigma = \frac{2}{3(1 - \nu)} \int_{T_1}^{T_2} E(\alpha_c - \alpha_a) \, dT$$

An evaluation of this expression for hexagonal zirconium is shown in Fig. 12.16: the yield stress of zirconium at 673 K is around 48 MN m^{-2}, so cycling in the range 293–673 K would be expected to produce plastic deformation. This has been confirmed by experiments.[58]

In random polycrystalline aggregates, thermal cycling leads to surface rumpling, but no dimensional changes. If, however, a preferred orientation is present in the material, repeated thermal cycling leads to changes in the specimen dimensions or growth, because the strains resulting from the deformation are dependent on crystal orientation. The role of texture is

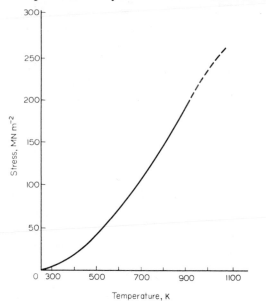

Fig. 12.16 Estimate of maximum stress induced in zirconium during thermal cycling. (Aitchison *et al.*, 1962, in *Properties of Reactor Materials and Effects of Radiation Damage*. Ed. by D. J. Littler, Butterworth, London)

particularly significant in the thermal cycling of α-uranium[59] where the growth rate during thermal cycling increases with the reduction in area given to a rod prior to cycling. If, on the other hand, the grain orientations are 'randomized' by annealing in the β-uranium region, the subsequent growth on thermal cycling is very small. The magnitude of the growth phenomenon in thermally cycled uranium rods is shown in Fig. 12.17, where the most severely cycled rods have grown 30 per cent. In practice, texture is by no means the only variable: in addition the temperature range, heating and cooling rates, time at the highest temperature and the grain size have all been shown to influence growth during cycling.[60] An interest-

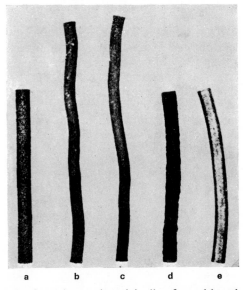

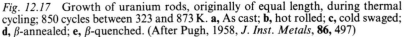

Fig. 12.17 Growth of uranium rods, originally of equal length, during thermal cycling; 850 cycles between 323 and 873 K. **a**, As cast; **b**, hot rolled; **c**, cold swaged; **d**, β-annealed; **e**, β-quenched. (After Pugh, 1958, *J. Inst. Metals*, **86**, 497)

ing side-effect of thermal cycling is the acceleration of creep as a result of the intergranular stresses which can augment the externally applied stress.[61, 62]

It is possible to study the growth process under simpler conditions by thermally cycling bi-crystals: in experiments with zinc, one crystal has been found to grow more rapidly than the other, but only when the maximum temperature was high enough to allow grain-boundary sliding.[63] This phenomenon is frequently referred to as 'thermal ratcheting'. On heating grain A tends to expand more rapidly than grain B in the direction parallel to the common boundary until it deforms plastically because of the restraints upon it. As soon as a sufficiently high temperature is reached, the stresses are relieved by grain-boundary sliding. On cooling sliding eventually ceases, and is replaced by slip when the stress builds up again in grain *A*. On reheating the stress will pass through zero and the whole process will be repeated, leading to further grain-boundary displacement and slip within the grains. Other theories do not invoke thermal ratcheting as a mechanism in polycrystalline metals. In uranium it has been concluded that two factors are essential for growth in addition to the anisotropy of thermal expansion within the grains: they are a double-fibre texture with each component having different coefficients of expansion along the rod axis, and a change in the mode of deformation as the temperature is raised.[64] The first of these factors seems not to be essential, although it is generally agreed that a texture of some kind is necessary for growth. On the other hand, there is support for the view that deformation must be by slip at low temperatures and by creep processes such as grain-boundary sliding and sub-structure formation at high temperatures.

Growth can also occur in cubic metals,[65] provided a texture is present, but the stresses must be induced by large temperature gradients which can occur when specimens of substantial dimensions are rapidly heated and cooled. It seems clear that to retain maximum dimensional stability under conditions which lead to stress generation, the degree of preferred orientation must be reduced to a minimum.

General references

1. UNDERWOOD, F. A. (1961). *Textures in Metal Sheets.* MacDonald, London.
2. BARRETT, C. S. and MASSALSKI, T. B. (1966). *The Structure of Metals.* John Wiley, New York and London.
3. BOAS, W. and MCKENZIE, J. K. (1950). "Anisotropy in Metals." *Prog. Metal Phys.*, **2**, 90.
4. WASSERMANN, G. and GREWEN, J. (1962) *Texturen metallischer Werkstoffe.* Springer Verlag, Berlin.
5. *Fundamentals of Deformation Processing* (1964). Sagamore Army Materials Conference, Vol. 1. W. A. Backofen *et al.* (Eds.). Syracuse University Press, New York.
6. *Recrystallization, Grain Growth and Textures* (1966). American Society for Metals.

References

7. SMALLMAN, R. E. and GREEN, D. (1964). *Acta metall.*, **12**, 145.
8. DILLAMORE, I. S. and ROBERTS, W. T. (1964). *Acta metall.*, **12**, 281.
9. HSUN HU CLINE, R. S. and GOODMAN, S. R. (1961). *J. appl. Phys.*, **32**, 1392.
10. SMALLMAN, R. E. (1955). *J. Inst. Metals*, **83**, 10.
11. GOODMAN, S. R. and HSUN HU (1964). *Trans. AIME*, **230**, 1413.
12. CALNAN, E. A. and CLEWS, C. J. B. (1950). *Phil. Mag.*, **41**, 1085.

13. CALNAN, E. A. (1954). *Acta metall.*, **2**, 865.
14. HAESSNER, F. (1963). *Z. Metallk.*, **54**, 98.
15. WASSERMANN, G. (1963). *Z. Metallk.*, **54**, 61.
16. COOK, M. and RICHARDS, T. L. (1940). *J. Inst. Metals*, **66**, 1.
17. READ, W. T. and SHOCKLEY, W. (1950). *Phys. Rev.*, **78**, 275.
18. HSUN HU, CLINE, R. S. and GOODMAN, S. R. (1962). *Trans. AIME*, **224**, 96.
19. BECK, P. A. and HSUN HU (1949). *Trans. AIME*, **185**, 627.
20. LIU, Y. C. and HIBBARD, W. R. (1953). *Trans. AIME*, **197**, 672.
21. BECK, P. A. and HSUN HU (1952). *Trans. AIME*, **194**, 83.
22. HSUN HU and GOODMAN, S. R. (1963). *Trans. AIME*, **227**, 627.
23. GOODMAN, S. R. and HSUN HU (1965). *Trans. AIME*, **233**, 103.
24. DUNN, C. G. and KOH, P. K. (1956). *Trans. AIME*, **206**, 1017.
25. CLINE, R. S. and HSUN HU (1965). *Trans. AIME*, **233**, 57.
26. RICKETT, R. L. and LESLIE, W. C. (1959). *Trans. ASM*, **51**, 310.
27. LESLIE, W. C. (1961). *Trans. AIME*, **221**, 753.
28. KRONBERG, M. L. and WILSON, F. H. (1949). *Trans. AIME*, **185**, 501.
29. BOWLES, J. S. and BOAS, W. (1948). *J. Inst. Metals*, **74**, 501.
30. MAY, J. E. and TURNBULL, D. (1958). *Trans. AIME*, **212**, 769.
31. FIEDLER, H. C. (1961). *Trans. AIME*, **221**, 1201.
32. FIEDLER, H. C. (1964). *Trans. AIME*, **230**, 95.
33. BECK, P. A., HOLSWORTH, M. L. and SPERRY, P. R. (1949). *Trans. AIME*, **180**, 163.
34. WALTER, J. L. and DUNN, C. G. (1960). *Trans. AIME*, **218**, 915.
35. DUNN, C. G. and WALTER, J. L. (1962). *Trans. AIME*, **224**, 518.
36. COOK, M. and RICHARDS, T. L. (1941). *J. Inst. Metals*, **67**, 203.
37. BALDWIN, W. M. (1946). *Trans. AIME*, **166**, 591.
38. WILSON, F. H. and BRICK, R. M. (1942). *Trans. AIME*, **161**, 173.
39. BURGHOFF, H. L. and BOHLEN, E. C. (1942). *Trans. AIME*, **147**, 144.
40. RICHARDS, T. L. (1949). *Prog. Metal Phys.*, **1**, 281.
41. BOURNE, L. and HILL, R. (1950). *Phil. Mag.*, **41**, 671.
42. HILL, R. (1950). *Mathematical Theory of Plasticity*. Oxford University Press, London, p. 315.
43. WHITELEY, R. L. (1960). *Trans. ASM*, **52**, 154.
44. ELIAS, J. A., HEYER, R. H. and SMITH, J. H. (1962). *Trans. AIME*, **224**, 678.
45. HOSFORD, W. F. and BACKOFEN, W. A. Reference 5.
46. GARBER, R. I., GRINDIN, J. A. and SHUBIN, Y. V. (1961). *Soviet Physics. Solid State*, **3**, 667.
47. HATCH, A. J. (1965). *Trans. AIME*, **233**, 45.
48. HONDA, H. and KAYA, S. (1926). *Sci. Reports*, Tohoku University, **15**, 721; KAYA, S. (1928), *loc. cit.*, **17**, 639.
49. GOSS, N. P. (1935). *Trans. ASM*, **23**, 515.
50. DUNN, C. G. (1949). In *Cold Working of Metals*. American Society for Metals, 113.
51. FAST, J. D. (1956). Philips Research Reports, **11**, 490.
52. ASSMUS, F., DIETERT, K. and IBE, G. (1957). *Z. Metallk.*, **48**, 344.
53. WIENER, G., ALBERT, P. A., TRAPP, R. H. and LITTMANN, M. F. (1958). *J. appl. Phys.*, **29**, 366.
54. DUNN, C. G. and WALTER, J. L. (1960). *Trans. AIME*, **218**, 449.
55. HOWARD, J. (1964). *Trans. AIME*, **230**, 588.

56. BOAS, W. and HONEYCOMBE, R. W. K. (1946). *Proc. R. Soc.*, **A186**, 57; (1947). *Ibid.*, **A188**, 427.
57. YOUNG, A. G., GARDINER, K. M. and ROTSEY, W. B. (1960). *J. nucl. Mat.*, **2**, 234.
58. AITCHISON, I., HONEYCOMBE, R. W. K. and JOHNSON, R. H. (1962). *Properties of Reactor Materials and Effects of Radiation Damage*. Ed. by D. J. LITTLER. Butterworth, London, p. 430.
59. CHISWICK, H. H. and KELMAN, L. R. (1956). *Proc. 1st International Conference on Peaceful Uses of Atomic Energy, Geneva*, **9**, 147.
60. MAYFIELD, R. M. (1958). *Trans. ASM*, **50**, 926.
61. MACINTOSH, A. B. and HEAL, T. J. (1958). *Proc. 2nd International Conference on the Peaceful Uses of Atomic Energy, Geneva*, **6**, 48.
62. ANDERSON, R. G. and BISHOP, J. F. W. (1962). Institute of Metals Symposium on Uranium and Graphite, p. 17. London.
63. BURKE, J. E. and TURKALO, A. M. (1952). *Trans. AIME*, **194**, 651.
64. PUGH, S. F. (1958). *J. Inst. Metals*, **86**, 497.
65. BOCKVAR, A. A., GULKOVA, A. A., KOLOBNEVA, L. J., SERGEEV, G. I. and TOMSON, G. I. (1958). *Proc. 2nd International Conference on Peaceful Uses of Atomic Energy, Geneva*, **5**, 288.
66. WARREN, B. E. (1959). *Prog. Metal Phys.*, **8**, 147.
67. CHRISTIAN, J. W. and SWANN, P. R. (1965). In *Alloying Behaviour and Effects in Concentrated Solid Solutions*, AIME, Ed. by T. B. Massalski. Gordon and Breach, New York, London, Paris.

Additional general reference

1. GREWEN, J. and WASSERMANN, G. (Eds.) (1969). *Textures in Theory and Practice*, Proc. International Symposium Clausthal 1968, Springer-Verlag Berlin.

Chapter 13

Creep in Pure Metals and Alloys

13.1 Introduction

In the deformation phenomena so far considered, the strain has been studied as a function of stress. If a stress in excess of the yield stress is suddenly applied to a specimen, it will give an instantaneous extension ϵ_1; however, if the stress is kept constant, the strain gradually increases with time until it approaches a steady value ϵ_2 (Fig. 13.1, Curve A). This time-dependent deformation process is known as creep and $(\epsilon_2 - \epsilon_1)$ is the creep strain. Creep also takes place at stresses substantially below the macroscopic yield stress. The attainment of a steady value of the strain ϵ_2 is usual at temperatures low with respect to the melting point; however, at elevated temperatures the creep curve does not become parallel to the time

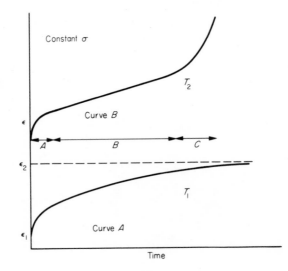

Fig. 13.1 Typical creep curves at two temperatures, $T_1 < T_2$

axis, but maintains a gradient and finally turns upward prior to failure, Fig. 13.1, Curve B. The first part of the curve obtained at both low and high temperatures is known as *primary* creep or *logarithmic* creep (*A*). It can occur in the absence of thermal activation, for example at 4 K. The second stage (*B*) is linear and is referred to as *secondary* or *steady-state* creep. The steady state represents a balance between work hardening and recovery processes which only occur if the temperature is high enough. This type of creep is very important in practice as it represents the majority of the creep strain resulting from the use of stressed components at high temperatures, e.g. in gas turbines or atomic reactors. The *third* stage or *tertiary* creep is best avoided in practice as it represents a region of rapidly increasing strain rate which ends finally in creep fracture. The fracture develops early in this stage as the growth of voids at the grain boundaries which extend, and open up as grain-boundary sliding takes place, leading finally to inter-granular creep fracture.

We shall first analyse the creep curves and discuss some of the empirical equations derived to describe creep behaviour. Then the structural changes which accompany creep will be discussed, and finally some modern theories of creep behaviour will be outlined.

13.2   Analysis of creep curves

Creep tests carried out under constant tensile load are not a fundamental way of studying the phenomenon, because as the specimen elongates, the stress on the cross-section increases. Andrade[7] pioneered the use of constant stress devices such as the hyperbolic weight, which enabled creep curves to be readily obtained under constant stress conditions. In these circumstances the behaviour is essentially that described by the curves in Fig. 13.1. These curves will tend to vary characteristically as the stress and temperature are changed (Fig. 13.2). There have been numerous attempts to describe, in the first instance, the basic shapes of creep curves by suitable equations, and also to take into account the effects of stress and temperature. One of the earlier relationships which provided an accurate description of many results was that due to Andrade:

$$\epsilon = \epsilon_0(1 + \beta t^{1/3}) . e^{\kappa t} \qquad\qquad \textbf{13.1}$$

where ϵ is the extension of the specimen at time t and ϵ_0, β and κ are constants.

The equation expresses the contribution of two types of flow to the increase in length of the specimen. Firstly, β flow which, when $\kappa = 0$, can be expressed as

$$\epsilon = \epsilon_0(1 + \beta t^{1/3}) \qquad\qquad \textbf{13.2}$$

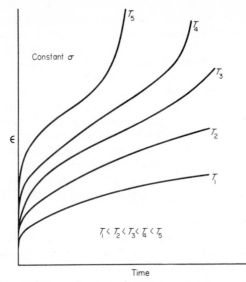

Fig. 13.2   Effect of increasing temperature on creep curves at constant stress.
(Increasing stress has a similar effect at constant temperature)

This contribution diminishes rapidly with time and describes the primary
creep (Fig. 13.3). Secondly, when $\beta = 0$

$$\epsilon = \epsilon_0 \, e^{\kappa t} \qquad\qquad \mathbf{13.3}$$

the κ creep component is shown as proceeding at a constant rate which

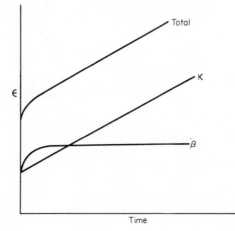

Fig. 13.3   The β and κ components of a creep curve. (After Andrade, 1910,
Proc. R. Soc., **A84**, 1)

corresponds to the steady-state or secondary creep (Fig. 13.3). Andrade took the view that the transient creep or β flow arose from slip processes within the grain, while the κ flow was caused by grain-boundary sliding. As we shall see, this picture is no longer fully valid, for the secondary creep arises not only from grain boundary flow, but more particularly from movement of dislocations within the grains.

The fact that any equation designed to fit a creep curve cannot apply over the whole experimental range is illustrated by the behaviour of many metals at low temperatures. Wyatt[8] tested polycrystalline copper over the range 77–443 K, and found that at the lower temperatures a logarithmic relationship held, but at higher temperatures an additional process caused the strain to be greater at a given time than that predicted by the logarithmic term (Fig. 13.4). He proposed an equation with three terms:

$$\epsilon = a \log t + bt^n + ct \qquad\qquad \textbf{13.4}$$

where a, b and c are constants and $n \simeq \frac{1}{3}$.

The fact that the equation has three terms of a different type, logarithmic, power law and linear, and three distinct constants means that it has a great deal of flexibility in describing mathematically the observed behaviour of metals under creep conditions. It is, however, as well to remember that the

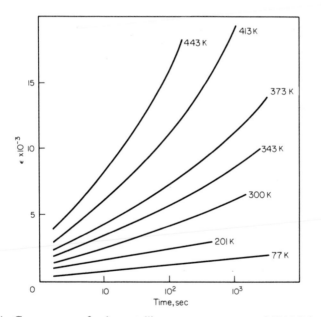

Fig. 13.4 Creep curves of polycrystalline copper at a stress of 176 MN m^{-2} over the temperature range 77–443 K. (After Wyatt, 1953, *Proc. Phys. Soc.*, **B66**, 495)

relationships referred to so far are empirical, and do not arise from any formal models of the creep process. Wyatt carried out temperature and stress changes during his tests on copper, which differentiated clearly between the creep behaviour in the low-temperature range and the higher temperature range. In the logarithmic creep range he found that a mechanical equation of state applied, in so far as the creep rate was uniquely defined as a function of the stress, strain and temperature. However, in the higher temperature range (above room temperature) a mechanical equation of state was no longer valid.

While the lower temperature creep phenomena are of fundamental interest, it is the behaviour at higher temperatures around which most interest centres, because new structural changes take place during deformation, and as this is the range of particular technological importance. Secondary creep, which occurs as a linear function of the time, is strongly temperature- and stress-dependent, the strain rate $\dot{\epsilon}$ conforming to a relation of the type

$$\dot{\epsilon} = A\,e^{-(Q - \alpha\sigma)/RT} \qquad\qquad \textbf{13.5}$$

where A and α are constants. Q is the activation energy, when σ the applied stress is zero.

Alternatively, the relationship

$$\dot{\epsilon} = \frac{B\sigma^n}{G}\,e^{-Q_{SD}/RT} \qquad\qquad \textbf{13.6}$$

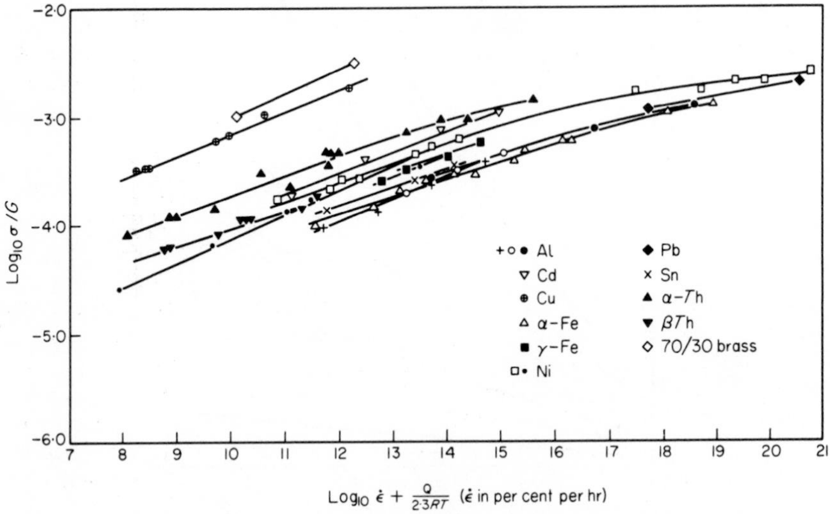

Fig. 13.5 Creep data on various metals. (After McLean, 1962, *Mechanical Properties of Metals and Alloys.* Wiley, New York and London)

where Q_{SD} is the activation energy for self-diffusion, has been advocated by Dorn[2] and Weertman.[28] This type of relationship allows the creep data from a number of different metals to be plotted on the one graph (Fig. 13.5). A straight line relationship would be expected, and is obtained on the left of the diagram, i.e. at low stresses and high temperatures when $n \simeq 5$; on the right-hand side the curves tend, however, to be concave towards the abscissa. It is concluded that the equation is relevant only at high temperatures and low stresses, where dislocation climb might be expected to be the rate-controlling process.

Much attention has been given to the determination of the activation energy Q, because this provides a way of distinguishing between several possible creep mechanisms. Moreover, a significant variation of Q as the temperature changes indicates that a change in mechanism is taking place (Section 13.5).

13.3   Structural changes during creep

In this section the structural changes which occur during secondary or steady-state creep and tertiary creep will be discussed. Two processes contribute to the deformation:

1. Slip within the grains.
2. Grain-boundary sliding.

Both these processes can be singled out experimentally for separate study. For example, slow deformation of single crystals at elevated temperatures will provide relevant information about slip, and experiments on bi-crystals with straight symmetrical boundaries will give basic information on grain-boundary sliding. However, to understand fully the behaviour of a polycrystalline aggregate, the relative contributions from both processes must be studied.

13.3.1.   *Deformation of single crystals under creep conditions*

If we limit consideration to steady-state creep conditions, the single crystals will be deforming almost entirely in Stage 3 of the stress-strain curve, with the added complication that slow straining will give time for recovery processes to take place and eliminate any work hardening. As the temperature of deformation is raised, the slip bands become coarser and wider spaced, until they are such infrequent events that they are unlikely to be important in fine-grained aggregates. However, between the coarse bands, very fine slip can be detected in the optical microscope[13] (Fig. 13.6). This fine slip completely fills in the space between the bands and must be taken into account. The use of an X-ray imaging technique to supplement optical examination has shown that the single crystals have become fragmented into sub-grains often strongly disoriented from each other (Fig. 13.7). In aluminium the presence of kink bands has a big influence on the disorientations within the crystal but more localized cell structures occur within regions between kink bands. Metals of lower stacking fault energy, e.g. copper, form a sub-structure much less readily because the dislocations must associate first over part of their length before climb is possible.

Fig. 13.6   Slip in an aluminium crystal deformed at a slow strain rate at 573 K. × 30

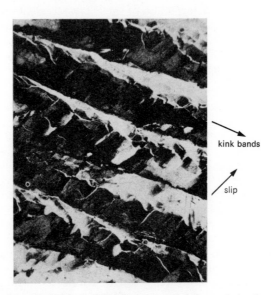

Fig. 13.7   X-ray micrograph of an aluminium crystal deformed slowly at 573 K. × 18. (Same orientation as Fig. 13.6)

Several metals are known to slip on new slip systems when the temperature of deformation is raised. For example, magnesium normally deforms on pyramidal planes above 473 K, and it has been confirmed that under creep conditions at 573 K, non-basal slip operates.[9] Zinc, too, at temperatures above 473 K deforms on $\{11\bar{2}0\}$ planes.[10] Moreover, aluminium, although it has a multiplicity of $\{111\}$ planes for slip, is thought to slip also on $\{100\}$ and $\{211\}$ planes[11] in $\langle110\rangle$ directions at high temperatures.

13.3.2. *Deformation of polycrystalline aggregates*

The work of Wood and Rachinger[12] on the slow deformation of aluminium over a range of temperature revealed significant structural changes by use of optical microscopy and X-ray Debye–Scherrer photographs. As

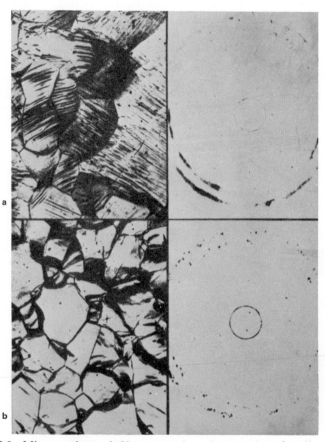

Fig. 13.8 Micrographs and X-ray powder photographs of polycrystalline aluminium. **a,** Deformed rapidly at 523 K; **b,** deformed slowly at 573 K (11·2 per cent in 56 hr at 3·65 MN m^{-2}). Micrographs ×75. (After Wood and Rachinger)

the rate of strain is reduced or the temperature increased, the slip bands become at first coarser and more widely spaced until they disappear completely. At the same time, the grain boundaries become more accentuated as grain-boundary sliding begins to take place. The X-ray diffraction rings which are blurred when taken from specimens subjected to high rates of strain and low temperatures (Fig. 13.8a), begin to sharpen as the strain rate is reduced until the rings are entirely composed of sharp spots (Fig. 13.8b). Wood attributed this to the gradual fragmentation of the grains in the absence of deformation by slip, but it is now known that this is a form of polygonization accentuated by the creep stress. This implies that slip must have taken place within the grains despite the absence of coarse slip bands, and that many of the dislocations have by a process of climb formed sub-boundaries or cell walls. This cell structure is one of the characteristic features of creep deformation in metals of high stacking fault energy, e.g. aluminium.

McLean[13] was able to show by the aid of phase contrast microscopy that the fine slip occurred within the grains of aluminium deforming under creep conditions, and he was further able to reveal the sub-structure by viewing in polarized light after anodic oxidation. Sub-structures have been found in a number of other metals, notably α-iron (Fig. 13.9), zinc, magnesium, cadmium and tin, which have rather high stacking fault energies so that the dislocations are only weakly dissociated, and thus are able to climb readily. On the other hand, copper[14] with a fairly low stacking fault

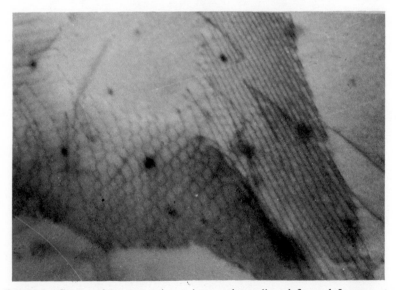

Fig. 13.9 Creep sub-structure in an iron-carbon alloy deformed 5 per cent at 823 K. Thin-foil electron micrograph. × 45,000. (After McLean, Crown copyright reserved)

energy develops a sub-structure more slowly during creep. Lead[15] shows very little sub-structure during creep. Both lead and copper tend to re-crystallize when the creep strain reaches a critical value. Similarly, nickel and austenitic steels recrystallize, although substructures form under some conditions.

In aluminium, the disorientations between sub-grains can become quite high (up to 40° tilt), particularly those on either side of kink bands, which can be quite pronounced in coarse-grained specimens at elevated tempera-tures. These bands divide the grains into disoriented lamellae, but then in the bands further smaller disorientations take place which are more characteristic of creep of fine-grained materials. The sub-grain disorienta-tion[13] is related to the strain, so that the disorientations increase as creep proceeds.

$$\theta \simeq \epsilon$$

where θ is the sub-grain disorientation in radians, and ϵ the creep strain.

The sub-grain size is inversely proportional to the stress and the strain rate. As the temperature is increased, the size of the sub-grains increases; however, the sub-grain size seems to be independent of the grain size, although the grain size of course represents the upper limit the sub-grains can achieve as the temperature is raised.

The thin-foil electron microscopy technique has enabled more precise observations to be made on the nature of sub-structures. In particular, observations have been made on α-iron,[16,17] which shows a marked sub-structure, with often well-defined dislocation networks (Fig. 13.9) as the specimen moves from the primary stage into secondary creep. Both tilt boundaries and hexagonal networks of dislocations are characteristic of the secondary creep stage. As the creep strain increases, the sub-boun-daries become better defined, and there are less residual dislocations within the sub-grains.[18] The sub-grain size in iron after creep at 773 K is much larger and sharper than in specimens deformed at room temperature. It is interesting to note that even a low stacking fault energy alloy such as austenitic Cr–Ni–Mn steel after creep at 973 K shows a sharply defined substructure[16] when examined by thin-foil electron microscopy.

Grain-boundary sliding is a prominent feature of polycrystalline metals deforming in the steady-state creep region. Microscopic observations show that boundaries inclined at 45° to the stress axis slide more than other orientations, suggesting that the shear stress along the boundary is signifi-cant. As in the case of bicrystals, individual boundaries do not move smoothly, but in small increments. However, the overall effect is a steady increase in strain originating from grain boundaries during a creep test.

A number of quantitative investigations have been made in which the magnitude of the grain-boundary strain has been determined. Two main methods have been used,

(a) Averaging the displacements measured at the individual boundaries.

The averaging methods vary, but all attempt to correct for the angles the boundary makes with the stress axis and with the surface of observation. So

$$\epsilon_{g.b.} = An\bar{l} \qquad\qquad \textbf{13.7}$$

where A = averaging factor, n = no. of boundaries/cm, $\bar{l}$ = mean value of the component of the g.b. displacement parallel to the stress axis.

(b) by measuring the change in grain shape in the interior of the specimen. It is assumed that the elongation of these grains is equal to the extension caused by crystallographic slip, and this is subtracted from the overall strain to get the grain-boundary strain.[74]

It has been shown[19] that both methods give comparable results for aluminium. Fig. 13.10, Curve 1, was determined by method (b), while curves 2–4 were obtained using method (a). These results indicate that the grain-boundary sliding for a given overall strain increases with temperature until a steady value is obtained. This plateau value then increases with decreasing grain size of the specimen, that is, it increases with increasing

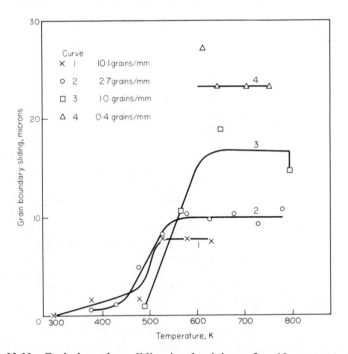

Fig. 13.10 Grain boundary sliding in aluminium after 10 per cent elongation, as a function of temperature. Curves 2, 3 and 4 were obtained from surface measurements, while curve 1 is from changes of grain shape in a section. (After McLean, 1962. *Mechanical Properties of Metals.* Wiley, New York and London)

grain-boundary area. There is an approximately linear relationship between grain-boundary sliding and elongation.

13.4   Contributions of slip and grain-boundary sliding to creep strain

McLean[13] has determined the relative contributions of slip and grain-
boundary sliding to creep in pure aluminium. He assumed that one of the
twelve possible {111} slip planes would be at approximately 45° to the
stress axis in all the grains of the specimen, then he calculated the strain
due to slip by measuring the spacing of the slip bands and the amount of
slip in each. From slip geometry

$$\epsilon_{\text{slip}} = (1 + \sqrt{2}np + n^2p^2) - 1 \qquad\qquad \textbf{13.8}$$

where n = no. of slip bands/unit length, p = displacement on each slip
band.

n and p were determined using optical and interference microscopy. The
grain-boundary contribution to the strain was determined by the method
(a) above, again making the assumption that all sliding boundaries are at
45° to the stress axis.

From the separate strain measurements, a total creep strain–time curve
was constructed and compared with that obtained by direct experimental
measurements (Fig. 13.11). This revealed that there was a substantial

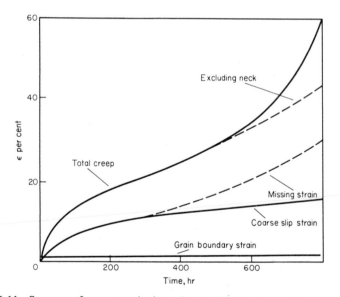

Fig. 13.11   Sources of creep strain in polycrystalline aluminium deformed at
473 K. (After McLean, 1952, *J. Inst. Metals*, **80**, 507)

contribution to the creep not accounted for by the coarse slip bands and the grain-boundary flow, which was attributed to fine slip between the coarse bands. This additional slip was revealed by phase contrast microscopy.

In Fig. 13.11 the grain-boundary contribution is small, about 10 per cent of the total, but this percentage depends very much on the metallurgical variables such as temperature, rate of straining, grain size. For example, in Fig. 13.12 it can be seen that the grain-boundary displacement increases with decreasing grain size and increasing stress for a given temperature. It has been found for several materials, that there is a linear relationship between sliding and the total strain. This implies that there is likewise a linear relationship between the strain due to slip and the total strain. The

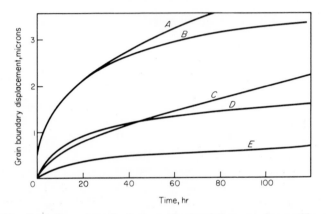

Fig. 13.12   Grain boundary displacement in aluminium as influenced by grain size and stress. **A,** 4·5 grains/mm: 11·4 MN m^{-2}; **B,** 1·0 grains/mm: 7·6 MN m^{-2}; **C,** 9·2 grains/mm: 7·6 MN m^{-2}; **D,** 4·5 grains/mm: 7·6 MN m^{-2}; **E,** 4·5 grains/mm: 5·2 MN m^{-2}. (After McLean, 1953, *J. Inst. Metals*, **81,** 293)

important question here is which process controls the other. McLean has pointed out that processes which slow down slip specifically also slow down grain-boundary sliding. For example, if a β-brass specimen is taken from the disordered state (above 673 K) to the ordered state (below 673 K) the creep rate diminishes by a factor of 10, and the prime effect of ordering is to make slip within the grains more difficult; however, the grain-boundary strain decreases in the same ratio. Again, if an alloy ages during creep and so makes slip more difficult, the sliding is again reduced in the same ratio. Clearly the slip within the grains is the process which controls the overall deformation. This is important, because if a metallurgical change is encouraged, which strengthens the grains, it will also lead to strengthening of the grain boundaries. However, this no longer holds in tertiary creep when grain-boundary failure develops.

13.5   Activation energy for steady-state creep

It is now clear that the structure of a metal under creep conditions undergoes substantial changes, and leads to the variations in creep rate which are characteristic of the creep curve. The completely empirical approach which seeks to obtain a relationship for strain in terms of time, temperature and stress for a given material is bound to be inadequate because the metallurgical changes will be different, or occur to different degrees as the variables are altered. As creep clearly depends on thermally activated processes it is thus necessary to examine in some detail the role of temperature on creep mechanisms.

The simplest assumption is that creep is due to a single activated process, in which circumstances the creep rate can be expressed in terms of an Arrhenius type rate equation:

$$\dot{\epsilon} = A\,e^{-\Delta H/RT} \qquad\qquad \textbf{13.9}$$

where $\dot{\epsilon}$ is the creep strain rate, and ΔH is the activation energy for the rate controlling process, T is the absolute temperature of the test, and R is the gas constant. The term A is complex, and includes not only the vibration frequency of the unit of flow, but also the change in entropy and a factor depending on structure, i.e. the distribution of dislocations, precipitate and grain boundaries. It is likely that in a complex process such as creep, more than one mechanism is operating at a given time. Assuming that these mechanisms are dependent on each other, the one which is slowest and requires the highest activation energy will be the rate-controlling process. If, however, the various processes in creep are independent of each other, the one with the lowest activation energy will be the dominant mechanism.

Dorn and co-workers[1, 20] assumed that the term A in eqn. **13.9** remained substantially constant over small temperature ranges and were thus able to develop a technique for determining the activation energy ΔH. A creep test is carried out at temperature T_1 at constant stress, until at a suitable strain, the temperature is suddenly changed by a small amount ΔT to T_2. This results in an immediate change in creep rate, which before and after the change is assumed to be associated with a similar metallurgical structure. It is thus considered that the change in creep rate arises entirely from the temperature change. If a single activation process is assumed, and the creep rates $\dot{\epsilon}_1$ and $\dot{\epsilon}_2$ before and just after the temperature change are measured, then applying eqn. **13.9**,

$$\dot{\epsilon}_1\,e^{\Delta H/RT_1} = \dot{\epsilon}_2\,e^{\Delta H/RT_2} \qquad\qquad \textbf{13.10}$$

It follows that

$$R \ln\left(\frac{\dot{\epsilon}_1}{\dot{\epsilon}_2}\right) = \Delta H \left(\frac{1}{T_2} - \frac{1}{T_1}\right) \qquad\qquad \textbf{13.11}$$

Typical results for polycrystalline aluminium are shown in Fig. 13.13

where the temperature has been changed downwards at A by 20 K and then upwards at B by 20 K. Fig. 13.13b plots the creep rate as a function of creep strain and reveals the two discontinuities in creep rate arising from the temperature changes. Using the above equations, values for ΔH between 126 and 132 kJ mol^{-1} were obtained in the temperature range 450–470 K.

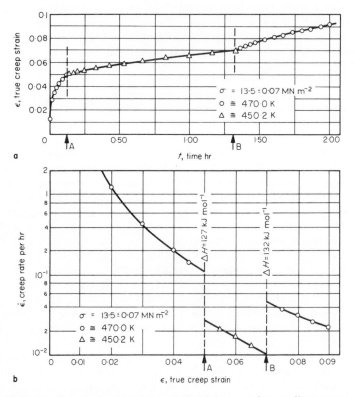

a

b

Fig. 13.13 **a,** Creep curve for pure aluminium under cyclic temperature changes. **b,** Creep rate–true creep strain curves. (After Dorn, 1957, *Creep and Recovery.* American Society for Metals)

Examination of the creep behaviour of aluminium over a wider temperature range showed that above 500 K the activation energy remained constant at about 149 kJ mol^{-1}, while between 240 and 370 K the activation energy was constant around 115 kJ mol^{-1}. This indicates that two distinct mechanisms of creep are rate controlling in the two temperature ranges. Between 370 and 500 K the activation energy increases from 115 kJ mol^{-1} to 149 kJ mol^{-1}, implying that the two processes are both operative and interdependent in this range.

An alternative method involves separate creep tests at the same stress at different temperatures, assuming that the structure will remain constant during steady-state creep, or better, that for a given strain the structure will be the same at T_1 and T_2. A power law is assumed to apply thus:

$$\epsilon = a[t\,e^{-\Delta H/RT}]^n \qquad \textbf{13.12}$$

where t = time, n and a are constants.

Creep tests are carried out at T_1 and T_2 and the times t_1 and t_2 to reach a chosen strain are determined. Then

$$t_1\,e^{-\Delta H/RT_1} = t_2\,e^{-\Delta H/RT_2}$$

and

$$\Delta H = \frac{RT_1 \cdot T_2}{T_2 - T_1}\,[\ln t_1 - \ln t_2] \qquad \textbf{13.13}$$

This method leads to difficulties if T_1 and T_2 are too widely separated, for structural differences will then arise; moreover, the shape of the creep curves can be different. However, it has been used satisfactorily with aluminium in the range 400–500 K, where the creep curves at several different temperatures can be represented on the same curve of creep strain plotted against $t\,e^{-\Delta H/RT}$, where ΔH is an activation energy of 142 kJ mol^{-1} (Fig. 13.14).

It is of some importance to know whether the magnitude of this activation energy is dependent on any of the metallurgical variables. The effect of

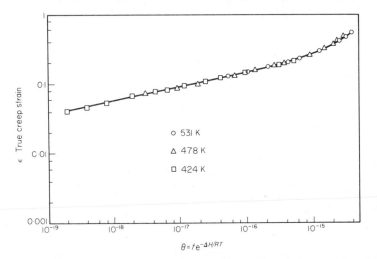

Fig. 13.14 Relation between log creep strain and log temperature compensated time for aluminium at a constant stress of 20·7 MN m^{-2}. (After Dorn, 1956. Some Fundamental Experiments on High Temperature Creep. Proceedings of Symposium at National Physical Laboratory, May/June 1954 on *Creep and Fracture of Metals at High Temperatures*. H.M.S.O.)

creep strain has been well investigated for aluminium, where ΔH is independent of strain within experimental error, both at high and low temperatures. The activation energy has also been shown to be independent of strain in copper,[21] thallium, magnesium, stainless steel[22] and several other alloys. Likewise, the effect of stress on ΔH appears to be fairly small for deformation at high temperatures when the activation energy for creep is very close to the activation energy for diffusion (Table 13.1). The activation energy is independent of temperature, provided this is high enough, but at lower temperatures significant changes take place.

13.6 Significance of the activation energy for creep

Many experiments have shown that there is not a unique activation energy for creep over the whole temperature range, but at high temperatures when steady-state creep predominates, the activation energy is in most cases very close to the activation energy for self-diffusion, which implies that the rate-controlling process is diffusion controlled. Some typical values are listed in Table 13.1.

Table 13.1 Typical values for activation energies for high-temperature creep (ΔH_C) and self-diffusion (Q_D)

	ΔH_C kJ mol^{-1}	Q_D kJ mol^{-1}
Aluminium	140	140
Copper	197	199
Indium	72	75
α-Iron	305	289
γ-Iron	297	293
Magnesium	117	134
Zinc	88	88

The conclusion reached is that vacancy formation and movement is a very significant phenomenon in steady-state creep. This, taken together with the ample evidence for sub-grain formation during secondary creep, leads to the view that dislocation climb is the rate-controlling process. Another generalization is that metals with large self-diffusion coefficients will show less creep resistance than metals with small coefficients.

Now

$$Q_D = Q_F + Q_M \qquad\qquad \textbf{13.14}$$

where Q_F and Q_M are the energies of formation and movement of a mole of vacancies. So metals in which vacancies move readily, i.e. Q_M is low, will have less creep resistance. Sherby and co-workers[23] have shown that this applies to body-centred cubic metals in which diffusion occurs more readily than in comparable face-centred cubic metals, and in particular α-iron is less creep-resistant than face-centred cubic austenite under comparable conditions.

13.7   Theories of creep

13.7.1. *Nabarro-Herring mechanism*

Since the importance of diffusion in steady-state creep has been shown, it is logical to examine the possibility of stress-directed diffusion taking place. It would be expected that if a rod specimen is under creep in tension, material would tend to move from the sides to the ends of the specimen to allow it to lengthen as the creep strain takes place. This could be achieved by a vacancy flow from the ends of a grain to its sides (Fig. 13.15), a type

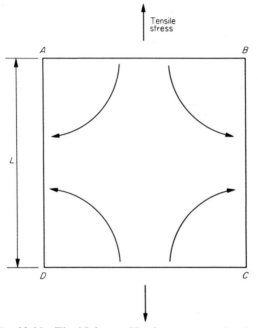

Fig. 13.15   The Nabarro–Herring creep mechanism

of viscous flow now referred to as the Nabarro-Herring mechanism.[24, 25] Vacancies are assumed both to originate and annihilate at grain boundaries, and the creep rate is controlled by the diffusion of the vacancies.

Along AB and CD the vacancy excess is

$$C - C_0 = C_0 \frac{\sigma b^3}{kT} \qquad \qquad \textbf{13.15}$$

where C = actual concentration, C_0 = equilibrium concentration, b^3 is the volume of a vacancy and σ is the applied stress. There is a corresponding deficit along AD and BC.

Now the rate of migration of atoms is then

$$\frac{dv}{dt} = \frac{2\sigma b^3 L D}{kT} \qquad \textbf{13.16}$$

where D = self-diffusion coefficient, and L = grain size. Then

$$\text{Creep rate } \dot{\epsilon} = \frac{1}{L^3}\frac{dv}{dt} = \frac{2\sigma b^3 D}{L^2 kT} \qquad \textbf{13.17}$$

from which it can be seen that the creep rate depends only linearly on the stress and should, therefore, be expected to be appreciable at low stresses, and at high temperatures, when vacancy formation and movement is adequate. It is interesting to note that at very high temperatures approaching T_M, experimentally a linear stress law is found, moreover, the creep rate varies with grain size in the way predicted by eqn. **13.17**.

13.7.2. Recovery theory

Cottrell and Aytekin[26] and Mott[27] developed a theory of steady-state creep which assumed that in this stage the rate of recovery was equal to the rate of work hardening:

$$-\left(\frac{\delta \tau_i}{\delta t}\right)_{\dot{\epsilon}=0} = \left(\frac{\delta \tau_i}{\delta t}\right)_{r=0} \qquad \textbf{13.18}$$

where r = rate of recovery, and τ_i = internal stress.

Each of these terms can be expressed thus:

$$\left(\frac{\delta \tau_i}{\delta t}\right)_{\dot{\epsilon}=0} = -A \exp\left[\frac{-(\Delta H_i - q\tau)}{RT}\right]$$

where A and q are constants and τ = yield stress, and

$$\left(\frac{\delta \tau_i}{\delta t}\right)_{r=0} = \frac{d\epsilon}{dt}\left(\frac{\delta \tau_i}{\delta \epsilon}\right)_{r=0} = \frac{d\epsilon}{dt} \cdot h$$

So in steady-state conditions

$$\dot{\epsilon}_s = \frac{A}{h} \exp\left[\frac{-(\Delta H_i - q\tau)}{RT}\right] \qquad \textbf{13.19}$$

Dislocation climb was regarded as the most likely recovery mechanism in steady-state creep, and in eqn. **13.19** ΔH_i was associated with climb and thus with volume diffusion. In practice this relationship does not appear to hold rigidly over the whole range of steady-state creep conditions, but does seem to have validity at high stress levels. The activation energy term, as has already been pointed out, is not stress-dependent in this form of creep.

13.7.3. *Weertman climb theory*

Weertman[28] explains the presence of fine slip during secondary creep by the operation of numerous Frank–Read sources on different parallel slip planes (Fig. 13.16). At low temperatures and stresses these loops would soon cease to operate as the dislocations from adjacent loops would tend to repel each other. However, at high temperatures, the dislocation loops are able to climb and annihilate each other, thus enabling further dislocation loops to be emitted from the sources. Clearly in a model of this type, the controlling process determining the rate of creep is that of climb. Weertman has derived the following equation for the rate of creep strain at low stresses:

$$\dot{\epsilon}_s = \frac{C\pi^2 \tau^{4\cdot 5} D}{b^{1/2} N^{1/2} G^{3\cdot 5} kT} \qquad\qquad \textbf{13.20}$$

where C is a constant $\simeq 0\cdot 25$, D is the coefficient for self-diffusion, τ is the applied stress, N is the density of dislocations taking part in climb, and G is the shear modulus.

A further equation, which is applicable at both high and low stresses, is

$$\dot{\epsilon}_s = \frac{C'\pi^2 \tau^2 D}{G^2 b^2} \sinh\left[\frac{\sqrt{3}\ \tau^{2\cdot 5} b^{1\cdot 5}}{8G^{1\cdot 5} N^{1/2} kT}\right] \qquad\qquad \textbf{13.21}$$

where $C' \simeq 1$.

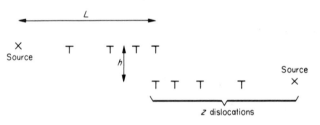

Fig. 13.16   Weertman model for secondary creep

Equation **13.20** requires the creep strain rate in steady-state creep to be proportional to the 4·5 power of stress, which appears to be very close to experimentally observed values provided the conditions are such that the creep process is diffusion controlled. The second equation shows that as the stress is reduced, the transition in behaviour is dependent on G, the shear modulus of the material. This transitional stress should increase with increasing G.

Strong support for this theory is obtained when diverse creep results for a number of metals are plotted in the form $\dot{\epsilon}/D$ against $\log \sigma/E$ where E is Young's modulus (Fig. 13.17), showing that the data are fairly consistent with a relationship of the general type indicated by eqn. **13.20**.

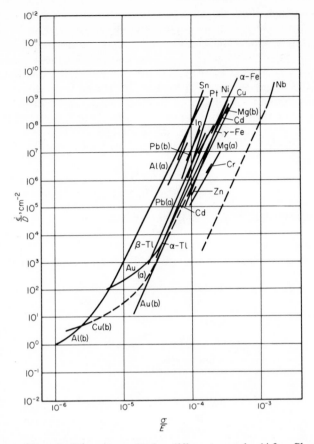

Fig. 13.17   Plot of $\dot{\epsilon}/D$ against σ/E for different metals. (After Sherby, 1962,
              Acta Metall., **10**, 135)

13.7.4. *Other possible dislocation mechanisms in high temperature creep*

So far the only dislocation mechanism considered apart from dislocation glide has been climb. There are, however, other possible processes which must be briefly considered. Firstly, the dislocation lines may become jogged by the glide dislocations intersecting each other or cutting through the forest dislocations. With screw dislocations, the jogs will have to move non-conservatively and generate vacancies. These vacancies will tend to restrain the movement of the jogged dislocations, unless they are free to move away,[29] so the rate controlling process is again the rate of diffusion of vacancies, and the activation energy is that for self-diffusion.

Diffusion down dislocations (pipe diffusion) is more rapid than is bulk diffusion, so it is reasonable to assume that vacancies will move rapidly along dislocations. It is then possible that any jogs which have been formed by intersections will be able to climb as the vacancies feed in. It has been suggested[30] that metals of low stacking fault energy, such as silver and copper, which do not normally polygonize in the absence of stress, may do so by this mechanism when the stress is superimposed at elevated temperatures. This would lead firstly to the formation of jogs, then subsequently, to climb of the jogs as the vacancies feed in along the dislocations.

13.7.5. Creep at intermediate temperatures

It has been pointed out that the activation energy for creep in aluminium changes between 400 and 500 K. This change in activation energy is illustrated in Fig. 13.18 for zinc,[31] where the activation energy for creep is

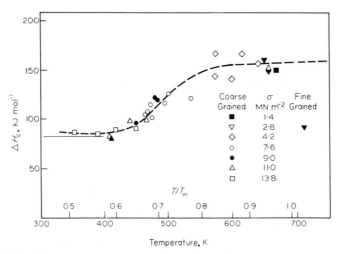

Fig. 13.18   Temperature dependence of the activation energy for creep in zinc. (After Tegart and Sherby, 1958, *Phil. Mag.*, **3**, 1287)

159 kJ mol^{-1} between 0.95 and 0.85T_M, then it decreases until 0.6T_M below which it becomes constant again, giving a value of 88 kJ mol^{-1} at 0.5T_M. Similar behaviour is found in a number of other metals including aluminium,[32] α-thallium,[33] copper[21] and steel,[34] where in all cases the activation energy falls when the temperature is below a certain limit.

The mechanism most likely to occur at these intermediate temperatures is cross-slip. One of the important properties of screw dislocations which has been previously referred to is the ability to move from one glide system to a new system provided that the slip direction is common to both slip systems. This process of cross-slip does not need diffusion to occur. Nevertheless, it can be thermally activated, as some energy is required to move a dislocation loop from its original slip plane to another somewhat less suitably oriented. Cross-slip is thus a type of recovery process, a dynamic recovery process which occurs during deformation, and is the dominant mechanism in Stage 3 work hardening of face-centred cubic single crystals (Chapter 4). It is significant that the range of this stage in aluminium is between 273 and 400 K ($0.20-0.3T_M$), when Stage 3 is the principal mechanism of deformation in aluminium single crystals. On the other hand, cross-slip in copper is much more difficult because of the more widely dissociated dislocations, so higher stresses and temperatures are needed for it to take place. Between 400 and 800 K ($0.3-0.6T_M$) copper has an activation energy for creep in the range 117–155 kJ mol^{-1}. It is considered that cross-slip operates in the creep of copper towards the upper end of this temperature range. These hypotheses have been confirmed by metallographic examination, which has revealed cross-slip in aluminium single crystals only at temperatures high enough for recovery creep[35] and the same phenomenon in copper at a much higher temperature, viz. about $0.5T_M$.[14]

A model for this process can be envisaged which involves the piling up or tangling of dislocations at obstacles such as Lomer–Cottrell sessile dislocations or at incipient cell walls. If then the stress and temperature are high enough, the screw components of the dislocations can escape from the obstacles by cross-slip, and thus produce further creep strain. However, as the temperature is insufficient for diffusion to occur, the edge components are unable to climb, so complete recovery is impossible. There will thus be an accumulation of edge dislocations on the slip plane which will result in at least some degree of work hardening. When the temperature is raised sufficiently for diffusion to occur, the edge components climb so that complete recovery is possible and steady-state creep results.

13.7.6. *Mechanism of creep at low temperatures*

The primary stage of creep at a sufficiently low temperature can be described by a logarithmic law which has been established in one form or another for a large number of metals and alloys. When a constant stress is applied to a specimen, it will give an instantaneous strain and as a result of work hardening the flow stress will be raised from τ_0 to τ. When the stress is maintained on the specimen, there will be some dislocations needing just a small impetus to make them move, while others are less favourably placed. Thermal energy fluctuations will in time cause the most favoured dislocations to move, but this process of thermally activated glide will diminish as the easier dislocations are moved, so the strain rate will gradually decrease. Thus at low temperatures creep occurs by normal glide which is blocked by obstacles sufficiently small to be overcome by thermal activation. In pure metals, as distinct from alloys, the obstacles which are likely to be overcome in this way are:

1. forest dislocations, which the primary slip dislocations have to intersect thus forming jogs;

2. nearby sessile dislocations;
3. the lattice friction (Peierls–Nabarro force).

The first attempt to develop a formal theory involving the thermal activation of glide dislocations to produce local deformations was made by Orowan,[36] who assumed that N, the number of local deformation processes per unit time, is given thus:

$$N = \nu \exp \left(\frac{-h^2 \epsilon^2 V}{2GkT} \right) \qquad \textbf{13.22}$$

where h is the slope of the stress–strain curve of the material, ϵ = shear strain, V = activation volume, and $h^2 \epsilon^2 = (\tau_i - \tau)^2$ where τ is the applied stress and τ_i is the local stress to cause a deformation. It follows that the activation stress to cause a unit of creep strain is $(\tau_i - \tau)$. Now the creep rate is defined in terms of N and s, where s is the average strain from one event, which is assumed by Orowan to be inversely proportional to the square of the total strain,

$$\frac{d\epsilon}{dt} = Ns = \frac{N}{\epsilon^2} \qquad \textbf{13.23}$$

so from eqns. **13.22** and **13.23**

$$\frac{d\epsilon}{dt} = \frac{V}{\epsilon^2} \exp \left(\frac{-h^2 \epsilon^2 V}{2GkT} \right) \qquad \textbf{13.24}$$

Assuming that the exponential term is independent of $1/\epsilon^2$, and on integrating, it follows that

$$\epsilon = \text{const. } t^{1/3} \qquad \textbf{13.25}$$

Many experimental results fit this equation, which corresponds to that for the Andrade β flow describing the transient component of creep. However, there are many other results where the index varies quite substantially from $\frac{1}{3}$.

The Orowan theory did not assume any detailed dislocation interactions which have to be thermally activated. One of the most likely interactions is that of glide dislocations with the forest dislocations threading through the slip planes (Fig. 13.19) where small sections of a dislocation line will first be activated over the stress hill, thus catalysing the movement of the rest of the dislocation.[37] If there are N dislocation segments per unit volume and each dislocation (Burgers vector b) when activated moves over an area A, then the creep rate is

$$\dot{\epsilon} = NAb\nu \, e^{-\Delta H/kT} \qquad \textbf{13.26}$$

ν = frequency of vibration of a dislocation loop of length l (Fig. 13.19).

We now have to find an expression for the activation energy ΔH of the process.[38] Physically before intersection, the two dislocations if extended must form constrictions, and then a jog is created in the dislocation; these processes need an energy ΔH_0. However, the applied stress helps the intersection process, and thus the actual energy ΔH is lower by a term which expresses the work done by the local stress τ_L over a distance $d/2$ (Fig. 13.19) where d is the diameter of the forest dislocation and b the Burgers vector, so the thermal energy needed is

$$\Delta H = \Delta H_0 - \sigma_L \cdot bdl \qquad\qquad 13.27$$

The applied stress σ is constant, but σ_L will change as the material work hardens, in the following way:

$$\sigma_L = \sigma - \frac{\delta\sigma}{\delta\epsilon} \cdot \epsilon \qquad\qquad 13.28$$

By using equations **13.27** and **13.28** in **13.26** a differential equation relating

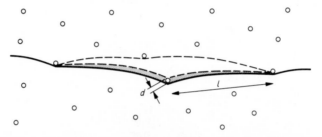

Fig. 13.19 Interaction of a glide dislocation with the dislocation forest. (After Schoeck, 1961, *Mechanical Behaviour of Materials at Elevated Temperatures*, ed. by Dorn. McGraw-Hill, New York)

ϵ the creep strain, and the rate of creep strain $\partial\epsilon/\partial t$ is obtained, the solution of which is

$$\epsilon = \frac{kT}{(\delta\sigma/\delta\epsilon) \cdot bdl} \cdot \ln(\gamma t + 1)$$

$$= \alpha \cdot \ln(\gamma t + 1) \qquad\qquad 13.29$$

where α and γ are constants.

The intersecting dislocation model thus gives a logarithmic time law for creep which is observed at low temperatures with many metals.

13.8 Tertiary creep—the onset of creep failure

Tertiary creep commences when the creep curve begins to deviate from linearity, and acquires a gradually increasing slope (Fig. 13.2). This can be due in part to the increase in stress due to change in the cross-section as a result of localized necking, but it persists when necking is avoided by the use of low stresses and high temperatures. In these cases, the increase in creep strain results primarily from the development of intergranular cavities and cracks which ultimately lead to intergranular creep failure, i.e. failure at the grain boundaries. Nevertheless, there is evidence that intergranular cracks can occur before the tertiary stage [39] and are well developed by the end of secondary creep. Density measurements [40] have revealed that in copper the density starts to fall in primary creep and decreases at a constant rate during secondary creep; in the tertiary stage, however, there is a marked decrease in density. The exact behaviour of a metal or alloy is very sensitive to the temperature of deformation and the strain rate, and in general the higher the temperature and the lower the strain rate, the earlier will be the stage at which cavities and cracks start to form.

Fracture appears to initiate in at least two ways, from wedge-type cracks, and from more or less spherical cavities. The wedge-shaped cracks (*W*-type) characteristically form at grain boundary triple points (Fig. 13.20),

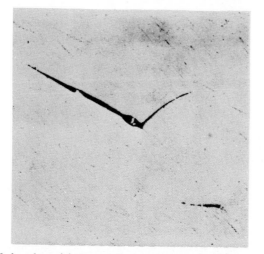

Fig. 13.20   Wedge-shaped intergranular cracks in Al–5 per cent Mg tested at 523 K. × 80. (After Grant and Mullendore, 1965, *Deformation and Fracture at Elevated Temperatures*. M.I.T. Press., Cambridge, Mass.)

and tend to grow along the grain boundaries which are normal to the applied stress. [41] This type of crack is usually encountered at lower creep temperatures, and at higher stress levels. These cracks are associated with grain boundary sliding which, as a result of stress concentrations, can lead to the opening up of cracks at the triple points in several ways (Fig. 13.21). Such *W*-type cracks have been observed in a large number of pure metals, and in commercial alloys such as Nimonic 80 and 90 and stainless steel.

The second type of grain boundary crack nucleation was first observed by Greenwood and co-workers[42] who, early in the tertiary creep range, found small spherical cavities in the grain boundaries of a number of metals

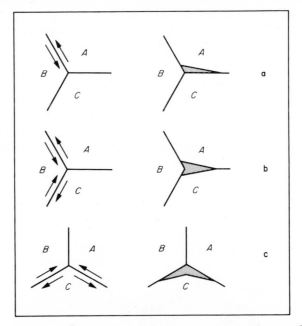

Fig. 13.21   Mechanisms for the formation of triple point cracks. (After Chang and Grant, 1956, *Trans. AIME*, **206**, 544)

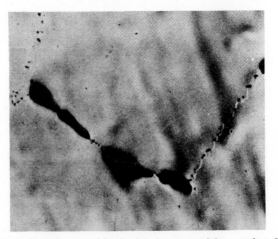

Fig. 13.22   Grain boundary cavities in A1–5 per cent Mg tested at 523 K × 100. (After Grant and Mullendore, *loc. cit.*)

and alloys. The cavities (r-type) are round or elliptical, and gradually increase in number and size within the tertiary range (Fig. 13.22). Typically at lower temperatures, the cavities are about 1μ apart and have been observed to be as small as $1\ \mu$ in diameter; they tend to form primarily on boundaries where substantial sliding is taking place. The influence of some important metallurgical variables on these two phenomena will now be considered.

13.8.1. *Effect of temperature*

As the extent of grain-boundary sliding increases with temperature at least to the point when grain-boundary migration occurs, it would be expected that the tendency toward intergranular crack formation would also increase. Moreover, at the higher temperatures, the mode of crack formation changes from that of W-type to that of r-type. As the temperature is raised the number of voids per unit length of boundary increases, and has been shown to be related to the increase in the amount of grain-boundary sliding[43] (Fig. 13.23). In some experiments the density of voids at boundaries decreases at very high temperatures, but this is because the boundaries tend to migrate and leave behind voids previously formed on them. The kinetics of void growth have been studied in silver,[44] where the mean void radius r increases linearly with time at a temperature T,

$$r = \alpha(T)t + r_0 \qquad\qquad \textbf{13.30}$$

The rate of void growth, $\mathrm{d}r/\mathrm{d}t = \alpha(T)$, increases with increasing T.

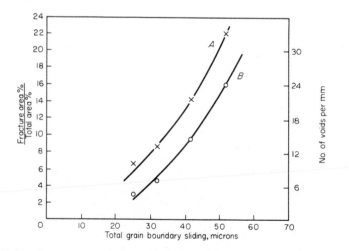

Fig. 13.23   Creep of copper bicrystals at $2 \cdot 1\ \mathrm{MN\,m^{-2}}$ in the range 923–1173 K. Both fracture area (curve A) and voids/unit length are plotted as a function of grain boundary sliding. (After Intrator and Machlin, 1959, *Acta metall.*, **7**, 149)

13.8.2. *Effect of stress*

It is often difficult to separate out the effects of stress and temperature, but in general at a constant temperature decreasing stress encourages void formation rather than W-crack nucleation. The higher the stress at a constant temperature the shorter the time to intergranular failure—this is a general pattern in creep behaviour. As high stress favours W-type failure the clear implication is that this type of crack propagates more rapidly than the cracks formed by coalescence of voids.

The type of stress has a large influence on void formation and on the localities where voids form. Bicrystal experiments show that stresses normal to a boundary do not produce cavities [45] because the boundary would not be subject to shear stress and thus sliding would not occur. On the other hand, if a grain-boundary is sliding, a superimposed tensile stress often encourages void formation. [46]

Void formation in tensile creep tests is predominantly in boundaries which make angles between 60° and 90° to the tension axis [47] but some occur in boundaries at lower angles, and some in boundaries parallel to the tensile axis. Compressive stresses generally help to minimize crack formation, but do not eliminate the phenomenon. When a hydrostatic pressure is imposed on a tensile specimen, void formation is very much reduced.

13.8.3. *Mechanisms of intergranular creep fracture*

The formation of W-type intergranular cracks was first explained by Zener. [48] This model assumes that grain-boundary sliding is blocked at triple points and a large stress concentration is built up. Superimposition of a tensile stress leads to the opening up of a wedge-shaped crack. The maximum tensile stress (σ_{max}) at the triple point is

$$\sigma_{max} = \left(\frac{L}{2r}\right)^{1/2} . \tau \quad \text{for } L \gg r \qquad \qquad \textbf{13.31}$$

where τ = shear stress along boundary, L = length of boundary, r = radius of curvature at tip of boundary.

If the grain boundary is firmly pinned at high temperatures by impurity particles, W-type cracking will take place if the stress is high enough. If, on the other hand, the boundary is free to migrate, some relief of the stresses at the triple points must take place [49] and the tendency to crack is reduced.

The first theory of cavity growth [42] assumed that the cavities formed by vacancy agglomeration, and that these grew gradually and eventually joined together to form cracks. Theoretical work has, however, shown that a very high supersaturation of vacancies is needed before a stable void can be formed. Another difficulty is that the Nabarro–Herring flow of vacancies is *away* from the transverse boundaries where cavities are most readily formed.

Analysis of void growth in copper by measuring the change in density has shown that void growth is very sensitive to creep strain and is also stress dependent.[40] An activation energy for growth of 121 kJ mol^{-1} is obtained, which is close to that for grain boundary diffusion in copper. It had earlier been proposed[50] that the grain boundaries were the source of the vacancies rather than diffusion within the grains; a model was set up for void growth on this basis, and tested by experiments on copper in tension with superimposed hydrostatic stress. From which

$$t_r \simeq \frac{KTa^3}{4(D_BS_B)(\sigma - P)\Omega}$$

13.32

where t_r = time to rupture; D_B = grain boundary diffusion coefficient; S_B = width of boundary; σ = applied stress; P = superimposed hydrostatic pressure; Ω = atomic volume; and a = spacing of voids.

Ideally, if the void nucleus size and distribution are constant, the time to rupture is controlled by $(\sigma - P)$. This is not strictly true, but when $(\sigma - P) \to 0$, no voids were detected in the specimen, and t_r was very large.

An alternative view is that the voids, like the W-type cracks, arise from grain-boundary sliding. Mechanisms have been proposed for the formation of cavities in the grain boundaries at jogs which can readily occur as a result of slip within the grains, or can be already present in boundaries prior to creep. One such model is shown in Fig. 13.24, where the jog in the

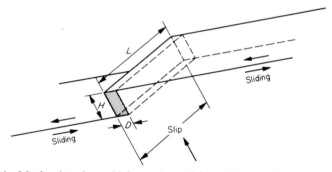

Fig. 13.24   Mechanism for void formation. (After Gifkins, 1959, in *Fracture*, ed. by Averbach *et al.* Wiley, New York)

boundary has occurred as a result of slip in the neighbouring grain triggered off by a dislocation pile-up. Subsequent sliding then opens up a hole at the jog. There is now experimental evidence for the existence of ledges in grain boundaries varying in height from 50 to 400 Å; the latter value was found in iron after creep at 823 K.[51] Ledges have also been observed by electron microscopy in specimens prior to creep.[18] Recent field ion microscopy studies[73] have clearly shown the existence of ledges which occur in the less organized regions of the boundary connecting regions of good atomic fit.

McLean[51] has pointed out that a cavity will probably not disappear if its diameter is greater than several atomic diameters. If surface tension causes the hole to become spherical then to obtain a stable crack, the following condition must be satisfied:

$$r > \frac{2\gamma}{\sigma} \qquad\qquad 13.33$$

where r = radius of hole, σ = tensile stress at ledge, and γ = surface energy.

As the stress is raised the stable void size becomes smaller, so that metals and alloys which can support a high stress will have smaller stable cavities. If we assume $\gamma = 1000$ mJ m^{-2}, a void 1 micron in diameter will be stable under a stress of 2·1 MN m^{-2}. Such values are consistent with those under which small voids ($\sim$ 1 micron diameter) are observed.

It is probable that voids grow both by a vacancy accretion method, and by formation and enlargement of a cavity by grain-boundary sliding. Experiments have shown that the cavity population is not only sensitive to stress and resulting grain-boundary sliding, but also to conditions which result in marked changes in the vacancy concentration. For example, by heating α-brass in vacuum to partially volatilize the zinc, the vacancy concentration is increased substantially, as is the tendency to form voids in a subsequent creep test.

Another cause of grain-boundary voids which undoubtedly plays an important part in creep ductility is the presence in the boundary of second-phase particles. Pure aluminium in creep can show substantial necking, but the addition of a small concentration of iron ($< 0·1$ per cent) to form particles of $FeAl_3$ at the grain boundaries results in intergranular fracture with no localized necking. It is also well known that the creep ductility of austenitic steels can be markedly reduced by the presence of $Cr_{23}C_6$ or NbC in the grain boundaries.[3] There is some evidence to suggest that these grain boundary particles nucleate cracks when the matrix-precipitate interface separates as a result of grain-boundary sliding. It would thus be expected that particles wetted by the matrix would be less detrimental to creep ductility than particles which were not wetted by the matrix. As second-phase particles are present even in relatively pure metals, it is likely that this is a real cause of intergranular fracture at elevated temperatures, as it is for ductile fracture at lower temperatures (Chapter 15).

13.9 Creep in alloys

The basic phenomena which occur during creep of pure metals have now been summarized. It is important to see how alloying influences this behaviour, and to study other relevant phenomena which are characteristic

of alloys. In general, pure metals are quite inappropriate in applications
for use at high temperatures where creep resistance is important, consequently strengthening is achieved both by alloying elements in solid solution, and by the formation of finely dispersed precipitates in the solid
solution. Another reaction which can influence creep resistance is the
order–disorder type of transformation.

13.9.1. *Solid solutions—effect of solute atoms on secondary creep*

Alloying elements in solid solution usually slow down the process of
recovery following cold work, and as the strain hardening rate is not
reduced the overall effect is to diminish the creep strain during steady-state
creep at a given stress level. Basic studies on the mechanical properties of
solid solutions have shown that solute atoms can increase strength by
several different mechanisms. These are

1. Segregation to dislocations to cause Cottrell locking.
2. Elastic interactions of solute atoms with moving dislocations to
   increase the Peierls–Nabarro stress or friction stress.
3. Segregation to stacking faults (Suzuki interaction).
4. Interaction with vacancies and dislocation jogs.
5. Formation of short or long range order.
6. Segregation to grain boundaries with consequent effects on grain
   boundary migration and sliding.

Each of these interactions can also be a possible source of strength under
creep conditions.

It has already been established that the rate-controlling process in
secondary creep is dislocation climb, so the solutes must in some way be
able to hinder this basic recovery process. In dilute solid solutions the
activation energy for creep ΔH_c is very close to that of the pure metal.
Moreover, the nickel–gold solid solutions have been examined over the
complete range[52] and the activation energy for creep shown to be between
167 and 209 kJ mol^{-1}. Fig. 13.25 plots these data together with the variation of $(\Delta H_d)_{Au}$ and $(\Delta H_d)_{Ni}$ with composition. It is clear that the rate of
self-diffusion in solid solutions is similar to that in pure metals. The solute
must, therefore, have its effect on creep rate by influencing the tendency of
the dislocations to climb rather than by hindering the movement of
vacancies. This could be done in two ways. Firstly, the solute atoms could
restrict movement of jogs by segregating to them,[53] and as the jog concentration need not be very high, a small concentration of solute atoms could
have a large effect. Such a marked retardation in climb has been observed
in polygonization studies on aluminium with small concentrations (0·02–
0·2 atomic per cent) of iron, lithium and sodium.[54] However, the various
solutes differ markedly in their effect on the recovery of metals, the biggest
influence being shown by solutes which show only a small solubility. Zinc

which is substantially soluble in aluminium has little effect on polygoniza-tion in concentrations up to 15 weight per cent.

The role of stacking fault energy in determining firstly the width of dissociated dislocations, and secondly their ability to climb has already been referred to in the case of pure metals. Substantial concentrations of solute can markedly lower the stacking fault energy of the solvent metal (Chapter 6), and so make the process of dislocation climb (and cross-slip) still more difficult. Reference has been made to the intrinsic superior creep resistance of γ-iron over α-iron; γ-iron stabilized at low temperatures usually contains chromium and nickel, and the high chromium content causes the stacking fault energy to be fairly low ($\sim$ 20–30 mJ m^{-2}), so the dislocations will climb less readily than those in α-iron under comparable creep conditions. Chromium presumably has a similar influence in nickel-base creep-resistant alloys, while the beneficial addition of cobalt in creep-resistant γ-iron alloys has been attributed to the same effect.[55]

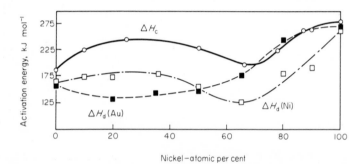

Fig. 13.25   Activation energy for creep of gold–nickel alloys as a function of composition, together with the activation energies for self-diffusion of gold and nickel. (After Sellars and Quarrell, 1961–2, J. Inst. Metals, **90**, 329)

The role of solute locking in steady-state creep can be expected to be small, and even in the fully strain-aged condition, alloys are unlikely to benefit above about $0.5T_M$ because the solute atoms will be able to diffuse readily, and so accompany the dislocations when they move. The situation is somewhat better if strain ageing leads to stable precipitation. It is also unlikely that a high Peierls–Nabarro stress or friction stress will persist at temperatures where secondary creep occurs, at least in dilute solid solu-tions. There is, however, the possibility that in concentrated solutions, the atomic bonding may be increased, particularly if the composition of the solid solution approaches that of a stable compound.[56]

13.9.2. Creep curves of solid solutions

The behaviour of a series of nickel–titanium solid solution alloys is shown in Fig. 13.26, where the marked effect of up to 2 per cent Ti on the creep strain at 1073 K is shown.[57] The creep strengthening follows the relationship

$$\epsilon_c = \epsilon_0 - \frac{d\epsilon_c}{dc} c \qquad\qquad \textbf{13.34}$$

where c = atomic concentration of solute, ϵ_c = creep strain, and ϵ_0 = instantaneous strain.

The rate of change of ϵ_c with concentration is dependent on the nature of the solute atom, atoms of some metals being much more effective than others. In general, one might expect similar rules to those which apply in

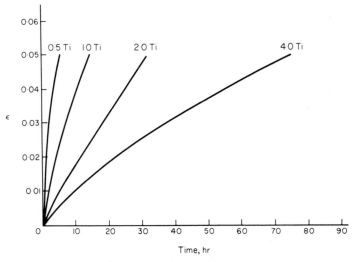

Fig. 13.26   Creep curves of nickel–titanium solid solutions at 1073 K, stress 2·76 MN m⁻², 0·5–4 atomic per cent titanium. (After Hazlett and Parker, 1954, *Trans. ASM*, **46**, 701)

solid solution strengthening at lower temperatures, but there is a lack of useful fundamental data. In solid solutions a decrease in the rate of secondary creep almost always means an increase in time until creep rupture occurs, but the rupture life could be shortened drastically by the presence of only a small concentration of grain-boundary precipitate. Dilute solid solutions of copper and magnesium in aluminium have substantially improved rupture lives when tested between 0·5 and $0\cdot8T_M$,[58] but significantly, in view of its effect on recovery of aluminium, zinc had little effect.

13.9.3. *Effect of order–disorder reactions on creep*

It is now well established that when a solid solution undergoes an ordering reaction, the creep strength is substantially raised. This has been found in β-brass,[59] where the creep rate drops markedly at 743 K (Fig. 13.27), the critical temperature T_c at which long-range order appears on cooling. Similar results have been obtained in iron–aluminium[60] and iron–nickel alloys.[61] In the iron–aluminium alloys it was found that creep in the ordered state also required a much higher activation energy, almost twice the activation energy for self-diffusion which corresponded with the activation energy for creep of the disordered alloy. The extra activation energy is probably needed to move a pair of dislocations which is necessary for each slip process, if order is to be preserved across the slip plane. The disorder created by the first dislocation of the pair is removed by the passage of the second dislocation. Such pairs of dislocations are similar to extended dislocations in metals of low stacking fault energy, in so far as they will not be able to cross-slip or climb readily.

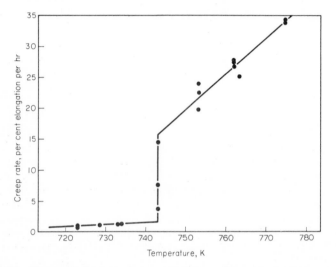

Fig. 13.27   Creep of β-brass above and below the critical temperature for ordering. (After Martin *et al.*, 1957, *Trans. AIME*, **209**, 78)

13.9.4. *Precipitation and dispersion hardened alloys*

The behaviour of alloys containing one or more types of precipitate or dispersion is of particular practical interest because into this category fall all the most important creep-resistant alloys. Unfortunately, their behaviour is extremely complex, and often difficult to predict without carrying out long-term creep tests. Part of the difficulty arises from the fact that metallurgical changes such as precipitation are taking place during the test, but it is also still not possible to predict satisfactorily the effect that changes in the nature of the precipitate or its distribution will have on the whole spectrum of creep properties over a range of temperatures, stresses and different environments. It is, however, possible to find some guiding principles, and to explain in fundamental terms the good creep resistance of certain types of alloy.

As in pure metals, the creep strain originates from dislocation movement within the grains and from grain-boundary sliding. The occurrence of a precipitate in the crystals and at the grain boundaries will obviously interfere with both these processes, so each will be considered. Under high-temperature creep conditions dislocation climb seems to be the rate-controlling process as it is in pure metals; however, there are not as many reliable data.[6] A number of steels give a comparable activation energy with that for α-iron; on the other hand, nickel-base Nimonic alloys give activation energies much larger than those for other nickel alloys. Aluminium alloys containing precipitate have the same activation energy for secondary creep as does pure aluminium.[62]

It is generally agreed that precipitates of the appropriate size and dispersion are much more effective in promoting creep resistance than solid solution hardening, although the two phenomena are additive in their effect. A distinction has been made[63] between inherent creep resistance arising from solid solution strengthening and precipitation present in the material prior to creep testing, and *latent* creep resistance which results from precipitation processes taking place during creep at elevated temperatures. A metallurgically unstable alloy can undergo marked structural changes during creep which are reflected in the creep curves. Precipitation processes can lead to a large reduction in secondary creep rate which can become zero or even negative; Fig. 13.28 illustrates the occurrence of this phenomenon in a complex creep-resisting steel tested over a range of stresses at 973 K. In these circumstances the newly formed precipitate frequently nucleates on the dislocations responsible for the creep strain and makes further movement of these dislocations difficult. In creep-resistant steels several alloy carbides, e.g. VC, TiC, NbC, Mo_2C, $Cr_{23}C_6$ which can be precipitated by prior heat treatment and during creep by appropriate choice of composition, are commonly found to be preferentially nucleated on dislocations.[3] In the early stages the particles are extremely small and closely spaced, so that these dislocations cannot move, but as coarsening occurs, the dislocations may become unpinned. The most effective precipitates are those which coarsen slowly, and thus provide a very fine dispersion to resist dislocation movement. The carbides NbC, TiC and VC are for this reason particularly effective in creep-resistant alloy steels, in which they can form dispersions with average particle sizes often much less than 500 Å diameter in the range 773–973 K.

Electron microscopy has provided ample evidence for the hindering of dislocation movement by precipitates, both at low and high temperatures. In the latter circumstances, dislocations are able to avoid the obstacles by dislocation climb, or by cross-slip.[64] It seems likely that at low stresses dislocation climb is the predominant mechanism, while higher stresses would encourage cross-slip. On the other hand, Ansell and Weert-man[65] have concluded that at high stress levels the Orowan mechanism would occur, the dislocations bowing around the particles leaving residual dislocation loops.

Ansell and Lenel[66] have carried out some basic experiments on the creep behaviour of coarse-grained aluminium–aluminium oxide dispersions (SAP type alloys) in the temperature range 700–900 K, and have determined activation energies by the Dorn method of temperature change. They obtained a value of 155 kJ mol⁻¹, which is close to the activation energy of self-diffusion in aluminium. This indicates that dislocation climb which is diffusion-controlled is the rate-controlling process for steady-state

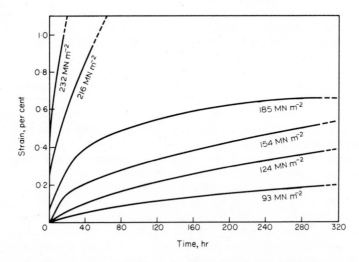

Fig. 13.28   Creep curves of G18B steel at 973 K (a complex austenitic steel with 0·4%C, 13 Ni, 13 Cr, 10 Co, 2·5 W, 2 Mo, 3 Nb and 1 Si). (After Oliver and Harris)

creep in this alloy. On the other hand, fine-grained SAP-type alloys behaved rather differently, giving a much higher activation energy (628 kJ mol⁻¹), which is compatible with a rate-controlling process involving the nucleation of dislocations in grain boundaries. This is only likely to occur when the normal dislocation sources within the grains are rendered completely inoperative.

A further strengthening mechanism which should be useful at high temperatures comprises precipitates in association with stacking faults in matrices of suitably low stacking fault energy.[67] For example, niobium, titanium, tantalum and vanadium carbides form fine dispersions in nickel–chromium austenitic steels, which can be primarily on extended dislocations, that is, thin discs of stacking fault. The precipitate is very fine and stable, and together with the intersecting faults forms a series of walls impermeable to creep dislocations. These walls can be regarded as a super-imposed array of boundaries which do not undergo boundary sliding.

13.10   Rules for development of creep resistance

While theoretical considerations cannot predict the creep behaviour of a complex alloy, it is nevertheless possible to state a series of rules which define ways of obtaining increased creep resistance.

1. The creep resistance at a given temperature is higher in metals and alloys of high melting point. This occurs because for any temperature a metal of high melting point has a lower rate of self-diffusion than a metal of lower melting point. As the rate of dislocation climb is proportional to the rate of self-diffusion, this process will be more difficult in a metal of higher melting point. The limiting temperature below which climb cannot occur readily is about $0 \cdot 5T_M$.

2. Creep resistance is greater in a matrix of low stacking fault energy, because the dislocations are dissociated, and thus find it more difficult to cross-slip and to climb, in order to avoid obstacles. The stacking fault energy of a pure metal can be lowered by solute additions. For this purpose solutes of high valency are best because they more readily increase the electron/atom ratio, and thus decrease the stacking fault energy. Fortunately, such solutes also tend to raise the flow stress more markedly than solutes of lower valence.

3. Solid solution hardening is a useful contribution. This is best achieved by use of solutes differing markedly in atomic size and valency from the parent metal. Unfortunately, these factors mitigate against extensive solid solubility.

4. Long-range order in solid solutions provides a further contribution to the creep strength of solid solutions, because the super-lattice disloca-tions are paired to preserve order across the slip plane, and are thus similar to extended dislocations.

5. Precipitates are essential to increase further the creep strength of a solid solution, and theory provides an estimate of the critical spacing of dispersion for optimum strength in terms of that just small enough to prevent dislocations bowing around the particles (approx. 10^{-6} cm). Un-fortunately, such fine dispersions are usually not stable at high tempera-tures because diffusion of the elements from the precipitates to the matrix allows coarsening. This can be minimized in several ways:

   (a) choice of elements in the precipitate which diffuse slowly,
   (b) use of a dispersed phase which is practically insoluble in the matrix, so that the re-solution of fine particles and growth of coarser particles is slow,
   (c) selection of a precipitate which is crystallographically closely matched to the matrix and so remains coherent longer.

6. Use of precipitates in association with crystal defects. Some precipitates form more readily than others on dislocations, and thus are important sources of strengthening, both at low and elevated temperatures. Precipitates which form *during* creep are particularly useful if they nucleate on dislocations. Nucleation in association with stacking faults is another form of strengthening. Precipitation at grain boundaries is useful in reducing grain-boundary sliding, but in many cases this leads to early cavity formation and premature intergranular cracking. High creep strength is thus often achieved, but at the expense of creep ductility. It is possible that grain boundary precipitates with low energy interfaces with the matrix are less likely to cause intergranular failure.

13.11 Deformation at high temperatures and high strain rates

Hot working is an important aspect of the fabrication of metals. Raising the temperature of deformation enables the deformation to be carried out with much less expenditure of energy and with much less risk of cracking. Moreover, the enhanced diffusion at elevated temperatures and the superimposed effects of deformation combine to minimize inhomogeneities due to segregation, and the coarse-grain structure typical of the cast state can be refined. Hot work involves high rates of strain and large strains which are not readily achieved in a laboratory test. For example, a hot tensile test can be carried out at a high rate of strain but only a limited amount of strain is obtained before fracture. For this reason, hot torsion tests have been widely used where the torque required to twist the specimen is a measure of the flow stress and the ductility is assessed by the number of revolutions to fracture.

It has been found[68,69] that the torque quickly rises to a maximum value then falls to a steady-state region where it is practically constant. Fig. 13.29 illustrates this point for aluminium, copper and nickel tested at equivalent temperatures, $T/T_M = 0.7$. In general, the effect of increasing strain rate is to raise the peak and move it towards higher strains. A similar effect is observed as the temperature of testing is reduced.

High strain rate data from hot torsion and also from tensile and compression tests have been found to follow the relationships

$$\Gamma = \Gamma_0 \theta^N \qquad\qquad\qquad 13.35$$

$$\sigma = \sigma_0 \dot{\epsilon}^N \qquad\qquad\qquad 13.36$$

where Γ = torque, θ = angular velocity, σ = flow stress, and $\dot{\epsilon}$ = strain rate. Γ_0, σ_0 and N are constants which are dependent on temperature. However, recent work[70] has shown that the data are much better correlated by a relationship containing constants which are independent of temperature, viz.

$$\theta = A\,(\sinh \alpha\Gamma)^{n'} \,.\, \exp\,(-Q/RT) \qquad\qquad 13.37$$

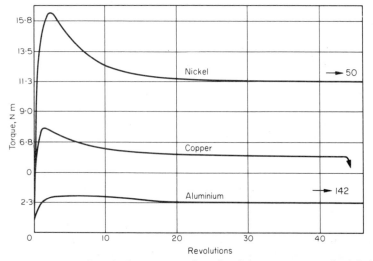

Fig. 13.29 Torque/revolutions curves for aluminium, copper and nickel at $T/T_M = 0.7$ (speed 66 rev/min). (After Hardwick and Tegart, 1961–2, *J. Inst. Metals*, **90**, 17)

where A, α and n' are constants. Figure 13.30 shows some data for a 0·25 per cent carbon steel subject to hot torsion in the range 1073–1473 K; the data now fall on a series of parallel straight lines over a wide range of strain rates. Similar results have been obtained for other carbon steels, austenitic stainless steel and copper. The empirical relationship is thus useful for the extrapolation of data and enables the determination of activation energies in an unambiguous way by plotting log sinh $\alpha\Gamma$ against $1/T$ for a constant strain rate $\dot{\theta}$, so that

$$\frac{d[\log \sinh (\alpha\Gamma)]}{d(1/T)} = \frac{Q}{2·3Rn'} \qquad \textbf{13.38}$$

The plot normally gives a straight line, and the activation energy Q is obtained simply from the slope. Alternatively, the activation energy of the deformation process can be obtained from a plot of log $\dot{\theta}$ against $1/T$ at constant Γ or constant sinh $\alpha\Gamma$.

The values for activation energies obtained from such plots are in some cases comparable with those obtained in creep, when the rate-controlling process is sub-grain formation by dislocation climb (Table 13.2). This is the case for aluminium and low carbon ferritic steels. However, with other metals such as copper and austenitic stainless steel, the activation energy in hot torsion is much higher than under creep conditions. In these cases the experimental evidence suggests that recrystallization is the rate-controlling process in hot torsion, while in creep it is usually recovery in the form of sub-grain formation.

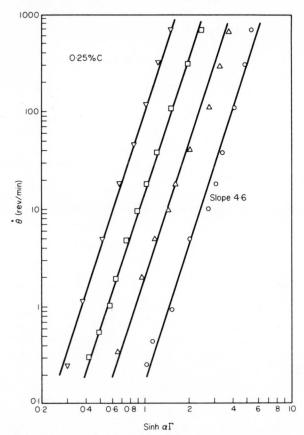

Fig. 13.30 Hot torsion data for 0·25 per cent C steel plotted using a sinh relationship. (After Sellars and Tegart, 1966, *Mem. Sci. Rev. Mét.*, **63**, 731)

Table 13.2 Activation energies for hot torsion and creep (after Sellars and Tegart[70])

Material	Activation energy kJ mol^{-1}	
	Hot torsion	Creep
Soft iron (0·05 per cent C)	280	256
0·25 per cent C steel	303	308
1·2 per cent C steel	389	257
18/8 stainless steel	414	314
Copper	301	201
Nickel	297	243
Aluminium	125–180	155

Structural observations on high stacking fault energy metals such as aluminium[71,72] have revealed that hot working causes the original grains to change to a network of sub-grains, the size and perfection of which depends on the strain, strain-rate and temperature. Aluminium under creep conditions behaves in a similar fashion, and thus it is not surprising that the activation energies for the two processes are the same, although the strain rates are so very different. Metals and alloys of lower stacking fault energy, where dislocation climb necessary to develop a sub-grain structure is more difficult, for example copper[71,72] and 18/8 stainless steel[68] recrystallize during hot working. The original grains become severely distorted and form an incomplete sub-structure which is followed by recrystallization, which entirely replaces the original distorted structure. These metals also can recrystallize either partly or wholly during creep,

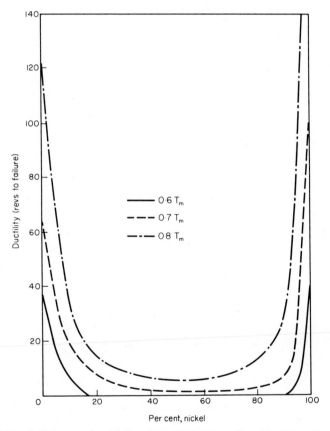

Fig. 13.31   Influence of nickel concentration on the ductility of copper–nickel alloys at various T/T_M ratios

but under the appropriate conditions of stress and strain-rate, recovery processes are rate controlling, and thus the lower activation energies listed in Table 13.2 are accounted for.

Nickel, although it has a substantially higher stacking fault energy, behaves rather like copper; however, during hot working a more strongly developed substructure is seen before recrystallization commences.[71] Under creep conditions sub-grain formation is the dominant structural change. When the whole range of copper–nickel solid solutions is examined, similar structural changes are observed to those occurring in the pure metals, but the alloys recrystallize at higher temperatures because solute atoms retard the grain boundary movement which is an essential part of recrystallization. This retardation of recrystallization is an effective contribution to superior strength at elevated temperatures of solid solutions compared to pure metals. The concentrated solid solutions are found to be much less ductile at elevated temperatures than the pure metals (Fig. 13.31) over a wide temperature range from 0.5 to $0.8T_M$; however, the detailed reasons for this striking effect have yet to be elucidated.

13.12 Superplasticity

A number of alloys when deformed in tension at elevated temperatures are capable of exhibiting neck-free plastic extensions which can exceed 1000%.[75] To achieve this condition the essential requirement is that the strain rate during deformation must not exceed the rate of recovery. Dislocations must be able to move and thus produce strain, then disappear at grain boundaries rather than remain in the grains. Consequently an elevated temperature at which climb can occur is necessary, and also a fine stable grain size. The requisite grain size can be achieved by use of a fine dispersion of a second phase, or by using a two phase aggregate such as a eutectic or eutectoid,[76] where the chemical dissimilarity of the phases prevents grain growth. There is still considerable speculation about the basic mechanism involved, but continuous recrystallization and grain boundary sliding have been suggested.

General references

1. *Creep and Recovery*. (1957). American Society for Metals.
2. DORN, J. E. (Ed.) (1961). *Mechanical Behaviour of Materials at Elevated Temperatures*. McGraw-Hill, New York and Maidenhead.
3. *Structural Processes in Creep*. (1961). Iron and Steel Institute Special Report No. 70.
4. Joint International Conference on Creep. Institute of Mechanical Engineers, London (1963).
5. KENNEDY, A. J. (1962). *Processes of Creep and Fatigue in Metals*. Oliver and Boyd, Edinburgh.
6. GAROFALO, F. (1965). *Fundamentals of Creep and Creep-Rupture in Metals*. MacMillan, New York and London.

References

7. ANDRADE, E. N. DA C. (1910). *Proc. R. Soc.*, **A84**, 1.
8. WYATT, O. H. (1953). *Proc. phys. Soc. Lond.*, **B66**, 495.
9. CHAUDHURI, A. R., CHANG, H. C. and GRANT, N. J. (1955). *Trans. AIME*, **203**, 682.
10. CAHN, R. W., BEAR, I. J. and BELL, R. L. (1953–4). *J. Inst. Metals*, **82**, 481.
11. SERVI, I. S., NORTON, J. T. and GRANT, N. J. (1952). *Trans. AIME*, **194**, 965.
12. WOOD, W. A. and RACHINGER, W. A. (1949–50). *J. Inst. Metals*, **76**, 237.
13. MCLEAN, D. (1951–52). *J. Inst. Metals*, **80**, 507; (1952–53). *Ibid.*, **81**, 133, 287, 293.
14. FELTHAM, P. and MEAKIN, I. D. (1959). *Acta metall.*, **7**, 614.
15. GIFKINS, R. C. (1953–4). *J. Inst. Metals*, **82**, 39.
16. GAROFALO, F., VON GEMMINGEN, F. and DOMIS, W. F. (1961). *Trans. ASM*, **54**, 430.
17. MCLEAN, D. and HALE, K. F. Reference 3.
18. KEH, A. S. and WEISSMANN, T. (1963). *Electron Microscopy and the Strength of Crystals*. John Wiley, New York and London. p. 261.
19. MCLEAN, D. and GIFKINS, R. C. (1960–1). *J. Inst. Metals*, **89**, 29.
20. DORN, J. E. (1956). Symposium on Creep and Fracture of Metals at High Temperatures. National Physical Laboratory, London. H.M.S.O.
21. LANDON, P. R., LYTTON, J. L., SHEPHERD, L. A. and DORN, J. E. (1959). *Trans. ASM*, **51**, 900.
22. GAROFALO, F., RICHMOND, O., DAVIES, W. F. and VON GEMMINGEN, P. Reference 4, pp. 1–31.
23. SHERBY, O. D. and SIMNAD, M. T. (1961). *Trans. ASM*, **54**, 227.
24. NABARRO, F. R. N. (1948). Report on Conference on Strength of Solids. Physical Society, London, p. 75.
25. HERRING, C. (1950). *J. appl. Phys.*, **21**, 437.
26. COTTRELL, A. H. and AYTEKIN, V. (1950). *J, Inst. Metals*, **77**, 389.
27. MOTT, N. F. (1953). *Phil. Mag.*, **44**, 742.
28. WEERTMAN, J. (1957). *J. appl. Phys.*, **28**, 362; (1960). *Trans. AIME*, **218**, 207.
29. MOTT, N. F. Reference 20, p. 21.
30. SCHOECK, G. Reference 2, p. 79.
31. TEGART, W. J. MC G. and SHERBY, O. D. (1958). *Phil. Mag.*, **3**, 1287.
32. SHERBY, O. D., LYTTON, J. L. and DORN, J. E. (1957). *Acta metall.*, **5**, 219.
33. SHERBY, O. D. (1958). *Trans. AIME*, **212**, 708.
34. JOSEFSSON, A. and LANGERBERG, G. (1958). *Jernkont. Annlr.*, **142**, 57.
35. LYTTON, J. L., SHEPHERD, L. A. and DORN, J. E. (1958). *Trans. AIME*, **212**, 220.
36. OROWAN, E. (1946–47). *Jl. W. Scotl. Iron Steel Inst.*, **54**, 45.
37. MOTT, N. F. (1953). *Phil. Mag.*, **44**, 742.
38. SEEGER, A. (1954). *Phil. Mag.*, **45**, 771.
39. NIELD, B. J. and QUARRELL, A. G. (1956–57). *J. Inst. Metals*, **85**, 480.
40. BOETTNER, R. C. and ROBERTSON, W. D. (1961). *Trans. AIME*, **221**, 613.
41. CHANG, H. C. and GRANT, N. J. (1956). *Trans. AIME*, **206**, 544.
42. GREENWOOD, J. N., MILLER, D. R. and SUITER, J. W. (1954). *Acta metall.*, **2**, 250.
43. INTRATER, J. and MACHLIN, E. S. (1959). *Acta metall.*, **7**, 149.

44. OLIVER, P. R. and GIRIFALCO, L. A. (1962). *Acta metall.*, **10**, 765.
45. DAVIES, P. W. and WILTSHIRE, B. (1961). *J. Inst. Metals*, **90**, 470.
46. CHEN, C. W. and MACHLIN, E. S. (1957). *Trans. AIME*, **209**, 829.
47. STACEY, R. D. (1958). *Metallurgia*, **58**, 125.
48. ZENER, C. (1948). *Elasticity and Anelasticity of Metals.* Chicago, University Press, p. 158.
49. GRANT, N. J. and MULLENDORE, A. W. (1965). *Deformation and Fracture at Elevated Temperatures.* M.I.T. Press.
50. HULL, D. and RIMMER, D. E. (1959). *Phil. Mag.*, **4**, 673.
51. MCLEAN, D. (1963). *J. Aust. Inst. Metals*, **8**, 45.
52. SELLARS, C. M. and QUARRELL, A. G. (1961–62). *J. Inst. Metals*, **90**, 329.
53. MOTT, N. F. (1957). *Dislocation and Mechanical Properties of Crystals.* John Wiley, New York and London, p. 350.
54. MONTUELLE, J. (1953). *Comp. Rend.*, **241**, 1304; (1959) Conference on Properties of Very Pure Metals, CNRS, Paris.
55. MCLEAN, D. (1962). *Metall. Rev.*, **7**, 481.
56. KORNILOV, I. J. (1956). Symposium on Creep and Fracture of Metals at High Temperatures. National Physical Laboratory, London. H.M.S.O.
57. HAZLETT, T. H. and PARKER, E. R. (1954). *Trans. ASM*, **46**, 701.
58. GEMMILL, G. D. and GRANT, N. J. (1957). *Trans. AIME*, **209**, 417.
59. MARTIN, J. A., HERMAN, M. and BROWN, N. (1957). *Trans. AIME*, **209**, 78.
60. LAWLEY, A., COLL, J. A. and CAHN, R. W. (1960). *Trans. AIME*, **218**, 166.
61. SUZUKI, T. and YAMAMOTO, M. (1959). *J. phys. Soc. Japan*, **14**, 463.
62. GIEDT, W. H., SHERBY, O. D. and DORN, J. E. (1955). *Trans. ASME*, **77**, 57
63. GLEN, J. (1958). *J. Iron Steel Inst.*, **189**, 333.
64. THOMAS, G. and NUTTING, J. (1957). *J. Inst. Metals*, **86**, 7.
65. ANSELL, G. S. and WEERTMAN, J. (1959). *Trans. AIME*, **215**, 838.
66. ANSELL, G. S. and LENEL, F. V. (1961). *Trans. AIME*, **221**, 452.
67. HONEYCOMBE, R. W. K., VAN ASWEGAN, J. S. T. and WARRINGTON, D. H. (1963). *Relation between Structure and Strength in Metals and Alloys.* National Physical Laboratory Conference, p. 380.
68. ROSSARD, C. and BLAIN, P. (1958). *Rev. Mét.*, **55**, 573; (1959), **56**, 285.
69. HARDWICK, D. and TEGART, W. J. MCG. (1961–62). *J. Inst. Metals.*, **90**, 17.
70. SELLARS, C. M. and TEGART, W. J. MCG. (1966). *Mem. Sci. Rev. Mét.*, **63**, 731.
71. HARDWICK, D. and TEGART, W. J. MCG. (1961). *Mem. Sci. Rev. Mét.*, **58**, 869.
72. LEGUET, R., WHITWHAM, D. and HERENGUEL, J. (1962). *Mem. Sci. Rev. Mét.*, **59**, 649.
73. BRANDON, D. G., RALPH, B., RANGANATHAN, S. and WALD, M. S. (1964). *Acta metall.*, **12**, 813.
74. RACHINGER, W. A. (1952–3). *J. Inst. Metals*, **81**, 33.
75. AVERY, D. H. and BACKOFEN, W. A. (1965). *Trans. ASM (Quart.)*, **58**, 551.
76. CHAUDHARI, P. (1967). *Acta metall.*. **15**. 1777.

Additional general references

1. SELLARS, C. M. and TEGART, W. J. MCG, (1972). *Hot Workability*, Int. Met. Reviews **17,** 1.
2. GITTUS, J. (1975). *Creep, Viscoelasticity and Creep Fracture in Solids.* Applied Science Publishers, London.
3. PADMANABHAN, K. A. and DAVIES. G. J. (1980). *Superplasticity.* Springer-Verlag, Berlin.

Chapter 14

Fatigue

14.1 General characteristics

If a metal is subjected to a large number of stress cycles (for example, alternating tension and compression) it will eventually fracture even when the maximum stress reached is well below the yield stress in a tensile test. This phenomenon, which is called fatigue, is of great importance where metal parts are subjected to fluctuating stresses. It has been estimated that 80–90 per cent of metal failures in practice arise from this cause.

The fatigue behaviour of a metal or alloy is best studied by subjecting standard-shaped specimens to large numbers of identical stress cycles until failure occurs. The tests are repeated for a number of different maximum stress levels, and the stress level in a given test is plotted against the logarithm of the number of cycles to failure (S–N curves). Figure 14.1 illustrates two typical types of behaviour. Both curves fall steeply at the higher stresses, but flatten out at lower stresses. Type A material shows a well-defined horizontal region below which failure will never occur even

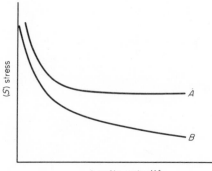

Fig. 14.1 Typical S–N curves for metals

in a prolonged test; this stress level is referred to as the *fatigue limit*. Such behaviour is typical of many steels. Most non-ferrous alloys on the other hand, behave in the way illustrated by curve *B*, where there is also a knee in the curve, but it continues to fall as the stress is lowered. Such materials do not have a fatigue limit, but an *endurance limit*, which is defined as the stress necessary to cause failure in a specified number of cycles (usually 10^8). The tendency towards fatigue is greatly enhanced by the presence of a stress concentration such as a sharp notch which, if suitably located, can lead to the initiation of the fatigue crack at its root. The growth of the crack occupies a large part of the fatigue life of the specimen. The fracture has two main zones, firstly a striated region with periodic markings growing from the point of initiation which represents the growth period when the crack moves a little during each cycle, and secondly a more uniform granular surface where the crack has finally propagated rapidly when the sound section has become too small to sustain the applied stress.

So far it has been assumed that the stress is symmetrically applied each side of zero, but in many cases in practice the mean stress may not be zero, for example in a tensile–compression test, there may be a superimposed tensile or compressive stress. To take this stress into account another type of diagram is used in which the alternating stress component σ_a and the static stress component σ_s form the two axes (Fig. 14.2). The boundaries of the plot are determined by the ultimate tensile stress $\sigma_{U.T.S.}$ which the combined stress $(\sigma_a + \sigma_s)$ cannot exceed. When the static stress is zero, values for σ_a can be obtained from an *S–N* curve for a series of fatigue

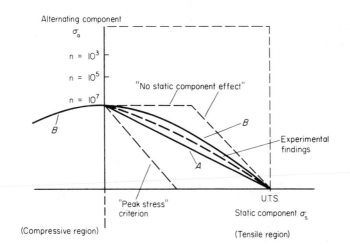

Fig. 14.2 Effect of alternating and static stress components on fatigue. (Kennedy, 1962, *Processes of Creep and Fatigue in Metals*, Oliver and Boyd, Edinburgh and London)

lives and plotted on the ordinate of Fig. 14.2. If the superimposed stress σ_s has no effect on the life, the curve would be parallel to the σ_s axis, but would eventually have to fall to the $\sigma_{U.T.S.}$ point. A more likely assumption is that the fatigue life would be represented by a relationship $(\sigma_a + \sigma_s) = C$ where C is a constant, which gives a straight line intersecting the abscissa at $\sigma_{U.T.S.}$ (curve A, Fig. 14.2). This line is usually referred to as Goodman's relation, which can be expressed in the general form

$$\sigma_a = -\frac{\alpha}{\sigma_{U.T.S.}} \sigma_s + \alpha \qquad\qquad 14.1$$

where α is the value of σ_a, when $\sigma_s = 0$ for a particular fatigue life. The experimental curves are not strictly linear and tend to lie above the Goodman curve. Another empirical relationship which gives a parabolic curve is due to Gerber (curve B, Fig. 14.2).

$$\sigma_a = \alpha\left[1 - \left(\frac{\sigma_s}{\sigma_{U.T.S.}}\right)^2\right] \qquad\qquad 14.2$$

The curve for this relationship lies above the experimentally determined curve, and is moreover symmetrical about the σ_a axis. This is not fully valid because it has been found frequently in practice that a *compressive* static stress has a beneficial effect on fatigue life.

There are many different types of fatigue test, but in the main we shall restrict consideration to uniaxial tests where the stress is symmetrically applied each side of zero and at a constant frequency, for example alternating compression and tension (push–pull), or a test in which a cylindrical specimen is rotated, and loaded transversely in a constant direction (rotating bend test). As in the case of unidirectional deformation, we shall firstly examine the behaviour of single crystals and then consider poly-crystalline aggregates.

14.2   Behaviour of single crystals

Gough and co-workers[9, 10] carried out a series of classical investigations on the behaviour of aluminium, silver, copper, iron and zinc in torsion. This type of test eliminated the need to test many differently oriented crystals, as different parts of the one crystal did not behave identically, because of variations in stress characteristics of the torsion test. Gough found that the slip planes and directions operating in fatigue were identical to the systems in unidirectional tests, and that the maximum resolved shear stress criterion was still valid.

Microscopic observations revealed that slip lines developed initially below the endurance limit in aluminium crystals,[10] but on continued cycling new slip lines eventually ceased to form, indicating that the crystals must have undergone work hardening at these low levels of stress. At higher stress levels similar hardening occurred, except that slip lines formed throughout the fatigue life but with decreasing frequency. Eventually fatigue cracks started within the slip bands.

Further evidence for hardening during fatigue comes from tensile tests on fatigued crystals. Patterson[11] made a comparison of the hardening of copper crystals in tensile deformation, and as a result of alternate tension and compression between shear (glide) strain limits ± 0.004. The hardening during fatigue was determined by measuring the maximum resolved shear stress in each cycle, which was plotted against the cumulative shear strain (Fig. 14.3a). Crystals of soft orientations exhibit marked Stage 1 hardening, whereas crystals likely to deform by multiple slip work harden much more rapidly; this is a close parallel with the behaviour of crystals during unidirectional deformation (Fig. 14.3b). The extended Stage 1 hardening

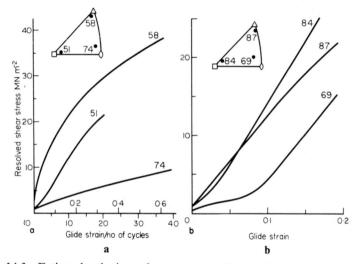

Fig. 14.3 Fatigue-hardening of copper crystals. **a**, Resolved shear stress plotted against total glide strain (upper scale) and number of cycles (lower scale); **b**, shear stress/shear strain curves in tensile tests. (After Patterson, 1955, *Acta metall.*, 3, 491)

during fatigue in soft crystals arises from the fact that no appreciable change of orientation occurs, so the crystal does not reach an orientation more favourable for the occurrence of sporadic secondary slip which initiates Stage 2 hardening in tensile tests. Further work on copper crystals[12] subjected to large numbers of stress cycles has confirmed that fatigue hardening takes place. Specimens fatigued at 78 K undergo further hardening if warmed to room temperature, which is interpreted as due to point defect movement which leads to jogs on dislocations.

Work hardening during fatigue has also been investigated by dynamic measurements of stress and strain[13, 14] which show that a hysteresis loop

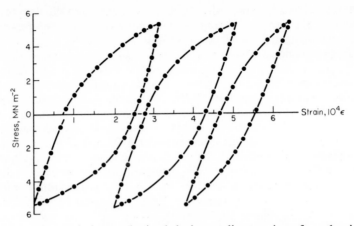

Fig. 14.4 Hysteresis loops obtained during cyclic stressing of an aluminium crystal. Cycles 8, 16 and 1280. (After Thompson *et al.*, 1955, *J. Inst. Metals*, **84**, 75)

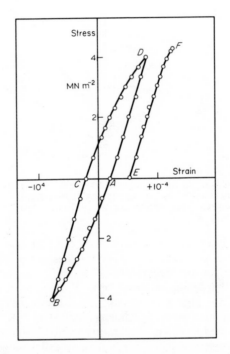

Fig. 14.5 A single hysteresis loop of an aluminium crystal, illustrating the Bauschinger effect (Thompson, 1958, *Phil. Mag.* Supplement 7 (25), 72)

is characteristic of each cycle (Fig. 14.4). In work on aluminium crystals at very low frequencies, the width W of the loop was found to decrease with increasing number of cycles, N, under conditions of constant stress amplitude, according to the empirical relationship

$$W = AN^{-q} \qquad\qquad \textbf{14.3}$$

where A is a constant and q is a measure of the rate of work hardening. q was found to be smaller for crystals deforming on one slip system than for crystals oriented for duplex or multiple slip.

If we look more closely at the typical fatigue hysteresis loop (Fig. 14.5) it will be seen that the behaviour is not symmetrical. If the crystal is first stressed in compression (AB) the curve soon departs from linearity to exhibit a substantial plastic strain on reaching B. On unloading to C, the curve is a straight line representing elastic behaviour, however, on imposing the tensile stress, plastic strain occurs very soon just above the zero stress level in the curve CD. This phenomenon, which occurs both in single crystals and polycrystalline aggregates, is known as the Bauschinger

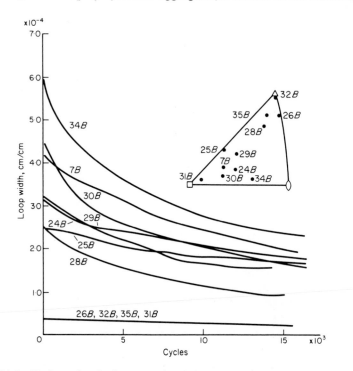

Fig. 14.6 Fatigue hardening on aluminium crystals of different orientations, shown by the change in hysteresis loop width with increasing cycles. Room temperature tests at $\pm 7\,\mathrm{MN\,m^{-2}}$ shear stress. (After Roberts)

effect after its discoverer. In the single crystal case, one explanation which has been put forward involves the operation of Frank-Read sources on the first leg of the cycle, the dislocations being locked in position by cross-slip so that on unloading they do not move until some stress is applied in the opposite sense. A detailed dislocation theory has, however, yet to be developed.

The broad hysteresis loop is a feature of the early stages of the fatigue life of a crystal. The width W falls rapidly in the first few per cent of the life, and attains a steady small value which represents almost elastic behaviour, although microscopic observations reveal that dislocation movements are still taking place. Figure 14.6 shows some results on hard and soft aluminum crystals tested at 800 cycles/min [15] and a fatigue shear stress of $\pm$ 7 MN m^{-2}. In most cases the loop width reached a steady value before 5 per cent of the fatigue life had passed.

Turning to hexagonal crystals, fatigue tests done on single crystals of cadmium have led to fatigue fracture, which was initiated at the boundaries of deformation twins with the matrix.[16] However, more extensive work

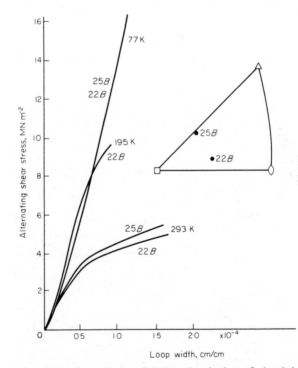

Fig. 14.7 Temperature dependence of fatigue hardening of aluminium crystals between 293 and 77 K revealed by loop width changes during stress build-up. (After Roberts)

at 77 K with zinc crystals favourably oriented for slip[17] revealed that fatigue failure did not take place below the normal tensile fracture stress at this temperature. This behaviour is in complete contrast to that of polycrystalline zinc, and of crystals where twinning is allowed to take place. It seems clear that under conditions of simple basal slip (*no* non-basal slip or twinning) fatigue crack nuclei are not formed, and the implication is that cross-slip or at least slip on intersecting systems is necessary for crack initiation.

Fatigue tests on aluminium[15] crystals over the range 77–273 K indicate that fatigue life at 77 K is much greater for a given stress than at room temperature and, moreover, that the dynamic work hardening as revealed by hysteresis loop measurements during gradual build-up to maximum stress is much more pronounced (Fig. 14.7). At room temperature a pure aluminium crystal will fail after about 500,000 cycles at 6 MN m^{-2}, while at 77 K a similar crystal will withstand a stress of 15 MN m^{-2} for a much larger number of cycles. Metallographic examination provides an explanation of the role of temperature in fatigue behaviour. At room temperature heavy slip bands or striations appear at an early stage (Fig. 14.8). These bands resemble slip in Stage 3 hardening in a tensile test, particularly that obtained at elevated temperatures where the slip bands are very broad and contain very many individual slip lines. It is well known that Stage 3 slip bands are formed by repeated cross-slip of dislocations which is thermally activated. The fatigue cracks are observed to form in the striations, so

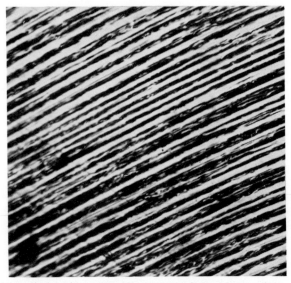

Fig. 14.8   Slip striations in an aluminium crystal fatigued 1·9 × 10^5 cycles at room temperature. Shear stress ± 10 MN m^{-2}. micrograph × 800

there is a direct correlation between the initiation of fatigue cracks and ease of cross-slip. At 77 K, striations are no longer observed in fatigued aluminium crystals even at stresses as high as $\pm$ 13 MN m^{-2}. If the stress is raised considerably ($>$30 MN m^{-2}) fine slip on more than one system is frequently observed but still no striations. In these circumstances failure does occur, but it does not resemble the typical fatigue failures at room temperature.

14.3   Basic experiments on polycrystalline aggregates

Many experiments with polycrystalline aggregates have shown that the mechanical effects of alternating and unidirectional stressing differ fundamentally. We have seen how hysteresis loops provide a way of measuring the energy dissipated in the specimen during the test. Haigh[18] devised an alternative method which measured the rate of heat production in the specimen during its life and on this basis divided the life into three stages.

1. First stage of initial hardening, which only occurs when a metal is stressed above its yield stress. It is accompanied by an initial heat evolution which usually only lasts a few thousand cycles. The hardening is very dependent on the exact conditions of the experiment, but can be lower than that in unidirectional tension, or substantially higher. For example, Broom and co-workers[19] found that polycrystalline aluminium gradually raised to the maximum cyclic stress at 90 K, hardened beyond the ultimate tensile strength at that temperature, but at room temperature the hardening was much less effective (cf. results on aluminium single crystals). The hardening has also been studied in polycrystalline copper by interrupting tests and carrying out tensile tests.[20] Within 100 cycles $\mp$ 77 MN m^{-2}, the proof stress had been raised from 22 to 72 MN m^{-2}, reaching 138 MN m^{-2} after 10^6 cycles (Fig. 14.9).

2. The second stage of a fatigue test is characterized by a gradual increase in the evolution of heat. An alternative way of expressing results is to plot the ratio of the energy dissipated per cycle E_c to the total vibrational energy E_v against the number of cycles. This ratio decreases rapidly in Stage 1, but in Stage 2 it increases slowly. There is a linear relationship between $\mathrm{d}(E_c/E_v)/\mathrm{d}N$ and the fatigue life; the greater the initial slope of the curve, the shorter the life. However, it should be emphasized that the dissipation of energy occurs throughout the whole of the specimen, whereas the fatigue failure is a localized crack propagation which is well advanced during Stage 2. It is thus unlikely that there is a direct relationship between the two. This is demonstrated by interrupting a test in Stage 2 and subjecting the specimen to an annealing treatment. If the temperature chosen is high enough, the specimen on re-cycling will again show high energy dissipation as at the beginning of the test; however, the fatigue life will be practically unaffected as the propagating cracks are not eliminated.

3. The third stage, in which the final propagation of the crack takes

place, occurs rapidly and with an increase in the dissipation of energy. This stage has many of the characteristics of ductile rupture.

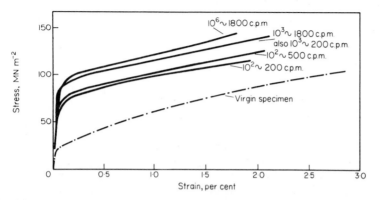

Fig. 14.9 Fatigue hardening of polycrystalline copper illustrated by stress–strain curves after cyclic stressing. (After Bullen, Head and Wood, 1953, *Proc. R. Soc.*, **A216**, 332)

14.4   Structural changes during fatigue

In the early stages of a fatigue test the slip band patterns are very similar to those occurring as a result of unidirectional deformation. Thompson

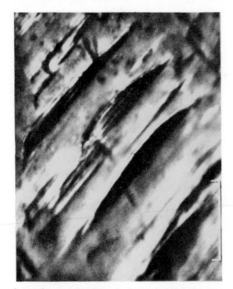

Fig. 14.10   Extrusions from slip bands in a cadmium-3 per cent zinc alloy. × 1300. (After Forsyth, 1957, *Proc. R. Soc.*, **A242**, 198)

and co-workers[13] by using a technique of electrolytic polishing to remove normal slip steps from the surface of copper polycrystals at various stages of a test, were able to show that after about 5 per cent of the life a few bands could not be removed in this way. These they called 'persistent slip bands'.

Controlled removal of the surface revealed that in the early stages the persistent slip bands were not more than 10 microns deep, increasing to 30 microns after 25 per cent of the specimen life. Eventually cracks were observed to start in the persistent slip bands which could thus be regarded as incipient cracks, so it is evident that crack propagation occurs during most of the fatigue life. Complete removal of the markings led to a pronounced extension of the fatigue life, and metallographic examination of the interior of the specimens revealed no further cracks. It is clear that fatigue crack nucleation is a surface phenomenon. Similar results have been obtained with aluminium.[21] Indeed, any physical or chemical treatment which minimizes fatigue crack initiation or removes incipient surface cracks will prolong the life of a component.

One of the most spectacular aspects of the metallography of fatigued specimens was reported by Forsyth[22] who observed the formation of very thin ribbons of metal on the surface of aluminium alloys. Microscopic observation indicated that the ribbons had extruded from well-defined slip bands. The extrusions are not normally regular but occur at intervals along the slip bands (Fig. 14.10). They have now been observed in a wide range of metals and alloys including copper, aluminium and also silver chloride.[23] The reverse effect, formation of intrusions or deep narrow clefts in the surface, has also been encountered on specimens which exhibit extrusions (Fig. 14.11). While most observations have been made at room temperature, extrusions and intrusions have been detected on copper fatigued as low as $2 \cdot 4$ K,[24] so a mechanism must be invoked which does not depend on a thermally activated process.

The experimental evidence indicates that for intrusions and extrusions to occur some cross-slip or slip on an alternative slip system must take place. In view of the difficulty of causing fatigue fracture in zinc crystals, it is of particular significance that intrusions and extrusions have not been observed in this metal.[25] This is consistent with the results on the plastic deformation of zinc which show that basal slip is predominant at room temperature, and that slip on non-basal systems occurs with considerable difficulty. In contrast, Partridge[25] has found extrusions in magnesium and cadmium. Recent thin-foil electron microscope studies on magnesium and cadmium have revealed the presence of non-basal slip, which is consistent with the observation of dipoles in these metals. On the other hand, dipoles are not observed in zinc.

The ease of occurrence of cross-slip does not seem to be an over-riding criterion for extrusion formation. For example, metals of low stacking fault energy, such as copper, in which cross-slip is difficult, readily develop intrusions during fatigue; moreover, this is also the case with copper-base solid solutions of very low stacking fault energies, e.g. Cu–7·5 atomic per cent Al ($\gamma \simeq 2 \ \mathrm{mJ\,m^{-2}}$) and 70/30 α-brass ($\gamma \simeq 15 \ \mathrm{mJ\,m^{-2}}$). On the other hand, pure aluminium, in which cross-slip is extremely easy, does not give rise to extrusions during fatigue, in contrast to the behaviour of many

Fig. 14.11   Fatigue striations in stainless steel showing both intrusions and extrusions. Thin-foil electron micrograph. (After Hirsch *et al.*, 1959, *Phil. Mag*, **4**, 721)

aluminium alloys. These results tend to diminish the significance of cross-slip operation, which is in any case temperature sensitive, and would not account for extrusions in copper at 2·4 K. It thus seems that mechanisms involving not only cross-slip but also slip on secondary systems must be considered. While extrusions have not been found under all conditions of fatigue, the process does normally lead to irregularities on the surface which are more complex than the slip steps formed during unidirectional deformation. Wood[26] and co-workers have applied the method of taper-sectioning to magnify irregularities in a direction to the surface. A taper angle of 2–3° allows the vertical magnification to be increased by 20 to 30 times. Results on copper and brass show that, after only 10 per cent of the fatigue life, many persistent slip bands are associated with notches on the surface (intrusions) or with peaks (extrusions). Fatigue cracks are then initiated at the roots of the intrusions, and propagate initially along the slip bands (Fig. 14.12), but at a later stage they tend to cross from one slip band to another (A and B). The crack is thus typically trans-crystalline

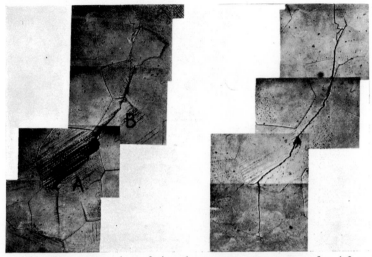

Fig. 14.12 Pure magnesium fatigued at room temperature for $1\cdot3 \times 10^6$ cycles (80 per cent life) at ± 20 MN m². **A,** As fatigued; **B,** after electropolishing to remove slip bands and reveal the crack more clearly. × 60 (After May and Honeycombe, 1963–4, *J. Inst. Metals,* **92,** 41)

(Fig. 14.12), and most evidence points to it commencing in the roots of surface notches produced on a fine scale by the movement of large numbers of dislocations in slip bands. Furthermore, the crack is initiated in the early stages, probably around 5–10 per cent, of the fatigue life.

Examination of fatigue fracture surfaces directly provides more information about the propagation stage. Forsyth[27] has referred to the crack formation on slip bands at intrusions as Stage 1, when the crack path is along the slip planes and thus in the plane of maximum shear stress. Stage 2 is used to describe a propagating stage when the general path of the crack is normal to the tensile axis of the specimen (in push–pull fatigue). The detailed metallography of the two stages is quite distinct. Stage 1 fractures normally show no striking features, but in Stage 2 pronounced bands are visible on the fracture surface (Fig. 14.13), which can in many cases be directly correlated with the cyclic treatment, in so far as they represent successive positions of the propagating crack. For example, an occasional increase in peak stress leads to the formation of a wider band. The propagating crack at this stage is behaving in a ductile manner, because during each cycle plastic deformation occurs at the head of the crack; moreover, each band does not lie on a plane but on a curved surface as distinct from Stage 1 propagation.

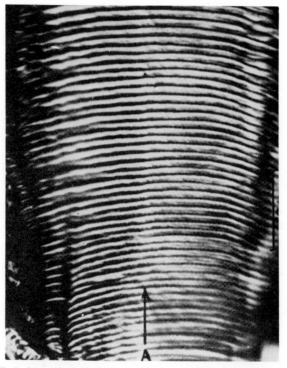

Fig. 14.13 Striations during the second stage of propagation of a fatigue crack. Direction of crack movement is shown by the arrow. (After Forsyth, 1963, *Acta metall.*, **11**, 703)

Thin-foil electron microscopy of fatigued copper crystals[28] has revealed that dislocation loops are a predominant feature even in the early part of the life, and that these remain the principle feature even when the stress amplitude exceeds that of τ_{III}. Similar observations have been made on aluminium[29] and aluminium–magnesium alloys[30] fatigued at fairly low stresses when clusters of jogs and loops gradually develop (Fig. 14.14). At a later stage in aluminium alloys, diffuse low angle boundaries form by linking up of the clusters, which gradually sharpen as new glide dislocations move into the boundaries.

Difficulty has been experienced in correlating these dislocation structures with the fatigue hardening behaviour.[31] For example, the loops
anneal out of fatigued copper in the range 573–673 K, the range in
which the energy stored in the metal during fatigue is released.[32] However,
the fatigue hardening is not completely removed at 873 K. It seems
clear that the general dislocation distribution, while it is typical of fatigue
deformation, does not represent the localized situation on slip striations
where cracking commences. As yet no detailed information on dislocation
distributions in these regions is available.

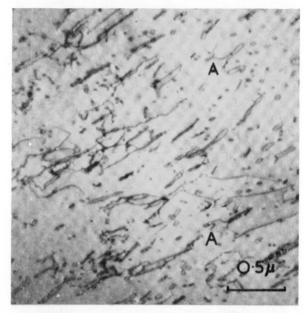

Fig. 14.14 Elongated dislocation loops and jogs in a fatigued Al–3 per
cent Mg alloy. Jogs forming at obstacles (A). Thin foil electron micrograph.
(After Waldron, 1965, *Acta metall.*, **13**, 897)

14.5   Theories of fatigue crack initiation

Since the experimental evidence supports the view that frequently
fatigue cracks originate in regions of slip which are associated with intrusions and extrusions on the surface, the relevant theories are those which
achieve crack nucleation by slip processes.

Mott[33] devised a mechanism involving the close approach of two edge
dislocations moving in opposite directions on closely spaced slip planes
(Fig. 14.15a), which leads to a cavity in the crystal when the dislocations
are abreast of each other—in the two-dimensional case, for example in the

bubble model, dislocations one slip plane apart will form a vacancy. Subsequent dislocations from the sources S_1 and S_2 will enlarge the cavity in a direction parallel to the Burgers vectors of the dislocations (Fig. 14.15b and c). If now one of the component screw dislocations AA' is free to move around the cavity circuit $ABCD$, it will extrude at the surface a similar slab $A'B'C'D'$ (Fig. 14.15d). This model requires the non-conservative movement of the jogged sections at J and J' (Fig. 14.15c), so is likely to be temperature dependent.

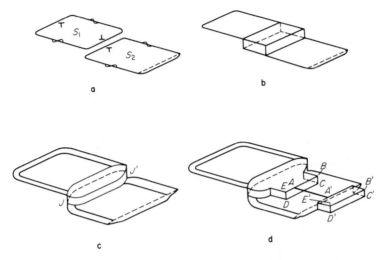

Fig. 14.15   Mott mechanism for formation of extrusions. (After Thompson, 1959, in *Fracture*, ed. by Averbach *et al.*, Wiley, New York)

Cottrell and Hull[34] have proposed a mechanism which is dependent on the operation of slip on two slip systems (Fig. 14.16a). During the tensile stage of the cycle both slip systems operate sequentially to produce two surface steps (Fig. 14.16b and c) while on the compressive stage the steps are converted to an intrusion and extrusion, the intrusion forming on the slip band which is first to operate (Fig. 14.16d and e). In this mechanism all the dislocation movements are conservative, so it should be able to take place at very low temperatures.

Another mechanism developed primarily to explain extrusions in hexagonal metals assumes the formation of jogs in screw dislocations[25] which develop into dipoles and then form loops of prismatic dislocations. Thin-foil electron micrographs give ample evidence of the existence of such loops. As the loops are entirely edge in character they are able to glide in the direction of the original Burgers vector b, and thus pinch out extrusions on the surface. Such movements of dislocation loops have not yet been observed directly under fatigue conditions, but they occur during the

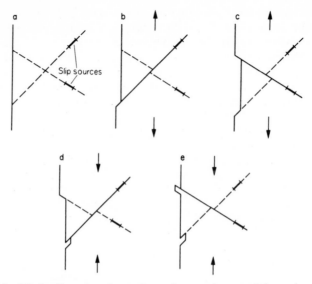

Fig. 14.16   Mechanism for formation of extrusions and intrusions. (After
Cottrell and Hull, 1957, *Proc. R. Soc.*, **A27**, 211)

deformation of face-centred cubic, body-centred cubic metals and in ionic
crystals such as NaCl and AgCl.

14.6   Metallurgical variables in fatigue

In attempting to explain the basic mechanisms of fatigue it is necessary
to eliminate many significant variables and to carry out experiments under
simple controlled conditions. However, it is also important to appreciate
the role of a number of variables which are encountered in practical
conditions.

14.6.1. *Stress amplitude*

The importance of the stress amplitude is implicit in the characteristic
form of the *S–N* curve, and in practice is the most important variable.
Steels are different from most other alloys in so far as they have an infinite
life below the stress corresponding to the fatigue limit, but this stress is
substantially below the elastic limit. It is tempting to associate the fatigue
limit in steels with a strain-ageing effect, where mobile dislocations which
could lead to the formation of a fatigue crack are locked by diffusion of
carbon atoms to form atmospheres or small precipitates. It is interesting
to note that some aluminium alloys which exhibit strain ageing also possess a definite fatigue limit. This viewpoint is supported by the large effect
which temperature is observed to have on the *S–N* curve.

The fatigue limit is substantially below the elastic limit; typically a mild steel with a lower yield stress of 210 MN m^{-2} will have a fatigue limit of 132 MN m^{-2}. This implies that substantial dislocation movement is taking place far below the point on the stress–strain curve which represents the macroscopic start of plastic deformation. There is plenty of evidence pointing to the movement of dislocations at stresses below the yield point. For examples, dislocation movement in silicon–iron has been detected at stresses as low as 0·75 σ_y.[35] These 'pre-yield point' dislocations are undoubtedly important in fatigue, and may also be significant in nucleating cracks in brittle fracture.

The fatigue limit of a steel will vary substantially depending on the exact heat treatment of the steel; however, there is an approximate correlation with the ultimate tensile strength (Fig. 14.17). The mean value of the fatigue limit/ultimate tensile strength ratio is about 0·5, but this is sensitive to the condition of the steel, and whether the specimens are notched or not. In martensitic steels the ratio can be as low as 0·25, while in ferritic steels it can reach 0·60.

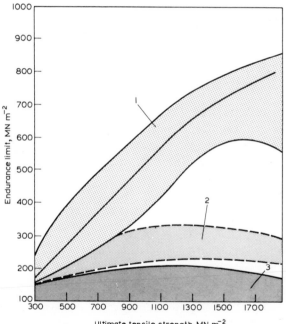

Fig. 14.17 Relation between endurance limit MN m^{-2} and ultimate tensile stress of steels. 1, Notch-free specimens; 2, notched specimens; 3, corroded specimens. (From *Prevention of the Failure of Metals under Repeated Stress*. Battelle Memorial Institute 1941. John Wiley, New York.)

14.6.2. *Stress system*

A normal laboratory fatigue test involves a constant stress amplitude, but in practice many variations are possible, several of which are shown in Fig. 14.18. Fatigue tests can be very complicated, involving combinations of tensile, compressive and shear stresses, but it is possible to make some simple generalizations. A tensile stress superimposed on a push–pull test

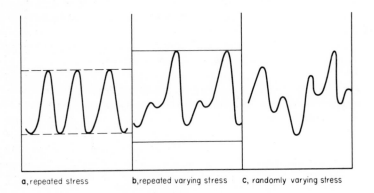

a, repeated stress   b, repeated varying stress   c, randomly varying stress

Fig. 14.18 Some modes of stressing in fatigue. (After Yokobori, *loc. cit.*)

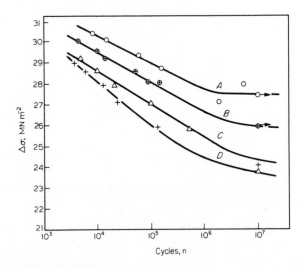

Fig. 14.19 Effect of superimposed mean tensile stress on *S–N* curves for an aluminium–zinc–magnesium alloy. **A,** Zero tensile stress; **B,** 69 MN m^{-2}; **C,** 138 MN m^{-2}; **D,** 207 MN m^{-2}. (After Kennedy, 1962, *Processes of Creep and Fatigue in Metals*, Oliver & Boyd, Edinburgh)

lowers the fatigue resistance, while a compressive stress increases it. Thus the effect of increasing the mean tensile stress is to displace the S–N curve to lower stress levels, while a mean compressive stress raises it.

Figure 14.19 shows some results for an aluminium–zinc–magnesium alloy, in which the mean tensile stress is increased from zero to 207 MN m^{-2}. Bearing in mind that fatigue failure commences at the surface of the specimen, introduction of compressive residual stress there would be expected to have a beneficial effect by retarding the opening of a crack. This is the reason why surface treatments such as shot peening* and skin rolling† improve performance under fatigue conditions. The adverse effect of mean tensile stresses can be understood in terms of their tendency to open up a propagating fatigue crack and so accelerate failure.

14.6.3. Stress concentrations

Most fatigue failures originate in stress concentrations produced by faulty design or manufacture. The large effect of a stress concentration in the form of a notch is best revealed in systematic tests using notched and unnotched material. The magnification of the applied stress at the root of a notch can be calculated from elasticity theory for a given notch section, and the ratio of this stress to the nominal stress is the *stress concentration factor* (q); likewise, the term *fatigue notch factor* (j) is applied to the ratio of fatigue strengths of unnotched and notched specimens. Another quantity, the notch sensitivity η is defined as

$$\eta = \left(\frac{j-1}{q-1}\right) \qquad\qquad \textbf{14.4}$$

and is used to express the notch sensitivity of the fatigue strength of materials. Thus a material which suffers no reduction in fatigue strength due to a notch has a notch sensitivity $\eta = 0$, while if a notch exerts its theoretical maximum effect, $\eta = 1$. The large effect of notching steel test specimens is shown in Fig. 14.20 where the fatigue limit/ultimate tensile strength ratio varies from about 0·5 for unnotched specimen to as low as 0·15 for notched material. In engineering components subject to fluctuating stresses, it is thus wise practice to avoid any design features which can act as a notch, for example sharp re-entrant angles, holes in highly stressed regions, etc. It is also desirable to give the component as good a surface finish as possible, and it is common practice to polish fatigue specimens to a high degree using a standard procedure so that reproducible results are obtained. Sometimes inhomogeneities in the materials, such as large inclusions or hard spots, act as stress concentrations, so in many applications, for example the manufacture of ball bearings, very clean homogeneous steels are preferred.

---

* Blasting the surface with small steel balls gives rise to compressive surface stresses, as does a light reduction pass † through a rolling mill.

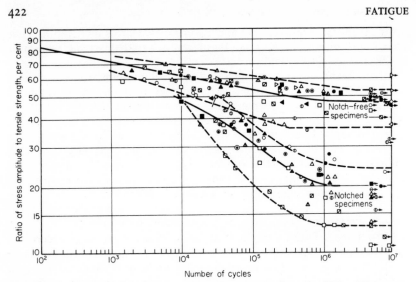

Fig. 14.20 Effect of notching on the *S–N* curves of steels. (After Weissman and Kaplan, 1950, *Trans. ASTM*, **50**, 649)

14.6.4. *Temperature*

As the temperature is lowered, the *S–N* curve moves to higher stress levels but does not change its shape[36] (Fig. 14.21). A thorough investigation of the behaviour of several face-centred cubic metals has been carried out over the range 4·2–293 K,[37] which shows that the fatigue strength/

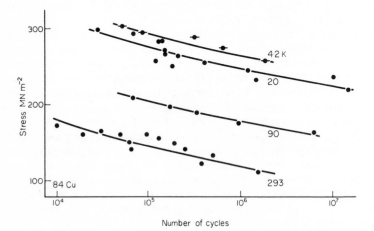

Fig. 14.21 *S–N* curves of copper in the range 4·2 K to 293 K. (After McCammon and Rosenberg, 1958, *Proc. R. Soc.*, **A242**, 203)

U.T.S. ratio falls between 293 and 100 K, then increases again. There is thus some indication that the mechanism of fatigue failure may change at very low temperatures.

With steels the fatigue limit is a maximum around 523-623 K, a phenomenon which appears to be connected with the tendency to strain-age. In this temperature range fatigue dislocations would suffer the greatest pinning, whereas at lower temperatures they would be free of solute atoms and precipitate. At higher temperatures the increased diffusivity of carbon leads to greater dislocation mobility.

14.6.5. *Frequency*

In general, as the frequency of stressing is increased, the longer is the fatigue life, but the main effect is at high frequencies, i.e. above 1 kc/sec. There are difficulties in the interpretation of some results, because considerable heating can occur at the higher frequencies.

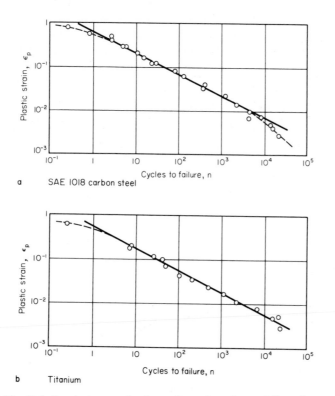

Fig. 14.22 Relation between plastic strain and cycles to failure for **a**, carbon steel (SAE 1018) and, **b**, titanium. (After Kennedy, 1962, *Processes of Creep and Fatigue in Metals*. Oliver and Boyd, Edinburgh)

14.6.6. *Strain amplitude*

Experiments using constant strain amplitude during fatigue have revealed that there is a simple relationship between the plastic strain ϵ_p and the number of cycles to failure (N_f).[38]

$$N_f = \gamma(\epsilon_p)^{-2} \qquad\qquad \textbf{14.5}$$

where γ is a constant. This relationship holds for a wide range of metals, a plot of log $\Delta\epsilon_p$ against log N_f (Fig. 14.22) giving straight lines of slope $-\frac{1}{2}$. While fatigue life can be improved by control of structure and surface condition, the wide validity of the above equation means that the most effective way of increasing fatigue life is to decrease the plastic strain amplitude.

14.6.7. *Chemical effects*

In normal circumstances fatigue cracks initiate at the surface, so it is not surprising that factors such as surface condition become important. Equally, the environment of the surface has been shown to have a large effect on fatigue properties. The classical work of Haigh and Jones[39] demonstrated that the fatigue life of lead could be increased tenfold by coating the surface with oil to exclude air. Gough and Sopwith[40] showed that copper and brass had higher endurance limits when tested in vacuum than in air. Later work[16] on copper in vacuum and air is illustrated in Fig. 14.23 where the beneficial effect of excluding oxygen is quite marked. It was found that crack initiation in vacuum was as rapid as in air, but that

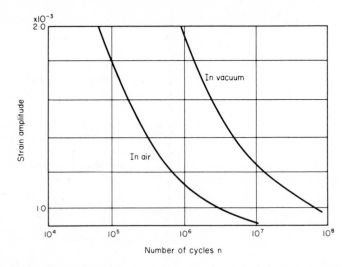

Fig. 14.23 Effect of atmosphere on the fatigue of copper. (After Thompson *et al.*, 1956, *Phil. Mag.*, **1**, 113)

the cracks took longer to propagate, so the role of oxygen must involve absorption on the crack surfaces and possibly the lowering of the surface energy. When the experimental conditions are broadened to include corrosive environments generally, the simultaneous application of alternating stress and a corrosive medium is more disastrous than their effect separately. This phenomenon is generally referred to as *corrosion fatigue*. It is hard to resist the conclusion that most practical cases of fatigue strictly fall into this category. However, some media are more effective than others, for example, sea-water has a very large effect on the *S–N* curves of various metals, including steels and light alloys. Corrosion fatigue is best reduced by the use of corrosion resistant alloys rather than alloys with a higher endurance limit, or alternatively the latter can be used if a fully effective corrosion resistant coating is employed.

Fretting is a related phenomenon to corrosion fatigue. It occurs between two surfaces which move periodically with respect to each other, and arises from the removal of metal from the surfaces by a mechanism akin to wear, that is by welding of asperities and tearing out of fragments which are usually associated with oxide and other corrosion products. Eventually fatigue cracks are often initiated in the damaged regions. Fretting can in many cases be eliminated by adequate lubrication.

14.6.8. *Grain size*

Generally the fatigue strength is raised as the grain size is reduced. It is assumed that this arises because the grain boundaries are good obstacles to the propagation of a fatigue crack as they are for the propagation of a brittle crack. While the brittle fracture grain size dependence has been studied in detail, the comparable fatigue case has received less systematic attention.

14.7   The propagation of fatigue cracks

It has been pointed out that the propagation of fatigue cracks is a discontinuous process which can be correlated to some degree with the stress cycles. There is also evidence that the mode of crack propagation is sensitive to the magnitude of the stress. At high stress levels on the upper arm of the *S–N* curve cracks can be more typical of ductile failure, nucleating in the interior, while at lower stresses the more 'typical' slip band nucleated cracks occur. This distinction has been supported by observations on the fatigue of magnesium oxide single crystals over a large stress range.[41] The total plastic strain to fracture in the high stress region is much smaller than that associated with cracking at low stresses.

There are many cases in practice where a fatigue crack occurs at a stress concentration such as a notch, and later becomes non-propagating.[42] This has led to the theory that material in the vicinity of a notch changes its state as a result of the high local stresses and develops enhanced resistance to cracking. Figure 14.24 plots the alternating stress against the notch sharpness (proportional to stress concentration factor), for mild steel plate. Curve 2 gives the stress to initiate the crack and Curve 1 the propagation stress.

The non-propagating cracks are thus found in the region to the right of the vertical line and bounded by curves 1 and 2. The fact that non-propagating cracks are found particularly in alloys prone to strain ageing, e.g. steels and aluminium alloys, has led to the view that the deformed region in the vicinity of the crack strain ages preferentially, and consequently offers greater resistance to propagation. The strain at the base of the notch is clearly an important factor, for it has been shown that temporary overloading of an already cracked specimen can render the crack propagation more difficult.[43]

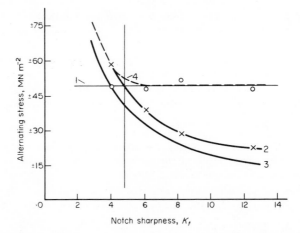

Fig. 14.24 Dependence of fatigue strength on notch sharpness for mild steel plate. Curve 1, Stress to propagate crack; Curve 2, stress to initiate crack; Curve 3, theoretical curve; Curve 4, fatigue limit. (After Frost and Phillips, 1957, *Proc. R. Soc.*, **A422**, 216)

Recent structural studies[44,45] on single crystals of aluminium and on polycrystalline copper and aluminium alloys indicate that two stages in the growth of the crack can be differentiated. The first is associated with cyclic shear on the slip planes in the presence of some secondary slip, while the second stage has a different mode of growth controlled by the normal stresses at the tip of the crack. Cross-slip is also important during this stage, and it has thus been suggested that the difficulty of cross-slip in lower stacking fault energy alloys accounts for their increased resistance to crack growth.[49]

The theory of crack propagation in fatigue is complex; however, there is general agreement that the rate of propagation of a fatigue crack can be expressed in the form

$$\frac{dx}{dN} = k \cdot c^n \qquad\qquad \textbf{14.6}$$

where N is the number of cycles, x is the distance moved, c is the half length of a central crack, and k is a constant depending on the stress level and the alloy. In some theories $n \simeq 1$,[46] while Head[47] has found that $n = \frac{3}{2}$, both of which predictions are borne out by some experimental results, but there is evidence that n can sometimes be as high as 3.

It has also been established that the rate of growth is a function of $\sigma_G \sqrt{c}$ where σ_G is the stress based on the whole specimen section.[45] So for two specimens in which cracks are growing at the same rate

$$k_1 c_1^n = k_2 c_2^n \qquad\qquad \textbf{14.7}$$

and

$$\sigma_{G_1}\sqrt{c_1} = \sigma_{G_2}\sqrt{c_2}$$

from which

$$A(\sigma_{G_1})^{2n} c_2^n = A(\sigma_{G_2})^{2n} c_2^n$$

so that equating coefficients

$$k = A\sigma_G^{2n}$$

and thus

$$\frac{dx}{dN} = A\sigma_G^{2n} \cdot c^n$$

so

$$\log\left(\frac{dx}{dN}\right) = \log A + 2n \log \sigma_G c^{1/2} \qquad\qquad \textbf{14.8}$$

Thus the experimental results should lie on a straight line of slope $2n$ if $\log(dx/dN)$ is plotted against $\log \sigma \cdot c^{1/2}$. However, experimental results do not completely follow this pattern if n is assumed to be 1 or 1·5. At low stresses $n \simeq 2$ while at high stresses it changes to approximately 3. Thus no simple expression holds over the whole range of experimental conditions, but the over-riding importance of stress and crack length is clear.

14.8   Fatigue at elevated temperatures

The fatigue strength of pure metals decreases progressively as the temperature is raised, a result which is scarcely surprising in view of the correlation between the endurance limit and the ultimate tensile strength. The situation in some alloys, particularly steels, is more complex, in so far as improved fatigue properties can occur between 523 and 623 K due to strain-ageing effects. At temperatures above this range, steels no longer exhibit a well-defined fatigue limit but behave in a similar way to most other metals and alloys.

We know that normal deformation at high temperatures produces different structural features from those occurring at lower temperatures, in so far as dislocations not only glide but climb; moreover, under creep conditions the grain boundaries play an important part both in deformation and fracture. Likewise, in fatigue as the temperature of deformation is raised, the mode of failure changes from transgranular initiated in striations, to intergranular nucleated at grain boundary triple points or cavities within the boundaries.[21,48] For example, pure magnesium fatigued in push–pull at 3000 cycles/minute fails transgranularly at room temperature, but intergranularly at 523 K[48] (Fig. 14.25). This transition is inhi-

Fig. 14.25 Pure magnesium fatigued at 523 K. 10^5 cycles (50 per cent life) at $\pm 5\cdot65$ MN m^{-2}. Intergranular cracks. $\times 200$ (May)

bited by coarse grain sizes or by alloying elements in solid solution. The failures bear a striking resemblance to creep failures at elevated temperatures. At low stress levels cavities form preferentially where intense slip bands meet the boundaries, lending support to the view that the creep cavity nucleation mechanism (Chapter 13) is taking place, at least in pure metals, although the presence of occasional inclusions which may act as nucleation points for cracks cannot be excluded. Formation of grain boundary cavities during fatigue has been detected in magnesium at temperatures as low as 423 K. At high stress levels, grain-boundary sliding leads to triple point cracking in a similar manner to that occurring in creep. It seems clear that the dislocation movements which lead to cracking on striations at low temperatures are modified by climb as the temperature is raised; this is substantiated by the detection of well-defined sub-grains. The grain boundary sliding then becomes the dominant process in nucleating cracks. It is thus not surprising that at high temperatures coarse-grained material has higher fatigue resistance than fine-grained material.

The role of frequency becomes more important as the temperature is raised, because the deformation is accompanied by processes which are diffusion controlled and thus time dependent, for example dislocation climb. Thus, we would expect the fatigue endurance to increase with increasing frequency at elevated temperatures. Figure 14.26 shows this trend for an aluminium alloy RR 58 tested in the frequency range 10–2000 cycles/minute at 473 K.

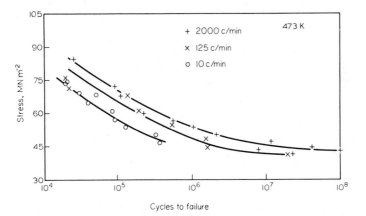

Fig. 14.26   *S–N* curves for aluminium alloy RR 58 at 473 K. Effect of frequency. (After Forrest and Tapsell)

14.9   Fatigue in practice

A phenomenon which leads to failure and involves maximum stresses substantially less than the macroscopic yield point must be cautiously considered by engineers who should be aware of the positive steps which can be taken to minimize the chances of failure. The endurance limit/ U.T.S. ratio is some guide, for most metals and alloys have values below 0·6, but this ratio can fall as low as 0·25. Perhaps the greatest hazard is that of stress concentrations in the form of notches, which may on one hand be simply a result of bad design, but may also arise from bad workmanship, e.g. rough surface finish, or from metallurgical faults such as inclusions, or cracks arising from heat treatment. Care must therefore be exercised at all stages of the manufacture of a part likely to be subjected to prolonged fatigue stresses. The metallurgist must play his part by providing clean material of uniform structure, but the engineer should realize that even the best alloys will fail by fatigue if the stress concentrations inherent in the design are severe enough. Not enough attention is paid to the likely environment of the part, and indeed many fatigue failures are really corrosion fatigue. This can be minimized by protective measures such as electroplating, other surface treatments or by cathodic protection. Once a fatigue crack has started it is too late usually for remedial action, and it is often unfortunately true that the cracks are not detected before catastrophic failure occurs.

General references

1. RASSWEILER, G. M. and GRUBE, W. L. (1958). *Internal Stresses and Fatigue in Metals*, Elsevier, Amsterdam.
2. CAZAUD, R. (1953). *Fatigue of Metals*. Chapman & Hall, London.
3. POPE, J. A. (Ed.) (1959). *Metal Fatigue*. Chapman & Hall, London.
4. THOMPSON, N. and WADSWORTH, N. J. (1958). 'Metal Fatigue,' *Phil. Mag.*, Supplement, *Advances in Physics*, **7** (25), 72.
5. Mechanisms of Fatigue in Crystalline Solids, International Conference, Florida, 1962; *Acta metall.*, 1963, **11**, 639–828.
6. KENNEDY, A. J. (1962). *Processes of Creep and Fatigue in Metals*. Oliver & Boyd, Edinburgh.
7. FORREST, P. G. (1963). *Fatigue of Metals*. Pergamon Press, Oxford.
8. YOKOBORI, T. (1965). *The Strength, Fracture and Fatigue of Materials*. Noordhoff, Groningen, The Netherlands.

References

9. GOUGH, H. J. (1933). *Proc. Am. Soc. Test. Mater.*, **33**, 3.
10. GOUGH, H. J., HANSON, D. and WRIGHT, S. J. (1927). *Phil. Trans. R. Soc.*, **A226**, 1.
11. PATTERSON, M. S. (1955). *Acta metall.*, **3**, 491.
12. BROOM, T. and HAM, R. K. (1959). *Proc. R. Soc.*, **A251**, 186.
13. THOMPSON, N., COOGAN, C. K. and RIDER, J. R. (1955). *J. Inst. Metals*, **84**, 75.
14. BUCKLEY, S. N. and ENTWISTLE, K. M. (1956). *Acta metall.*, **4**, 352.
15. ROBERTS, E. and HONEYCOMBE, R. W. K. (1962–63). *J. Inst. Metals*, **91**, 134.
16. THOMPSON, N., WADSWORTH, N. J. and LOUAT, N. (1956). *Phil. Mag.*, **1**, 113.
17. FEGREDO, D. M. and GREENOUGH, G. B. (1958–59). *J. Inst. Metals*, **87**, 1.
18. HAIGH, B. P. (1928). *Trans. Faraday Soc.*, **24**, 125.
19. BROOM, T., MOLINEUX, J. H. and WHITTAKER, V. N. (1956). *J. Inst. Metals*, **84**, 356.
20. BULLEN, F. P., HEAD, A. K. and WOOD, W. A. (1953). *Proc. R. Soc.*, **A216**, 332.
21. SMITH, G. C. (1957). 'A Discussion on Work Hardening and Fatigue.' *Proc. R. Soc.*, **A242**, 189.
22. FORSYTH, P. J. E. (1954). *J. Inst. Metals*, **82**, 449.
23. FORSYTH, P. J. E. (1957). 'A Discussion on Work Hardening and Fatigue.' *Proc. R. Soc.*, **A242**, 198

24. HULL, D. (1957). *J. Inst. Metals*, **86**, 425.
25. PARTRIDGE, P. G. (1965). *Acta metall.*, **13**, 517.
26. WOOD, W. A., COUSLAND, S. MCK. and SARGANT, K. R. (1963). *Acta Metall.* **11**, 643
27. FORSYTH, P. J. E. (1963). *Acta metall.*, **11**, 703.
28. SEGALL, R. L., PARTRIDGE, P. G. and HIRSCH, P. B. (1961). *Phil. Mag.*, **6**, 1493.
29. WILSON, R. N. and FORSYTH, P. J. E. (1959). *J. Inst. Metals*, **87**, 336.
30. WALDRON, G. W. J. (1965). *Acta metall.*, **13**, 897.
31. SEGALL, R. L. and FINNEY, J. M. (1963). *Acta metall.*, **11**, 685.
32. CLAREBROUGH, L. M., HARGREAVES, M. E., WEST, G. W. and HEAD, A. K. (1957). 'A Discussion on Work Hardening and Fatigue,' *Proc. R. Soc.*, **A242**, 160.
33. MOTT, N. F. (1958). *Acta metall.*, **6**, 195.
34. COTTRELL, A. H. and HULL, D. (1957). 'A Discussion on Work Hardening and Fatigue,' *Proc. R. Soc.*, **A242**, 211.
35. STEIN, D. F. and LOW, J. R. (1960). *J. appl. Phys.*, **31**, 362.
36. ALLEN, N. P. and FORREST, P. G. (1956). ASME International Conference on the Fatigue of Metals, Inst. Mech. Eng., 327.
37. MCCAMMON, R. D. and ROSENBERG, K. M. (1957). 'A Discussion on Work Hardening and Fatigue.' *Proc. R. Soc.*, **A242**, 203.
38. COFFIN, L. F. (1959). Symposium on Internal Stresses and Fatigue of Metals. Elsevier Publishing Co., Amsterdam.
39. HAIGH, B. P. and JONES, B. (1930). *J. Inst. Metals*, **43**, 271.
40. GOUGH, H. J. and SOPWITH, D. G. (1935). *J. Inst. Metals*, **56**, 55.
41. CORNET, I. and GORUM, A. E. (1960). *Trans. AIME*, **218**, 491.
42. FROST, N. E. and PHILLIPS, C. E. (1957). 'A Discussion on Work Hardening and Fatigue.' *Proc. R. Soc.*, **A422**, 216.
43. FROST, N. E. and PHILLIPS, C. E. (1956). *J. Inst. Mech. Engrs.*, **170**, 713.
44. FORSYTH, P. J. E. (1962). Proceedings of Symposium on Crack Propagation, Cranfield.
45. MCEVILY, A. J. and BOETTNER, R. C. (1963). *Acta metall.*, **11**, 725.
46. FROST, N. E. and DUGDALE, D. S. (1958). *J. Mech. Phys. Solids*, **6**, 92.
47. HEAD, A. K. (1953). *Phil. Mag.*, **44**, 925; (1956). *J. appl. Mech.*, **78**, 407.
48. MAY, M. J. and HONEYCOMBE, R. W. K. (1963–64). *J. Inst. Metals*, **92**, 41.
49. HAASEN, P. (1965). *Alloying Behaviour and Effects in Concentrated Solid Solutions*, ed. by Massalski, AIME Met. Soc. Conf. Vol. 29. Gordon & Breach, New York, London & Paris.

Additional general references

1. FROST, N. E. MARSH, K. J. and POOK, L. P. (1974). *Metal Fatigue*. Clarendon Press, Oxford.
2. THOMPSON, A. W. (Ed.) (1977). *Work Hardening in Tension and Fatigue*. Metallurgical Society of AIME, New York.
3. *Fatigue and Microstructure* (1979). ASM Materials Science Seminar 1978, American Society for Metals.
4. FONG, J. T. (Ed.) (1979). *Fatigue Mechanisms*. A Symposium at Kansas City 1978, American Society for Testing and Materials, Philadelphia.
5. KLESNIL, M. and LUCAŠ, P., (1980). *Fatigue of Metallic Materials*. Elsevier Amsterdam.

Chapter 15

Fracture

15.1 Introduction

A very ductile metal will neck down in the central region of a tensile specimen, and finally break with up to 100 per cent reduction in the cross-sectional area of the specimen. This type of fracture is known as ductile fracture. We shall examine the mechanism by which it occurs and indicate the factors which determine the reduction in area prior to failure. In contrast, many solids, particularly body-centred cubic metals and ionic solids, fracture in a brittle fashion at low temperatures. This is brittle or cleavage fracture, which is often preceded by only a small amount of plastic deformation, and the crack propagates rapidly along well-defined crystallographic planes which possess low surface energies.

Amongst the other important modes of fracture is intergranular creep fracture, which has already been considered because of its relevance to tertiary creep. Fatigue fracture too has special characteristics which have been considered in Chapter 14. Finally, another type of fracture of practical importance is intergranular brittle fracture which, like cleavage fracture, occurs at low temperatures, and is a result of a brittle grain boundary phase or the segregation of solute atoms to the boundaries.

15.2 Brittle fracture in amorphous materials

Non-crystalline materials such as glass are completely brittle at normal temperatures, and the macroscopic fracture stress is well below the calculated values of the theoretical fracture stress. This stress can be determined in a similar way to the yield stress of an ideal solid (Chapter 3) and likewise gives a stress of approximately 10^5 MN m^{-2}. If the strength of freshly drawn glass or silica fibres is measured, it approaches the theoretical strength; however, rubbing with a finger is sufficient often for the fracture stress to drop catastrophically. It is now known that the decrease in strength is due to the creation of tiny surface cracks which have been revealed in various ways, e.g. by decoration with evaporated sodium.[6]

Griffith[7] was the first to assume that the discrepancy between the fracture strengths of ideal brittle solids and real solids was due to the presence of small elliptical cracks (Fig. 15.1). At the tip of a crack of length $2c$ in a thin plate under plane stress, there is a stress concentration such that σ_m the maximum stress at the head of the crack is

$$\sigma_m = 2\sigma\left(\frac{c}{\rho}\right)^{1/2} \qquad \textbf{15.1}$$

where σ = applied tensile stress, and ρ = radius of curvature of crack root. This result was first calculated by Inglis in 1913.[8] The point is that the stress concentration at the root of the crack allows the theoretical strength to be achieved locally, while the material as a whole is under a

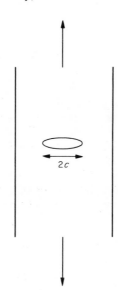

Fig. 15.1   An elliptical crack

relatively small stress. When the applied stress is high enough, the crack will start to propagate and release elastic energy; however, some energy is needed to create the new crack surfaces.

The elastic energy available for the case of thin plates (plane stress) is

$$U_E = -\frac{\pi c^2 \sigma^2}{E} \qquad \textbf{15.2}$$

per unit plate thickness. This term is negative because the growth of the crack removes or releases strain energy. The crack creates two new surfaces, the surface energy of each being $(2c)\gamma$ thus,

$$U_S = 4c\gamma \qquad \textbf{15.3}$$

where U_S = total new surface energy, and γ = surface energy/unit area.

The Griffith criterion is that the crack will spread if the increase in surface energy is less than the decrease in strain energy. So the equilibrium condition is defined as that in which the change in energy with crack length is zero.

So

$$\frac{\mathrm{d}U}{\mathrm{d}c} = \frac{\mathrm{d}\,(U_E + U_S)}{\mathrm{d}c}$$

$$= \left(-\frac{2\pi c\sigma^2}{E} + 4\gamma\right) = 0$$

Giving

$$\sigma = \left(\frac{2\gamma E}{\pi c}\right)^{1/2} \qquad\qquad \textbf{15.4}$$

One feature of this equation is that the stress is inversely related to crack length, so that as the crack grows, the stress needed drops, thus the crack accelerates rapidly. One can put typical values of σ, γ and E for brittle solids into eqn. **15.4** to get an indication of likely crack sizes. For glass, values of the order of 2×10^{-4} mm are obtained, which are close to observed crack sizes. However, when the same is done for zinc which undergoes cleavage fracture at low temperatures, the length of crack required to initiate cleavage is several millimetres. This is clearly not true, as even much smaller cracks have not been seen in zinc, so that in crystalline solids other types of fracture nucleating centres must exist.

15.3   Crack propagation criterion in crystalline solids

The Griffith theory is entirely concerned with elastic cracks in a material which does not undergo plastic deformation prior to fracture. It is now well established that plastic deformation is needed in crystals to nucleate a crack and, in addition, more plastic deformation occurs during crack propagation. This deformation tends to blunt the crack by increasing the radius of curvature at the root, which means that more energy is needed to continue the propagation.

In these circumstances Orowan[9] has shown that the Griffith relationship is modified as follows:

$$\sigma = \left[\frac{2E\gamma\rho}{\pi ca}\right]^{1/2} \qquad\qquad \textbf{15.5}$$

where ρ = radius of curvature of crack root, and a = interatomic spacing across fracture plane.

This relationship implies that the fracture stress increases with the degree of blunting of the crack root which was originally of atomic dimensions (a). The relationship is identical with the Griffith formula when $\rho = a$.

15.4   Crack nucleation in crystals

There is now much experimental evidence that in crystalline solids plastic deformation precedes brittle fracture, although the extent of dislocation movement prior to propagation of the crack can be very small. It is, therefore, logical to conclude that dislocation interactions are responsible for the formation of crack nuclei, and a number of possible mechanisms have been proposed, for some of which there is direct experimental support.

The simplest model involves the pile-up of a series of edge dislocations at a grain boundary or another strong obstacle,[10] in which a crack nucleus is formed when several of the dislocations at the head of the pile-up coalesce (Fig. 15.2). This is made possible by the stress concentration

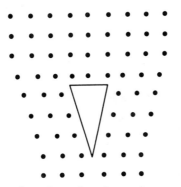

Fig. 15.2   Formation of crack nucleus by coalescence of edge dislocations. (After Yokobori)

which occurs at the head of the pile up (Chapter 3), and which leads to a tensile stress on a plane normal to the slip plane,[11] which is assumed to reach the ideal fracture stress. This theory thus envisages a crack forming normal to the slip plane. Stroh[12] made a further study of this type of model (Fig. 15.3b) and found that the tensile stress σ is a maximum at 70·5° to the slip plane and the following expression holds:

$$\sigma_{\max} = 2 \left(\frac{l_0}{3c}\right)^{1/2} . \tau \qquad\qquad \textbf{15.6}$$

where τ = applied shear stress, l_0 = length of slip line occupied by pile-up, $2c$ = crack length.

Now we use the Eshelby, Frank, Nabarro equation for a dislocation pile-up (Chapter 3, Section 3.7):

$$n = \frac{\pi l_0 \tau (1 - \nu)}{Gb}$$

where G is the shear modulus, and n is the number of dislocations in the pile-up and finally obtain a condition for the formation of a crack, which is

$$n\tau > \frac{12\alpha G}{(1 + \nu)} \simeq 0 \cdot 7G \qquad \textbf{15.7}$$

α is a constant $\simeq 0 \cdot 06$.

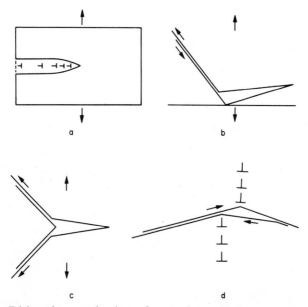

Fig. 15.3 Dislocation mechanisms for crack nucleation. **a**, Elastic crack regarded as a pile-up of edge dislocations; **b**, pile-up against a boundary forming a crack; **c**, crack forming by movement of dislocations on two slip planes; **d**, crack formation at tilt boundary. (After Cottrell, 1959, in *Fracture*, ed. by Averbach *et al.* John Wiley, New York and London)

So a crack is initiated when the local magnified shear stress $n\tau$ reaches a value of about three-quarters the shear modulus. Substitution of appropriate values for τ and G into eqn. **15.7** give values for n between 10^2 and 10^3, which are feasible numbers of dislocations to be associated in pile-ups. It is interesting to note that cracks at the ends of pile-ups have been observed in magnesium oxide crystals[13] in approximately the predicted orientation relative to the slip plane.

Another process involves the movement of dislocation pile-ups on two intersecting planes in a body-centred cubic metal, in which two glide dislocations on $\{110\}$ $\langle 110 \rangle$ slip systems can interact to form a cavity dislocation.[14]

$$\tfrac{1}{2}a[\bar{1}\bar{1}1] + \tfrac{1}{2}a[111] \rightarrow a[001]$$

A crack is nucleated as further dislocations combine (Fig. 15.3c) and the resultant dislocations feed in to enlarge the cavity. This model produces a crack on the $\{001\}$ plane, which is the observed cleavage plane for iron, and some other brittle cubic crystals. Microscopic evidence for this mechanism has been obtained in iron[15] and in magnesium oxide.[16] In the latter material, etch pits reveal dislocation pile-ups on intersecting slip bands, and cracks in the $\{001\}$ orientation are observed in the region of the intersections as the theory predicts.

Hexagonal metals frequently cleave on the basal plane which is the slip plane, so a model is needed for crack nucleation in the same plane as the glide dislocations. Gilman[17] has pointed out that kink planes normal to the basal plane, which can be assumed to be walls of edge dislocations, can provide crack nuclei (Fig. 15.3d). Slip is assumed to occur on the basal plane on one side of the kink boundary, and cause a displacement which leads to the formation of a crack along the slip plane[18] (Fig. 15.4).

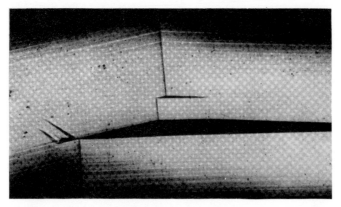

Fig. 15.4 Basal plane crack in zinc near sub-boundary (cf. Fig. 15.3**d**). (After Gilman, 1957, in *Dislocations and Mechanical Properties of Crystals*, ed. Fisher, *et al.*, John Wiley, New York and London)

Deformation twinning can also cause the nucleation of brittle cracks in body-centred cubic metals[19] (Fig. 15.5). Recent work[20,55,56] has shown that crack nuclei can be formed in association with twins when deformation twin lamellae intersect, or when a twin lamella meets a grain boundary. Cracks are more likely to occur when the resolved normal stress on the cleavage plane is high, and the line of intersection of the twins is close to this cleavage plane. Stroh[1] has analysed the initiation of cracks by piling up of dislocations at other dislocations acting as obstacles, and reached the conclusion that such barriers are normally insufficiently strong for the nucleation of cracks. It is thus concluded that twin and grain boundaries are more favourable places for the initiation of cracks.

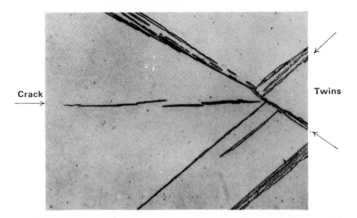

Fig. 15.5 Nucleation of a cleavage crack by intersecting twins in a molybdenum crystal compressed rapidly at room temperature × 200. (After Reid, 1965, *J. Less Common Metals*, **9**, 105)

15.5 Brittle fracture of single crystals

Many crystalline solids undergo brittle or cleavage fracture at low temperatures, and ionic and covalent solids are more prone to exhibit brittleness than undergo substantial plastic deformation. True cleavage fracture is unknown amongst the face-centred cubic metals, but is found in a number of body-centred cubic metals, notably iron, molybdenum, chromium, vanadium, tungsten and tantalum, where the occurrence is associated with the presence of impurity atoms in interstitial solid solution. A number of hexagonal metals notably zinc, magnesium, titanium, zirconium and beryllium all undergo brittle fracture along the basal plane, if the temperature is low enough. While the presence of impurities tends to make these metals brittle at higher temperatures, it is still not clear whether reduction of the impurities to a very low level will eliminate the phenomenon, for example zinc of very high purity still cleaves readily at low temperatures. Metals of other crystal structures are frequently brittle, for example bismuth, antimony and uranium. Some typical cleavage planes are listed in Table 15.1. It is not possible to adopt a single criterion for the fracture plane, e.g. closest packed plane or that of lowest surface energy, but metals of the same crystal structure normally behave in a similar way.

Early work on the fracture of single crystals indicated that fracture

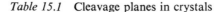

Table 15.1   Cleavage planes in crystals

Metal	Crystal structure	Fracture plane
Iron	b.c.c.	{001}
Molybdenum	b.c.c.	{001}
Chromium	b.c.c.	{001}
Tungsten	b.c.c.	{001}
Tantalum	b.c.c.	{110}
Vanadium	b.c.c.	{110}
Zinc	c.p.h.	{0001}
Beryllium	c.p.h.	{0001}
Bismuth	Rhombohedral	{111}
NaCl	Rocksalt cubic	{001}
MgO	Rocksalt cubic	{001}
LiF	Rocksalt cubic	{001}
Diamond	Cubic	{111}

occurred at a constant normal stress, that is, the stress normal to the cleavage plane at fracture was constant. More recently workers have not

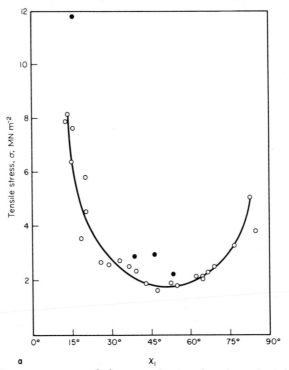

Fig. 15.6   Fracture stresses of zinc crystals as a function of angle between slip plane and axis. **a**, Tensile stress. The closed circles are for specimens with two (as distinct from one) favourably oriented slip directions.

confirmed this criterion in zinc[21] and iron,[22] and have not been able to find a single alternative criterion for fracture. The tensile stress at fracture varies very greatly with the crystal orientation (Fig. 15.6a) with a minimum when the fracture plane is at about 45–50° to the tensile axis. Figure 15.6b shows the results plotted in terms of stress normal to the cleavage plane, and shear stress in the cleavage plane in the slip direction. It is clear that a shear stress criterion is better than a normal stress one, but neither provides an adequate basis for the effect of orientation on fracture stress.

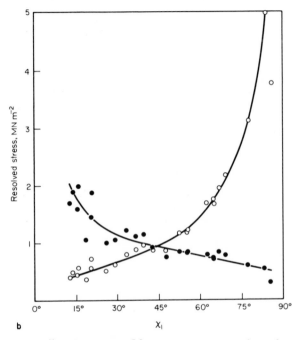

Fig. 15.6 **b**.   o tensile component of fracture stress normal to cleavage plane. ●, resolved shear stress component along slip plane in slip direction. (After Deruyterre and Greenough, 1956, *J. Inst. Metals*, **84**, 337)

Iron crystals of normal purity exhibit a well-defined transition from ductile to cleavage fracture when the temperature of testing is lowered from 120 to 20 K. At 77 K, the orientation of the crystal determines whether it will behave in a brittle or ductile manner.[23] Soft orientations towards [110] tend to be ductile, while harder orientations are brittle, suggesting that brittle crack nuclei are more likely to form when secondary slip is more probable. This gives support to the crack nucleation mechanism involving coalescence of dislocations on two different slip planes, a mechanism likely to be favoured in the absence of grain boundaries and twins. More recent work has interpreted the orientation dependence in terms of the occurrence of twins.[55]

15.6 General characteristics of brittle fracture—the transition temperature

We now have some dislocation mechanisms for the formation of crack nuclei, and a criterion which determines whether or not the crack will propagate, but we must also explain why some metals exhibit brittle fracture and others do not. To do this, several metallurgical aspects of the problem have to be considered.

Firstly, the cleavage type of brittle failure is essentially a low temperature phenomenon, which can be entirely eliminated if a sufficiently high deformation temperature is employed. There is thus a transition from ductile to brittle behaviour which usually occurs over a very narrow temperature range, and the temperature is known as the transition temperature. The brittle behaviour is revealed by impact tests carried out over a range of temperatures. At the higher temperatures the specimens absorb much energy as ductile failure occurs, but at the transition temperature, the energy needed to break the specimens suddenly becomes very much smaller (Fig. 15.7). In the case of steel, the energy absorbed in a Charpy impact test can fall from well over 136 J down to less than 7 J through the transition temperature. It should be emphasized that there is no one constant temperature at which this ductile brittle transformation occurs, for a given metal. The transition temperature is sensitive to a number of metallurgical variables of which purity and grain size are the most impor-

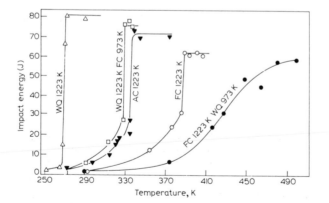

Fig. 15.7 Impact transition curves for a high purity iron–0·12% carbon alloy, showing effect of heat treatment. W.Q. = water quenched; A.C. = air cooled; F.C. = furnace cooled. (After Allen *et al.*, 1953, *J. Iron Steel Inst.*, **174**, 108)

tant for a relatively pure metal, but with steels the exact heat treatment (Fig. 15.7) and even the steel-making practice are very significant variables.

Secondly, the occurrence of brittle fracture is associated with certain crystal structures, in particular the body-centred cubic structure, and is more pronounced in the presence of impurities which form interstitial solid solutions in metals of this structure (Chapter 9, Section 9.7.6). The most familiar example is carbon in α-iron which, even if present to the extent of only a few parts per million, will cause the metal to undergo a ductile-brittle transition. The attainment of very low carbon contents by zone purification leads to iron which has substantial ductility (reductions in area up to 90 per cent) even at 4·2 K.[24] The body-centred cubic Group 5A metals vanadium, niobium[25] and tantalum[26] all exhibit ductile brittle transitions to a cleavage type of fracture on {001} planes, if the temperature is sufficiently low, in all cases below 150 K, and with very pure materials much lower. The body-centred cubic Group 6A metals tungsten, molybdenum and chromium are more susceptible to brittle fracture than the Group 5A metals, and have correspondingly higher transition temperatures. However, with the less pure grades of these metals the brittle fracture goes along the grain boundaries (intergranular). As the purity is increased by zone refining, the mode of fracture changes to transgranular on {001} planes in the case of tungsten and molybdenum. The cleavage path appears to be determined primarily by the surface energy which can be a significant part of the energy in the modified Griffith equation. For pure metals the {001} plane fulfils this requirement, but at higher impurity contents, the crack path changes to the grain boundaries, because segregation of impurities lowers the surface energy. The surface energy term γ is then defined:

$$\gamma = 2\gamma_s - \gamma_b \qquad\qquad \textbf{15.8}$$

The surface energy requirement for fracture $2\gamma_s$, is reduced by the amount of the grain boundary energy γ_b. The γ_s term is further reduced by the presence of impurities until $(\gamma_s)_{boundary} < (\gamma_s)_{\{001\}}$ and intergranular failure occurs. A good example of this transition from transgranular to intergranular brittle fracture as a result of impurity segregation to the boundaries is shown by iron containing small concentrations of phosphorus[27] or oxygen.[28]

The main role of the interstitial impurity atoms is the introduction of a sharp yield point in the stress–strain curve. As we have seen, this is a result of the strong asymmetric distortion which these atoms produce in the body-centred cubic structures (Chapter 6), which causes them to have a large interaction energy with the stress fields of dislocations. This means that strong dislocation locking will take place, so that when plastic deformation occurs it will be either by the sudden avalanche of dislocations which are torn from their solute atmospheres (Cottrell–Bilby theory), or by the rapid movement of newly generated dislocations, which have no solute atmosphere (Gilman–Johnson theory). In either case, the conditions are suitable for the coalescence of dislocations to form crack nuclei, for the first dislocations to move are acted on by a high stress, and thus will have a high velocity. As the temperature is lowered, the yield stress rises steeply (Fig. 9.17, Chapter 9); this is a general characteristic of body-centred cubic metals, and appears to be true even when the interstitial impurity level is very low[29] (Fig. 15.8). In terms of dislocations this means that the Peierls–Nabarro stress in b.c.c. metals is strongly temperature dependent. Consequently, the first dislocations to be mobile will move more rapidly at lower temperatures, and the chances of forming crack

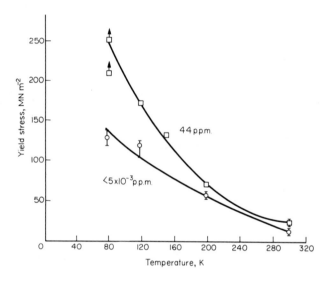

Fig. 15.8 Temperature dependence of yield stress of iron (approx. 44 p.p.m. carbon) and very pure iron (0·005 p.p.m. carbon). (After Stein, 1966, *Acta metall.*, **14**, 99)

nuclei will increase. If the critical flow stress becomes sufficiently high, twinning frequently becomes an important feature of the deformation process. This, too, leads to sudden yield drops as twin lamellae are formed, so that conditions are suitable for crack nucleation by one of the mechanisms involving twins. A further important point to be considered is that the marked temperature dependence of the flow stress makes plastic deformation at the tip of the crack more difficult as the temperature is lowered. Thus, at low temperatures there will be less plastic blunting of the crack, and propagation will occur more readily.

15.7   Theory of the ductile brittle transition

Recent theories have attempted to determine the most difficult part of the fracture process to achieve, from the three stages which can be differentiated. Firstly, slip dislocations are unlocked or generated, then these form crack nuclei, some of which propagate to give the typical transgranular brittle cleavage crack. There have been different views on which of these three stages is the most difficult to carry out. The original dislocation crack nucleation theory developed by Stroh[30] assumed that crack nucleation was the major obstacle, and that once a nucleus formed it would propagate readily. On the other hand, Cottrell[1] and Petch[1] decided that the propagation stage of fracture was likely to be the significant one, because microcracks which do not propagate are frequently observed.

When a notched iron specimen is tested in impact within the transition temperature range, the first crack to form is ductile and involves a large amount of localized plastic deformation, but as it propagates it changes to a cleavage-type crack. True ductile failure occurs by the linking up of a large number of small ductile crack regions, whereas one brittle crack can be responsible for the failure of the whole specimen, and is referred to as a Griffith-type crack propagation, because the modified Griffith criterion can then be applied. If the yield stress of the material is higher than the stress needed to propagate the brittle crack, the specimen will fail in a brittle manner, but if not, it will be ductile.

Stroh has shown that the stress σ_f necessary to propagate a brittle crack formed by the coalescence of n dislocations is defined thus:

$$\sigma_f = \frac{4\gamma'}{nb\left(1 + \dfrac{1}{\sqrt{2}}\right)}$$ **15.9**

where γ' is the effective surface energy of the growing crack, and b is the Burgers vector.

If now σ_i is the friction stress of an unlocked dislocation, then the effective stress operating on the dislocation is $(\sigma_f - \sigma_i)$ once the friction stress is overcome. These dislocations are generated in grains of diameter d which gives a maximum value for the length of a slip band formed by sudden release of dislocations at the yield point. Once the crack nucleus is formed at the end of a slip band, there will be a back-stress which is defined as follows:

$$(\sigma_f - \sigma_i) = \frac{4Gnb}{(1 - v)d}$$ **15.10**

Using eqns. **15.9** and **15.10**

$$\sigma_f = \frac{4G\gamma'}{\alpha(\sigma_f - \sigma_i)d} \quad (\alpha = \text{constant})$$ **15.11**

which defines the condition for the propagation of a brittle crack.

Now if we assume that the fracture occurs at the yield stress, the familiar relationship between yield stress and grain diameter

$$\sigma_f = \sigma_i + kd^{-1/2} \qquad \textbf{15.12}$$

can be introduced into **15.11** giving a criterion for the change from ductile fracture to cleavage, i.e. when σ_f reaches a critical value,

$$\sigma_f \simeq \frac{4G\gamma'}{\alpha k d^{1/2}} \qquad \textbf{15.13}$$

This equation defines the stress σ_f above which brittle fracture occurs; when σ_f becomes less than the right-hand side of the equation, ductile fracture takes place. The relationship embodies many of the experimentally determined aspects of brittle fracture. If σ_f is increased, brittle fracture is favoured. This will occur when the temperature is lowered, largely as a result of an increase in the frictional stress σ_i (eqn. **15.12**). So eqn. **15.13** is strongly temperature dependent, as would be expected from the sharpness of the transition temperature range. The frictional stress is essentially the stress necessary to move an unlocked dislocation through the crystal, and this is strongly temperature dependent even when the interstitial impurity content is low.

In the Stroh model of fracture, the fracture stress is determined by a nucleation process involving the coalescence of dislocations under a shear stress. This is not a valid hypothesis because it is known that hydrostatic tensile stresses increase the brittleness of mild steel, a result well established by experiments on thick and thin plates where the brittle behaviour is more extreme in the former. Again, if eqn. **15.12** is equally valid for yield and fracture stresses it suggests that the friction stress σ_i does not affect brittle behaviour, a deduction which is far from the truth. If on the other hand it is assumed that the growth or propagation of the crack is the critical stage, the applied stress on a crack nucleus would be higher for a metal with a high σ_i.

The Cottrell–Petch [14, 31] modification of the Stroh brittle fracture theory assumes that when the yield stress exceeds the stress for growth of the crack the material is brittle and vice versa. The stress for coalescence of groups of dislocations into crack nuclei is lower than the growth stress, so when the yield stress exceeds the coalescence stress, microcracks, or non-propagating crack nuclei will form.

Cottrell rewrites the Griffith equation (**15.4**) using a group of n dislocations to form the crack nucleus. The quantity na then represents the displacement between the faces of the dislocation crack, where a is the interatomic distance in the plane of the crack. The new form of the equation is

$$na\sigma \simeq 2\gamma \qquad \textbf{15.14}$$

where $na \simeq \sigma_c/E$. This equation represents the transition from a stable

crack nucleus to a continuously growing crack. The left-hand side of
15.14 is an energy term comprising the product of the applied stress σ and
the total Burgers vector which is equated to the surface energy term. The
total Burgers vector can be expressed in terms of the length d of the dislocation pile-up, and the effective shear stress on the dislocations namely
$(\sigma - \sigma_i)$

$$na = \frac{\sigma - \sigma_i}{G} d \qquad\qquad \textbf{15.15}$$

This equation relates the shear strain in the crack, na/d, to the effective
shear stress.

In a brittle material σ can be identified with the yield stress σ_y which is
defined in eqn. **15.12**, so substitution in this equation gives

$$na \simeq \frac{k_y d^{1/2}}{G} \qquad\qquad \textbf{15.16}$$

which is now put into eqn. **15.14**,

$$\sigma_y k_y d^{1/2} = \beta G \gamma \qquad\qquad \textbf{15.17}$$

or alternatively

$$k_y(\sigma_i d^{1/2} + k_y) = \beta G \gamma \qquad\qquad \textbf{15.18}$$

Equations **15.17** and **15.18** provide a criterion for fracture to just take
place at the yield point. The material is ductile when the left-hand side
becomes smaller than the right. The constant β will vary with the mode of
deformation and the presence of any stress concentrators such as notches.
These equations bring out the effect of such variables as grain size, friction
stress and the type of stress system imposed.

The temperature dependence of the friction stress σ_i is easily determined
from plots of the lower yield stress against $d^{-1/2}$ for several temperatures by
measuring the intercepts on the ordinate. There is, however, some evidence
that the σ_i term can contain a temperature dependent component σ_i' and a
temperature independent component σ_i''.

The temperature dependence of σ_i' is as follows:

$$\sigma_i' = k\, e^{-\phi T} \qquad\qquad \textbf{15.19}$$

where k and ϕ are constants, and is apparently still so even when the interstitial content is very low. The temperature independent term is thought
to arise from temperature-insensitive dislocation interactions with precipitates and lattice defects.

Assuming that σ_i as a whole has a similar temperature dependence to
σ_i' then

$$\ln \sigma_i = \ln B - \alpha T \qquad\qquad \textbf{15.20}$$

where B and α are constants.

Using eqns. **15.12** and **15.13**, we get an expression for the transition temperature, T_c

$$\ln B - \alpha T_c = \ln \left(\frac{4G\gamma'}{k} - k\right) - \ln d^{-1/2} \qquad \textbf{15.21}$$

This type of equation predicts a linear dependence of T_c on $\ln d^{-1/2}$ which has been found experimentally by Petch (Fig. 15.9).

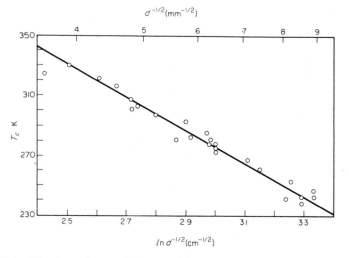

Fig. 15.9   The dependence of the transition temperature of mild steel on the grain size. (After Petch, 1959, in *Fracture*, ed. by Averbach *et al.* John Wiley, New York and London)

The friction stress at a particular temperature is substantially increased by various treatments.

(a) Increased interstitial solid solution: σ_i for iron increases by approximately 46 MN m^{-2} for each 0·01 per cent carbon or nitrogen in solid solution, and the transition temperature is appreciably raised. This would be expected from the substantial effect of interstitial elements on the initial flow stress (Chapter 6).

(b) σ_i is raised by quench ageing and strain ageing. Fine precipitates formed during these treatments tend to pin down or restrict dislocations.

(c) Radiation damage raises σ_i and the transition temperature substantially.[32] Aggregates of point defects produced by irradiation tend to act in a similar way to fine precipitates.

(d) σ_i is strain-rate dependent: slow strain rates encourage ductile behaviour and thus a lower transition temperature, by utilizing thermal energy to help overcome obstacles to dislocation movement.

It should now be clear that the actual transition temperature in a steel is likely to be dependent on a number of interacting variables.

15.8 Practical aspects of brittle fracture

When a fracture commences, potential energy stored as elastic energy in the stressed material is used partly to provide the surface energy of the new crack surfaces, and partly to provide the kinetic energy of the crack. Using the Griffith model it can be shown[33] that the velocity of the crack v is

$$v = \sqrt{\frac{2\pi}{k}} \sqrt{\frac{E}{\rho}} \left(1 - \frac{c_0}{c}\right)^{1/2} \qquad\qquad 15.22$$

where c_0 = critical crack size, $c = \frac{1}{2}$ actual crack size at a given instant, and k = constant. This equation shows that the velocity of crack propagation increases as the crack length increases, and reaches a limiting velocity for large values of c. This limiting velocity is between 0·4 and 0·5 of the speed of sound, so that the process of brittle fracture can occur with catastrophic rapidity.

The phenomenon became prominent on the introduction of all-welded constructions, such as ships and bridges, where the welding provided a continuous path for the propagation of the crack which would otherwise have stopped at joints between plates or girders. However, brittle failures in steel structures of many types have been experienced since the middle of the nineteenth century when steel began to be used widely for structural work. The most spectacular failures occurred in many of the all-welded Liberty ships produced during the second World War, when nearly 1500 incidents involving serious brittle failure occurred.[3] Nineteen ships broke completely in two without warning; one in particular was just at the wharf-side having completed its acceptance trials (Fig. 15.10). Brittle fracture was in fact one of the most serious metallurgical problems of the 1939–45 war, and regrettably is still far from being eliminated as a cause of structural failure. Other important structures such as bridges, pressure vessels, gas transmission lines, penstocks, storage vessels for liquid gases at low temperatures have all suffered similar failures, often catastrophic, without warning of the event.

A common feature of all failures is that they have occurred in cold weather, but in many cases the recorded temperature was not lower than 273 K. The failures reflect the basic behaviour of mild steel tested over a range of low temperatures in the laboratory when a sharp transition from ductile to brittle failure is observed. It is further clear that the transition temperatures of steels can vary over a wide range of temperatures, and are dependent on a large number of interdependent variables, of which the grain size is only one. Certainly, other things being equal, a fine-grained steel will have a lower transition temperature than a coarse-grained steel, and thus provide some protection from brittle failure.

The composition of the steel is a complex variable about which much remains to be understood. Carbon raises the transition temperature, and if the carbon content of iron is reduced to very low levels the transition temperature can be very low indeed. The effect is not limited to carbon in solid solution, for the room temperature solubility in ferrite is about 0·003 per cent and the transition temperature is substantiatially raised as the carbon content is increased from 0·04 to 0·44 per cent, although the

Fig. 15.10 Brittle fracture in a welded tanker. (After Parker, 1957, *Brittle Behaviour of Engineering Structures*. John Wiley, New York & London)

transition range is less sharp (Fig. 15.11a). Nitrogen has a similar effect to that of carbon.

Many metallic alloying elements have a beneficial effect on the transition,[3] for example manganese (Fig. 15.11b) and nickel lower the transition temperature, although their influence is not yet fully understood. One difficulty is that alloying additions in steels often have more than one role, for example manganese is a ferrite strengthener, but it is also a mild deoxidizer, and combines with most of the sulphur in the steel to form manganese sulphide inclusions. Aluminium is a useful addition as it is a deoxidizer and produces a fine grain size mainly by forming fine particles of aluminium oxide and nitride which restrain grain growth. The resulting lower transition temperatures are thus primarily a matter of finer grain sizes. Phosphorus and sulphur provide an interesting contrast. Phosphorus is soluble in α-iron and raises the transition temperature, while sulphur has a low solubility and is usually present as inert manganese sulphide particles with little effect on the transition temperature. It is clear, therefore, that changes in steel-making practice which influence not only composition but other factors such as de-oxidation and final grain size to mention only two, can be significant in this very complex problem.

It should be emphasized that the fabrication process employed is often the crucial point.[3] The Liberty ships failed primarily because they were welded, and allowed continuity of cracks throughout the whole structure,

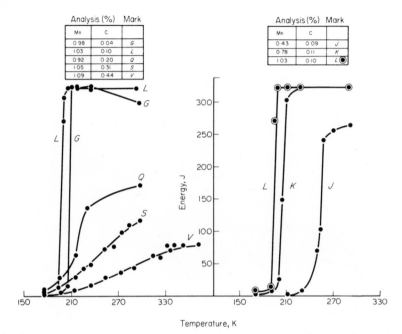

Fig. 15.11 Effect of, **a**, carbon and, **b**, manganese on the impact transition curves of iron–carbon alloys. (After Armstrong and Warner, 1955, Symposium on Impact Testing. Amer. Soc. Test. Mat. Special Publ. 176)

but equally important is the fact that welds and the adjacent heat-affected zones are often sources of stress concentrations, or even small cracks which help to initiate brittle fracture in material which otherwise would not fail. Bad welding practice or steel with too high hardenability* leads to martensitic areas near the welds which are brittle and very prone to microcracks, as well as high stress concentrations which under the appropriate conditions encourage brittle crack propagation. Proper attention to welding

* Steel which transforms slowly because of the presence of certain alloying elements, thus allowing martensite to form at relatively slow rates of cooling.

practice, including careful heating and cooling of the welded zone, can in many cases eliminate weld cracking.

The role of stress concentrators such as notches is of the greatest importance, as the presence of a notch raises the transition temperature. In engineering structures, it is often impossible to eliminate such stress concentrations, so the characteristics of the material in the presence of a notch must be known.

15.9 Metallography of brittle fracture

Single crystals of zinc provide one of the best means of studying ideal cleavage fracture surfaces, for by cleaving zinc at 77 K very smooth surfaces can be obtained, almost free from evidence of plastic deformation. However, it is usual to see fine steps or 'river markings' running roughly in the direction of propagation of the crack (Fig. 15.12). These markings indi-

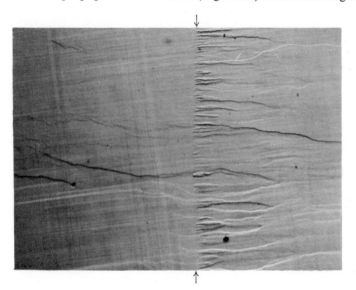

Fig. 15.12 Zinc crystal partly cleaved, then fully fractured, in liquid nitrogen. River markings or fracture surface. × 100. (After Gilman)

cate that the cleavage surface is not all on the same plane but stepped, and that the crack has been deflected at the steps to other parallel planes. The most likely obstacles to cause deflection of the crack are screw dislocations which emerge on the fracture surface, so that as the crack passes the dislocation it will move on two different levels.[34] Some screw dislocations are likely to pre-exist in the material, but another significant source of dislocations is the stress concentration at the root of the crack which causes plastic flow ahead of the crack. A fast-moving crack will initiate little deformation ahead, while a slow-moving crack will allow adequate time for plastic flow to occur. This deformation ahead of a crack has been nicely illustrated in lithium fluoride crystals where dislocations ahead of a stopped crack can be revealed by etch pitting.[35] The correlation between river markings and the velocity of the crack can be shown by interrupted cleavage tests on zinc, silicon iron or lithium fluoride where part of the fracture is initiated at, say, 77 K, while the remainder is carried out at room temperature. In such experiments, the river markings are much more numerous at the higher temperature when the crack moves at a lower velocity. Similarly, when a crack is stopped at a particular temperature, the river markings become more intense as it slows down and decrease when the crack is re-started (Fig. 15.12).

Turning to polycrystalline metals, it is interesting to relate the microstructure to the transition range. Below the transition temperature, the fracture is 100 per cent cleavage but micro-cleavage cracks usually one grain long have been detected above this temperature.[36] The cracks occur in not more than about 2 per cent of the grains, and then only in those which have undergone some prior deformation. These microcracks do not lead to brittle fracture above the transition temperature, so it can be assumed that the conditions for crack nucleation do not necessarily satisfy the requirement for crack propagation through the whole of the section. It is thus concluded that nucleation and propagation are distinct physical processes. Examination of the fractures clearly reveals the grain boundaries as major obstacles to crack propagation and, in fact, some grains may be so badly oriented with respect to the propagating crack that they fail by ductile rupture. In general, microscopic observations tend to support the view that the increased fracture strength of fine-grained materials arises not only from the greater difficulty of nucleating the crack, but also from the improved resistance to propagation.

15.10   Fracture toughness

In recent years emphasis has been placed on the conditions leading to crack propagation in engineering materials such as structural steels. The stress needed to cause fracture in a conventional test specimen is not a direct or reliable guide to the behaviour of a crack in an engineering structure, where crack propagation can often occur at stresses below the macroscopic yield stress of the materials as measured in a tensile test. Different materials show different crack propagation characteristics, so it is important to know for a given type of crack what the stress for propagation is. The crack propagation stress is influenced not only by the material, but by the design of the structure, for this determines the crack shape and the stress distribution in its vicinity.

As we have seen, a modified Griffith criterion for crack propagation can be obtained for crystalline solids, assuming that the surface energy term is augmented by the work done in plastic deformation as the crack propagates. Irwin[37] and Orowan[9] demonstrated that the deformation term was in fact the dominant one. So a stressed material in which a crack is propagating provides a strain energy rate, i.e. energy per unit area of crack formed, which is usually referred to as $\mathscr{G}$. Irwin[38] has shown that the energy rate can be expressed as

$$\mathscr{G} = \frac{P}{2}\frac{\delta\epsilon}{\delta c} \qquad\qquad 15.23$$

where P = load, and ϵ = strain. This equation can be used for the experimental determination of $\mathscr{G}$ by measuring ϵ as a function of the crack size c. The transition from slow to rapid fracture is characterized by a critical value $\mathscr{G}_c$, which is a measure of fracture toughness.

Irwin[38] has devised another fracture toughness criterion, a stress intensity criterion which he has shown to be equivalent to the energy criterion. Fracture takes place when a critical stress distribution is reached, the stress intensity factor K being determined by the applied stress and the crack geometry. It is defined for an infinite plate as

$$K = \sigma(\pi c)^{1/2} \qquad\qquad 15.24$$

Like $\mathscr{G}$, K can be determined experimentally using a stressed specimen in which a crack is slowly propagating as the load is increased. The transition to rapid fracture is characterized by the critical stress intensity factor, K_c, which thus is an alternative measure of fracture toughness.

15.11 Ductile fracture

Ductile failure by deformation implies a fracture process which takes place after substantial deformation, but this is by no means a satisfactory definition because brittle fracture can also be preceded by a marked degree of plastic deformation (delayed cleavage fracture). Perhaps the most familiar example of ductile fracture follows the development of a neck during tensile testing when a cup-and-cone fracture eventually occurs (Fig. 15.13b). This is a complex process which arises primarily from the stress distribution occurring in a tensile test, and from the work hardening characteristics of the material.

Many pure metals exhibit instead a double cup-and-cone fracture at room temperature (Fig. 15.13a), while the single cup-and-cone fracture is typical of a number of familiar alloys such as brass, duralumin and medium carbon steels. A third type of ductile fracture (Fig. 15.13c), which is planar in character, is frequently associated with high carbon steels at room temperature and some pure metals fractured at very low temperatures.

The main difference between ductile and brittle fracture, which is of profound practical significance, is that the propagation of a ductile crack involves substantial plastic flow, whereas in a brittle crack the fracture proceeds with a minimum of further plastic deformation. With cleavage the fracture energy is needed mainly to overcome the cohesive forces between atoms on each side of the crack path, whereas in ductile fracture this contribution is greatly outweighed by that needed for plastic deformation associated with the crack process.

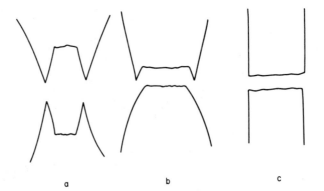

Fig. 15.13 Modes of ductile failure: **a**, double cup-and-cone; **b**, cup-and-cone; **c**, planar.

15.12 Ductile fracture in single crystals

The simplest form of ductile rupture occurs in some hexagonal and tetragonal metals deformed at elevated temperatures, when prolonged slip occurs on a few widely spaced slip bands leading to failure by glide. Zinc, cadmium and tin have been observed to behave in this way. However, as the deformation temperature is lowered, the slip band spacing becomes much closer, and necking down of the crystals precedes rupture. This is frequently associated with slip on more than one system and with twinning, so that the geometrical behaviour of the specimen becomes complex.

In face-centred metal crystals, ductile failure is also preceded by necking which, in pure metals may lead to a reduction in area of 100 per cent; however, in other cases a real fracture zone occurs at some stage during the necking process. For example, copper crystals deformed at 4·2 K fracture with only limited necking after substantial elongations.[39] Micro-examination of the necked regions of aluminium and copper single crystals has revealed[40] that a narrow zone of heavy slip bands occurs either on the primary or a secondary system, and further localizes the deformation until it is apparently occurring on one plane (macroscopically). Two stages of this process are shown in Fig. 15.14. In pure aluminium failure at room temperature is almost completely by glide, but in the case of copper there is evidence for the formation of ductile cracks. Large channel-like holes sometimes develop in the final stages of necking,[41,42] which have been explained as a result of heavy slip operating alternately on two non-parallel slip systems.[9]

Alloy crystals behave somewhat differently in so far as necking is not always severe, and planar ductile fractures can be obtained across the whole crystal section. Elam[41] found that solid-solution crystals of alu-

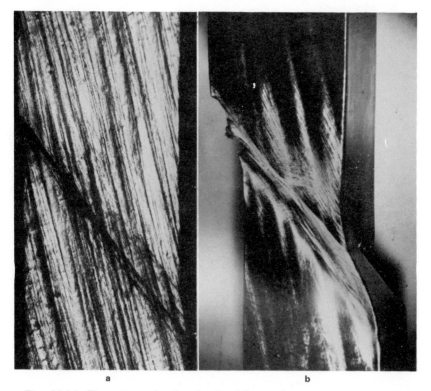

| a | b |

Fig. 15.14 Two stages in the ductile failure of a copper crystal. × 6
a, primary slip and localized slip on secondary system; **b,** severe shear on secondary plane and tearing.

minium–zinc with low zinc contents give fractures typical of pure aluminium, but with 10 per cent zinc or more the fracture becomes planar, and inclined at approximately 45° to the tension axis. This work showed that the fracture plane was the {111} slip plane, as also was the case with crystals of aged aluminium–5·5 wt per cent copper alloy[43] (Fig. 15.15). The fracture stresses for a wide range of crystal orientations have been determined[44] by using an aluminium–5·5 per cent copper alloy in several heat

Fig. 15.15   Planar fracture of an aged aluminium–5·5 wt per cent copper
crystal at room temperature. × 10.

treated conditions, which minimized the formation of a neck by raising
the capacity of the alloy for work hardening. In these circumstances, the
fracture occurred along a slip plane and a constant resolved shear stress
criterion was established. Fig. 15.16 compares the normal and resolved
shear stresses at 293 K obtained from crystals aged at 438 K, and illustrates that the resolved shear stress at fracture is nearly constant.

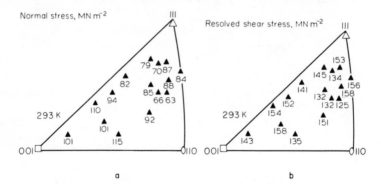

Fig. 15.16   Fracture stresses of aluminium–5·5 per cent copper crystals. **a,** Normal
stress; **b,** resolved shear stress.

15.13   Ductile fracture in polycrystalline aggregates

In tensile deformation of polycrystalline aggregates, necking is usually
the precursor to ductile fracture so it is necessary first to understand the
criterion for the commencement of neck formation.

Neck formation in practice commences at or near the point of maximum
load, i.e. $L = L_{max}$, and if it were not for the phenomenon of work
hardening, necking would commence immediately the metal yields. Work

hardening raises the yield stress; however, this effect is offset to some extent by the gradual elongation and consequent thinning of the specimen as deformation proceeds. When the increase in stress due to the decrease in cross-sectional area exceeds that due to work hardening, an unstable condition exists and necking begins.

Now
$$L = \sigma A$$

where σ is the true stress and A the actual cross-sectional area of the specimen.

$$dL = A \, d\sigma + \sigma \, dA \qquad\qquad \textbf{15.25}$$

$$= 0 \text{ at the maximum load point.}$$

Assuming that the volume V of the specimen does not change during deformation, then

$$V = A_0 l_0 = A l$$

$$A_0 = A(1 + e)$$

where A_0 is the initial cross-sectional area, l_0 the initial length, and e is the conventional strain.

It follows that

$$0 = A \, de + dA(1 + e) \qquad\qquad \textbf{15.26}$$

Combining eqns. **15.25** and **15.26** we obtain the following

$$d\sigma = \frac{\sigma \, de}{1 + e} = \sigma \, d\epsilon, \text{ where } \epsilon \text{ is the natural strain}$$

$$\therefore \frac{d\sigma}{d\epsilon} = \sigma \qquad\qquad \textbf{15.27}$$

Consequently, necking will take place at a strain when the slope of the true stress–strain curve equals the true stress at that strain.

We have now to decide at what stage of the necking process ductile fracture commences. One difficulty is that metals and alloys vary very greatly in the reduction in area they undergo in the necked region prior to failure. Moreover, the behaviour of one metal or alloy is dependent on the exact conditions of stressing. For example, if a hydrostatic pressure is superimposed on a tensile test, the degree of necking prior to failure is increased greatly[45] and the material is effectively more ductile. Again, if a metal is deformed by cold rolling, in which circumstances substantial compressive stresses are present, it will sustain much higher strains before fracture than in a tensile test. Furthermore, if subsequently tested in tension, the cold worked metal will still neck substantially prior to fracture. This and other similar evidence indicates that a strain criterion for ductile fracture is not valid,[46] and that regardless of the amount of cold work a material has received, it will not crack in a ductile manner, unless further plastic strain is imposed under favourable stress conditions.

The question then arises whether there is a stress criterion for the ductile fracture of polycrystalline materials. When the necking phenomenon is eliminated in single crystals a shear stress criterion for ductile fracture is found, but it is unlikely that this is of general application to the fracture zones of polycrystals where the specimen is under a complex triaxial stress system.

15.14   The initiation and propagation of ductile cracks

The examination of polished sections taken through the necks of deformed metals has revealed that in the early stages of necking only the effects of deformation by slip can be seen. Slight straining after sectioning reveals the very coarse slip bands characteristic of this stage of the deformation process. At a later stage small pores are found which gradually multiply and eventually nucleate a crack by joining together[47] (Fig. 15.17). Ludwik[48] first pointed out that ductile failure commenced by the

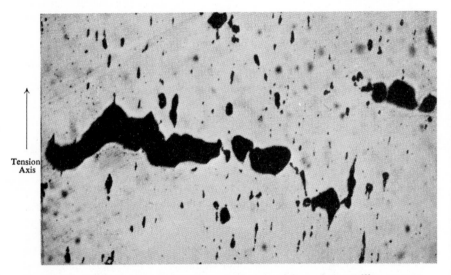

Tension Axis

Fig. 15.17   Voids linking up in the necked region of a polycrystalline copper specimen. (After Puttick, 1959, *Phil. Mag.* **4,** 164)

growth of such a crack in the centre of the specimen outwards. The pores mainly originate at small inclusions or particles of a second phase; Puttick,[49] as a result of a study of the nucleation of pores in iron and copper, has concluded that the pores are created by the metal flowing away from the interfaces of small inclusions not wetted by the matrix. In particular, small cavities were found in association with copper oxide inclusions in tough pitch copper, while in O.F.H.C. copper a similar result was obtained, except that the particles were too fine to be seen in the optical microscope.[50]

The link between inclusions and the initiation of ductile fracture is thus now well established, and explains very adequately the increase in reduction in area as the purity of a metal is raised. Ideally, a pure metal free of inclusions should thus exhibit 100 per cent reduction in area in the necked zone, but there is some evidence to support the view that ductile cracks can initiate also in the absence of inclusions.[51] In these circumstances, crack nuclei would be expected to form by dislocation coalescence as in the nucleation of brittle cracks, but the subsequent propagation would be different. The propagation of the ductile crack occurs by the joining up of pores which, because of the axial component of the stress, have become elongated. The resulting coalescence has been called a process of internal necking which is a description of the breaking of bridges between adjacent cavities elongated in the tensile direction (Fig. 15.17). This process continues until a wide crack normal to the tension axis occupies the centre of the necked region. There is the tendency to widen in the direction of tension, and in the case of metals which do not work harden greatly, for example pure metals, this leads to a double cup-and-cone fracture. By a change in the work-hardening characteristics of the material, or by restraint of the flow in the tensile direction, for example by having a wafer specimen comprising a soft metal in a hard metal sandwich, it is possible to keep the crack path roughly normal to the tension axis to give a fracture of type shown in Fig. 15.13c.

The final cone region of the fracture is a region of intense shear occurring at approximately 45° to the tension axis. Metallographic observation confirms this general picture by revealing marked shear distortion in this zone, for example the pores in the vicinity of the zone are elongated in the direction of shear. When one attempts to fit together the two halves of a cup–cone fracture it is obvious that they do not coincide, so the conclusion must be drawn that the surfaces of the cone not only slide under the influence of the shear stress component, but move apart as a result of the tensile stress. The cone region of the fracture can be eliminated by straining the specimen in a rigid frame which enables the rapid shear deformation to be suppressed.

15.15 Metallographic examination of ductile fractures

Fracture replica techniques* have enabled rough fracture surfaces to receive detailed examination in the electron microscope. Crussard and co-workers[52] have shown that the main characteristic of the fibrous or cup region of a ductile fracture is the formation of a continuous pattern of

* A very thin carbon film is evaporated on to the fracture and is subsequently removed by electrolytic polishing or etching for examination in the electron microscope.

dimples or shallow depression on each surface of the fracture. These vary in size from 0·5 to 20 microns in diameter, and are clearly the result of the linkage of the cavities formed in the necked region (Fig. 15.18) They probably first arise ahead of the propagating ductile crack as a result of stress concentrations at inclusions and, as they grow, concentrate the crack path in their direction, so producing a continuous propagation. This interpretation is strikingly confirmed by the fact that in many materials the dimples can nearly all be associated with individual inclusions (Fig. 15.18). There is also evidence that similar fracture patterns can be obtained in the

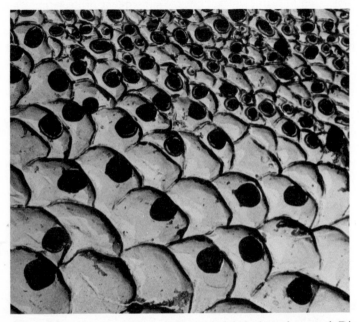

Fig. 15.18   Fracture surface of an overheated nickel–chromium steel. Dimples associated with sulphide particles. Fracture replica. × 5000.

absence of visible inclusions, in which cases the cracks may arise from dislocation interactions.

The appearance of the cone or shear zone of the fracture surfaces is strikingly different. It contains regions of elongated dimples which arise from the entrapment of voids in the shear zone, but there are also zones which are practically featureless, the proportion of these regions being about 1 : 1. The featureless zones have been interpreted as due to decohesion along the path of severe shear. Similar observations have been made on single crystals of aluminium–copper,[44] where a shear stress criterion for fracture was obtained, and in polycrystalline aluminium alloys which show 45° fractures after suitable ageing treatments.[53]

The general picture of ductile fracture is thus confirmed by direct observation of the fracture surfaces which offer a remarkable contrast to those resulting from brittle fracture. The 'plastic work factor' in the propagation of a ductile crack is clearly revealed, as is the important role of inclusions in acting as nucleating centres for cracks. However, although these points are well established, the overall mode of failure is very susceptible to the method of testing, the resulting stress distribution in the neck and work hardening characteristics of the material. These factors lead to a wide variety of fracture patterns. Perhaps the most important implication is the dominant role of inclusions in controlling the degree of ductility as expressed in terms of reduction in area of the neck prior to failure. There is evidence that the type of distribution of the inclusions has a decisive effect on the final mode of fracture. Normally inclusions are strung along the axis of a rolled or drawn rod, as a consequence of the method of manufacture, and the axes of tensile specimens usually coincide with the rod axis, with the result that a cup-and-cone fracture is commonly observed. However, a transverse tensile specimen from the same material often shows little necking and much lower reductions in area. In the latter case, the inclusions drawn out by the working processes present the maximum surface area to the tensile stress, while in the former they present the minimum area.

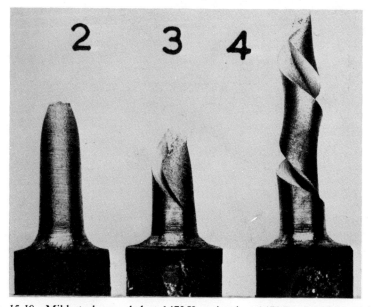

Fig. 15.19 Mild steel annealed at 1473 K, twisted at 1473 K, then fractured at room temperature. **2,** No revolutions: **3,** 1·5 revolutions; **4,** 4 revolutions. (After Hardwick)

A spectacular illustration of the role of inclusion distribution in determining the ductile crack path is revealed by hot torsion experiments on steel and aluminium.[54] Steel specimens given several revolutions in torsion at 1473 K, when subsequently fractured in tension at room temperature, exhibit a spiral fracture tracing out the path the inclusions were drawn into during the hot torsion experiment (Fig. 15.19). Other experiments with aluminium of two grades of purity led to the similar result that the impure inclusion-containing material showed a spiral fracture on deformation at room temperature, while the pure material exhibited a normal ductile fracture.

General references

1. AVERBACH, B. L., FELBECK, D. K., HAHN, G. T. and THOMAS, D. A. (Eds.) (1959). *Fracture*. Swampscott Conferences. John Wiley, New York and London.
2. *Fracture of Solids*, AIME, Met. Soc. Conference, Vol. 20. John Wiley, New York and London (1962).
3. PARKER, E. R. (1957). *Brittle Behaviour of Engineering Structures*. John Wiley, New York and London.
4. TIPPER, C. F. (1962). *The Brittle Fracture Story*. University Press, Cambridge.
5. TETELMAN, A. S. and MCEVILY, A. J. JNR. (1967). *Fracture of Structural Materials*. John Wiley, New York and London.

References

6. ANDRADE, E. N. DA C. and TSIEN, L. C. (1937). *Proc. R. Soc.*, **A159**, 346; GORDON, J. E., MARSH, M. D. and PARRATT, M. E. M. L. (1959). *Proc. R. Soc.*, **A249**, 65.
7. GRIFFITH, A. A. (1920–21). *Phil. Trans. R. Soc.*, **A221**, 163.
8. INGLIS, C. E. (1913). *Trans. Inst. Naval Arch.*, **55**, 219.
9. OROWAN, E. (1948). *Rep. Prog. Phys.*, **12**, 185.
10. ZENER, C. (1948). In *Fracturing of Metals*, American Society for Metals. P. 3.
11. MOTT, N. F. (1953). *Proc. R. Soc.*, **A220**, 1.
12. STROH, A. N. (1954). *Proc. R. Soc.*, **A233**, 404.
13. STOKES, R. J., JOHNSON, T. L. and LI, C. H. (1958). *Phil. Mag.*, **3**, 718.
14. COTTRELL, A. H. (1958). *Trans. AIME*, **212**, 192.
15. HONDA, R. (1961). *Acta metall.*, **9**, 969.
16. PARKER, E. R. Reference 1, p. 181.
17. GILMAN, J. J. (1954). *Trans. AIME*, **200**, 621.
18. STROH, A. N. (1958). *Phil. Mag.*, **3**, 597.
19. CAHN, R. W. (1955). *J. Inst. Metals*, **83**, 493.
20. HULL, D. (1960). *Acta. metall.*, **8**, 16.
21. DERUYTERRE, A. and GREENOUGH, G. B. (1956). *J. Inst. Metals*, **84**, 337.
22. BIGGS, W. D. and PRATT, P. L. (1958). *Acta metall.*, **6**, 694.
23. ALLEN, N. P., HOPKINS, B. E. and MCLENNAN, J. E. (1956). *Proc. R. Soc.*, **A234**, 221.
24. SMITH, R. L. and J. L. RUTHERFORD (1957). *Trans. AIME*, **209**, 857.

25. ADAMS, M. A., ROBERTS, A. C. and SMALLMAN, R. E. (1960). *Acta metall.*, **8**, 328.
26. MORDIKE, B. L. (1961). *Z. Metallk.*, **52**, 587.
27. HOPKINS, B. E. and TIPLER, H. R. (1958). *J. Iron Steel Inst.*, **188**, 218.
28. REES, W. P. and HOPKINS, B. E. (1952). *J. Iron Steel Inst.*, **172**, 403.
29. STEIN, D. F. (1966). *Acta metall.*, **14**, 99.
30. STROH, A. N. (1955). *Phil. Mag.*, **46**, 968; (1957). *Adv. Phys.*, **6**, 418.
31. PETCH, N. J. (1958). *Phil. Mag.*, **3**, 1089; HESLOP, J. and PETCH, N. J. (1958). *Ibid.*, **3**, 1128.
32. CHURCHMAN, A. T., MOGFORD, I. and COTTRELL, A. H. (1957). *Phil. Mag.*, **2**, 1271.
33. ANDERSON, O. L. Reference 1, p. 331.
34. GILMAN, J. J. (1955). *Trans. AIME*, **203**, 1252; (1958). *Ibid.*, **212**, 310.
35. GILMAN, J. J. (1957). *Trans. AIME*, **209**, 449.
36. HAHN, G. T., AVERBACH, B. L., OWEN, W. S. and COHEN, M. Reference 1, p. 91.
37. IRWIN, G. R. (1948). In *Fracturing of Metals*, American Society for Metals. p. 147.
38. IRWIN, G. R. (1964). *Applied Materials Research*, **3**, 65.
39. BLEWITT, T. H., COLTMAN, R. R. and REDMAN, J. K. (1957). In *Dislocations and the Mechanical Properties of Crystals*. John Wiley, New York and London. p. 179.
40. BEEVERS, C. J. and HONEYCOMBE, R. W. K. Reference 1, p. 474.
41. ELAM, C. F. (1925). *Proc. R. Soc.*, **A109**, 143.
42. ROSI, F. D. and ABRAHAMS, M. S. (1960). *Acta metall.*, **8**, 807.
43. KARNOP, R. and SACHS, G. (1928). *Z. Phys.*, **49**, 486.
44. BEEVERS, C. J. and HONEYCOMBE, R. W. K. (1962). *Acta metall.*, **10**, 17.
45. BRIDGMAN, P. W. (1952). *Studies in Large Plastic Flow and Fracture.* McGraw-Hill, New York and Maidenhead.
46. COTTRELL, A. H. Reference 1, p. 20.
47. TIPPER, C. F. (1949). *Metallurgia*, **39**, 133.
48. LUDWICK, P. (1926). *Z. Metallk.*, **18**, 269.
49. PUTTICK, K. E. (1959). *Phil. Mag.*, **4**. 164; (1960). *Ibid.*, **5**, 759.
50. RODGERS, H. C. (1960). *Trans. AIME*, **216**, 498.
51. BEEVERS, C. J. and HONEYCOMBE, R. W. K. (1962). *Phil. Mag.*, **7**, 763.
52. CRUSSARD, C., PLATEAU, J., TAMHANKAR, R., HENRY, G. and LAJEUNESSE, D. Reference 1, p. 524.
53. RYDER, D. A. and VIAN, R. E. (1962). *J. Inst. Metals*, **90**, 383.
54. HARDWICK, D. (1960). Ph.D. thesis. University of Sheffield.
55. HARDING, J. (1967). *Proc. R. Soc.*, **A299**, 464.
56. REID, C. N. (1965). *J. Less Common Metals*, **9**, 105.

Additional general references

1. *Ductility* (1968). Papers at an American Society for Metals Seminar 1967, American Society for Metals.
2. KNOTT, J. F. (1973). *Fundamentals of Fracture Mechanics*. Butterworths, London.
3. TAPLIN, D. M. R. (Ed.) (1978). *Advances in Research on the Strength and Fracture of Materials* (Fracture 77) Vols. 1–4, Proc. 4th International Conference on Fracture, Waterloo, Canada 1977, Pergamon, Oxford.
4. FRANCOIS, D. (Ed.) (1982). *Advances in Fracture Research* (Fracture 81) Vols. 1–6, Proc. 5th International Conference on Fracture, Cannes 1981, Pergamon, Oxford.

Author Index

Subject Index